MOTORBOOKS POWERTECH SERIES ™

How to Tune & Modify

CHRYSLER *FUEL INJECTION*

BEN WATSON

First published in 1997 by Motorbooks International Publishers & Wholesalers, 729 Prospect Avenue, PO Box 1, Osceola, WI 54020-0001 USA

Library of Congress Cataloging-in-Publication Data Available

ISBN 0-7603-0371-1

On the front cover: Gone are the carburetors and induction systems of the past. These are the major components used today on Chrysler's fuel-injected vehicles. The air plenum, control module, and injectors work together to provide superb performance, reliability, and low emissions. All parts courtesy of Chrysler Corporation. *David Gooley*

Printed in the United States of America

CONTENTS

INTRODUCTION

I have been working with electronic fuel-injected engines for over twenty years. My experience began with the introduction of the Bosch D-Jetronic-equipped engines in the early 1970s. Over the course of time I have striven to find the ultimate method for troubleshooting fuel injection and driveability problems. In seeking this ultimate answer I have literally journeyed from the U.S. Virgin Islands to Saipan; from 13 degrees above the equator to 23 degrees below the North Pole seeking a master. I have reached a conclusion: there is no ultimate method.

Driveability problems are the bane of both the consumer and the technician. Every skilled technician has had the "possessed" automobile to diagnose. I have had several. The most memorable was a 1990 Dodge Caravan. This minivan was equipped with a 3.0-liter fuel-injected engine. When the vehicle was put into reverse and the steering wheel turned to the left, the engine would die. I cannot even begin to tell you everything I did to repair this vehicle. Nothing worked. After a couple of days of hair pulling I called the zone office for Chrysler in Portland. One of its top people came to help. We worked together on the problem for two more days. On the morning of the third day, the car was gone, never to be seen again. Fortunately for everyone concerned, this car had yet to be sold.

In or around Fairbanks, Alaska, there is a car that the automotive repair community of that northernmost output of shopping malls and Denny's restaurants has dubbed Christine. "Christine" is, for those of you not familiar with Steven King, the title character in one of his books. Christine was a possessed

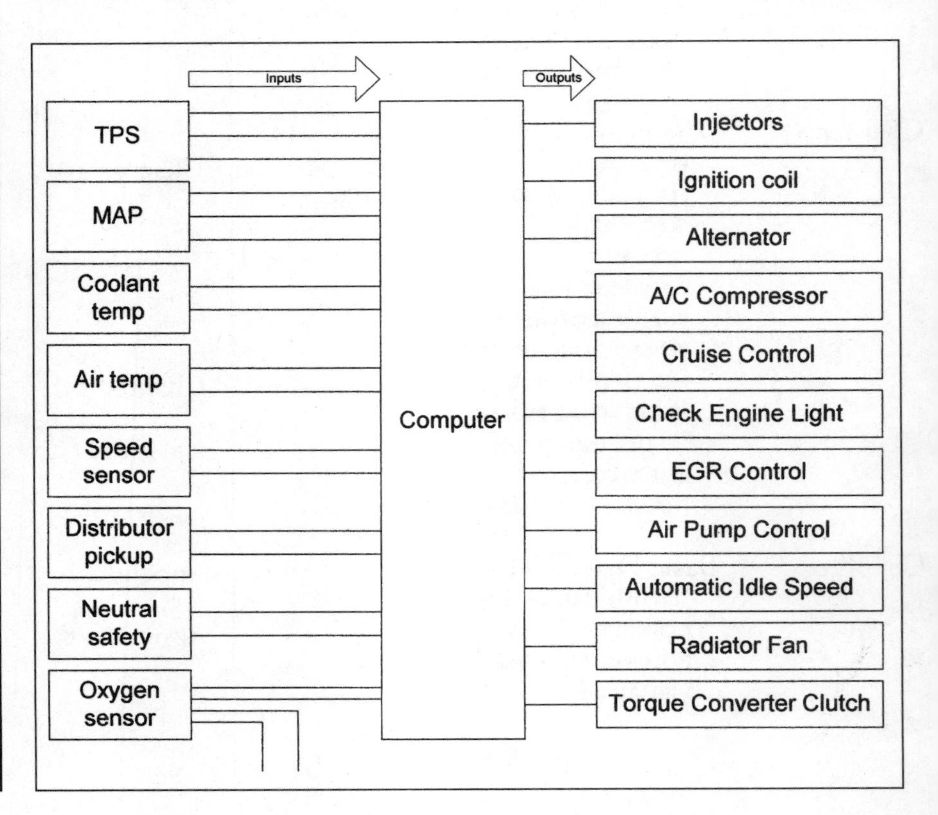

Chrysler began installing its electronic fuel injection system in the 1984 model year. From the beginning, Chrysler incorporated a wide range of abilities into the computer. While competitors' systems were controlling just the injectors with the fuel injection computer, Chrysler was controlling the injectors, the ignition system, the charging system, the air conditioning system, the cruise control, the idle speed, the engine cooling system, the emission control devices, and other items. The Chrysler computer basically has been designed to control everything under the hood.

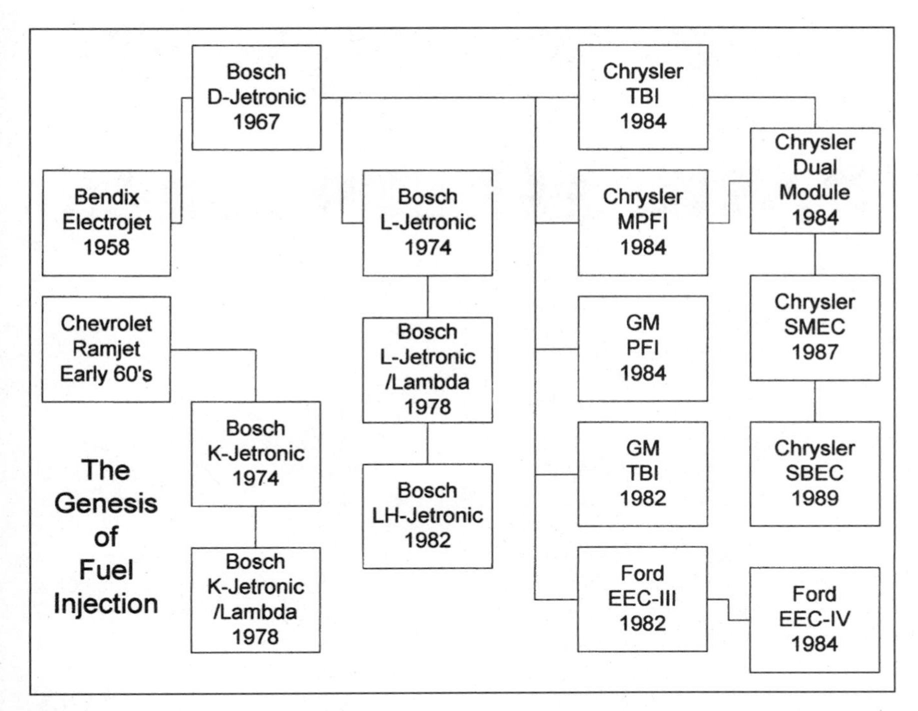

Mass production use of electronic fuel injection dates from the 1960s on European- and Japanese-built applications. The domestic manufacturers were slow to begin using fuel injection. Although Chrysler experimented with the use of electronic fuel injection in the late 1950s, it waited almost 20 years after the Europeans to begin using its modern generation of fuel injection. The wait was worth it. The injection systems of the 1980s not only control virtually every engine function, but also have the ability to diagnose themselves.

No matter how skilled a technician is, no matter how much experience he/she has, there will always be the car and the driveability problem that he/she cannot solve. Many is the time that I would work myself to extreme frustration then ask an associate to look at the car. Quite often the associate would find the problem in a matter of minutes. Fortunately, for my rather frail male ego, just as often I would find the source of problems for my associates in a matter of minutes. When you spend too much time looking at a problem, you become "snow-blind." It is then time for others to take a look.

1958 Plymouth. The Christine of Fairbanks is a little 2.2-liter Dodge. I first became aware of its existence when one of my students brought it for the class to troubleshoot. It had a dead number 4 cylinder. After all their repairs it still had a dead number 4 cylinder. Even after replacing the cylinder head and the engine again it still had a dead number 4 cylinder. We began, as all good diagnosis should, with a visual inspection. Everything was fine. We did a compression test and checked the ignition system on an engine analyzer. We checked the fuel pressure, we analyzed the serial data, we checked injector flow, we checked for intake and exhaust restriction, we tried everything that made sense and things that made no sense. The car is still driven around Fairbanks and still has the dead number 4 cylinder.

The message in these two depressing stories is that no one knows everything about troubleshooting fuel injection. Even with nearly 25 years of experience there are still those cars that refuse to be fixed by me. At this point you might be saying, "Gee, if he has all that experience and still cannot fix some cars, what chance do I, an amateur, have?" There is an advantage that the consumer or enthusiast always has over the professional—time. Time is the best troubleshooting tool there is. Time to analyze the problem, to test, to retest, to test again under other conditions. The professional technician's primary handicap is a lack of time. During the days I spent working on the aforementioned Caravan I made $0.00 Even if I had found the problem, warranty policy was only to pay for the time it took to make the actual repair. Can you imagine what the time I spent on that vehicle would have cost a paying customer if I had been adequately compensated for my time? I worked on it for about 32 hours. At the labor rate that shop charged that would have been $1,920! With Christine I had only four hours that the car owner was willing to give me to make the diagnosis. Time in each case would have allowed me to find the problem.

Now on the positive side. These are two unsolved mysteries in almost 25 years of working with these systems. The rest were resolved. The primary purpose of this book is to help you resolve Chrysler driveability problems, either your own or those of others.

A great deal of emphasis is placed on the importance of Ohm's Law in the understanding of automotive electronic systems. The most important piece of information about Ohm's Law is that when the resistance in a circuit increases, the current-carrying capability of that circuit decreases. Conversely, when the resistance of a circuit decreases, the current-carrying capability increases.

CHAPTER ONE

BASIC ELECTRONICS

Ohm's Law

Should you ever find yourself speaking before a group of auto mechanics, and should you desire to elicit a group groan, and induce sleep in the group, you should say that you are going to discuss Ohm's law. We are now going to discuss Ohm's law.

Volts is the amount of electrical pressure there is in the circuit. It is possible to have volts even when no work is being demanded of the circuit or its electrical devices. Batteries have voltage when they are sitting on the shelf, so does a lemon and so does a human. Voltage is the amount of electrical pressure pushing the current through a circuit.

If the pressure pushing electricity through a conductor is known as voltage, then the volume at which the electricity is flowing is called amps. A common analogy used to explain electrical circuits is city water pressure. Volts is like pounds per square inch (psi). Amps is like gallons per minute (gpm).

The measurement and control of amps has always been important in automotive electrical systems. The principle tests of battery performance are current capacity tests. One of the principle tests of starter performance is the amperage draw test. In both of these cases the volume of current being measured is extremely large. The automotive service industry has traditionally dealt with currents of 1 to 500 amps and even more.

The advent of automotive electronics has brought currents much smaller than the auto technician has ever dealt with before. For example, the oxygen sensor circuit at times carries a current of less than .0000000005 (one-half trillionth) amp! Although currents in these ranges behave exactly the same way as larger currents, their perceived behavior can be different.

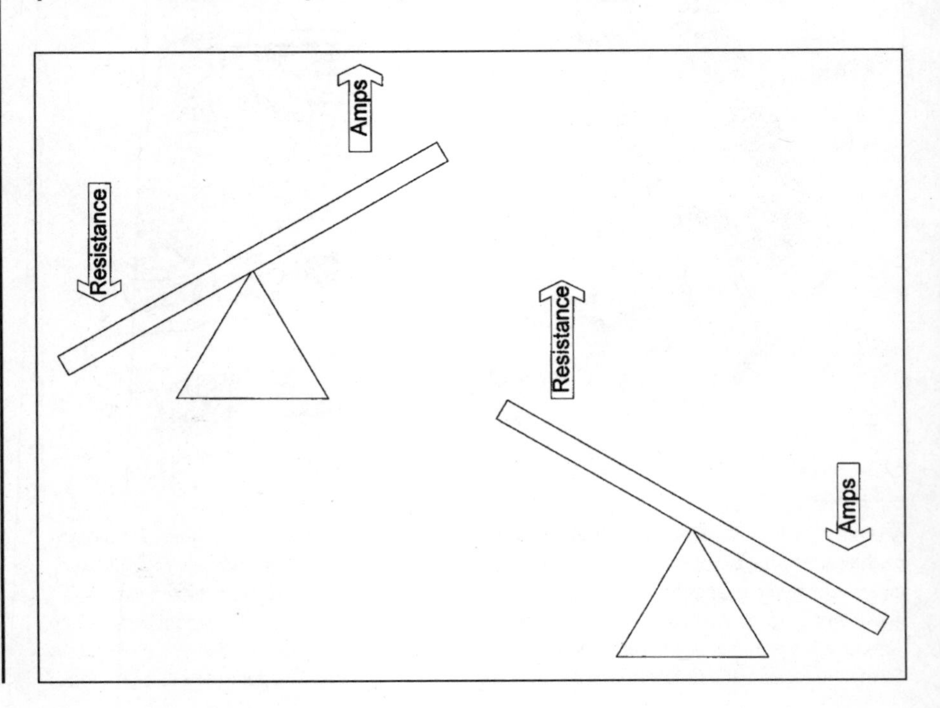

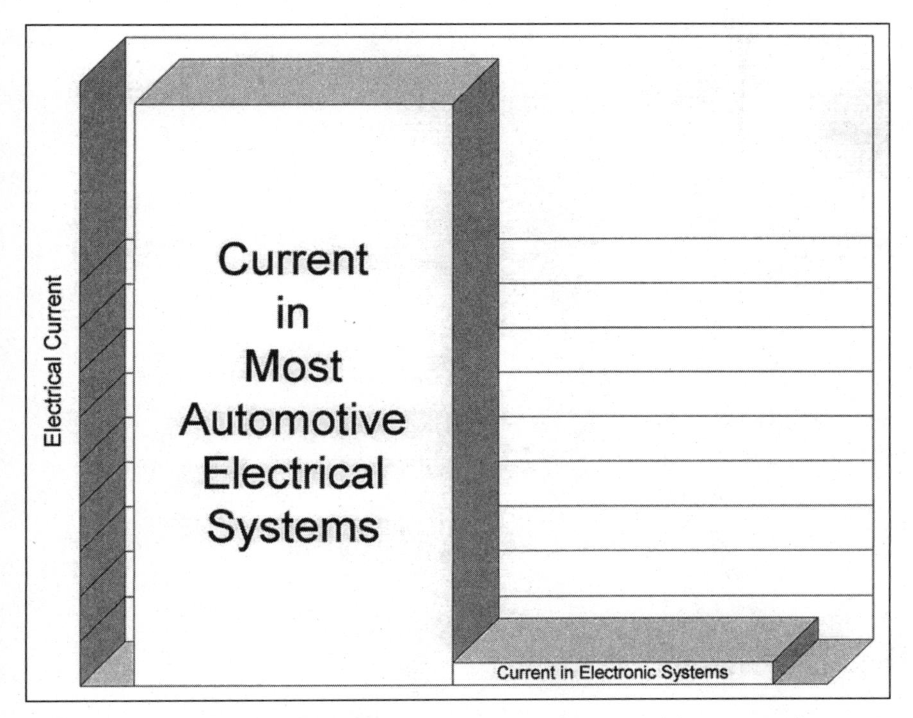

Historically, the electrical current flowing through the circuits of an automobile is measured in tens and even hundreds of amps. Circuits in electronic fuel injection systems have very high resistance. This means that the current in these circuits is always extremely low.

Normally when a conductor carrying a current accidentally touches ground prematurely, the result is very dramatic. Since the power supply for most circuits on the automobile is either the battery or the alternator, a conductor shorting to ground can easily be overloaded. Extreme heat builds up in an overloading circuit, and the conductor can actually melt. Most of the electronic circuits on the automobile have power supplies with a maximum potential far less than what the conductor can carry. When a low current circuit becomes grounded, the conductor is not affected. The effect on the power supply, however, can be devastating. The power supply could be overloaded to the point of destruction.

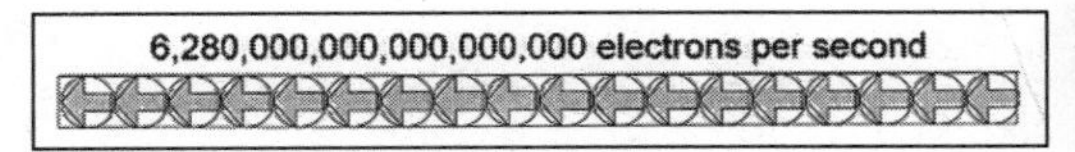

The current, or amps, flowing in a circuit actually represents the movement of electrons through the conductor.

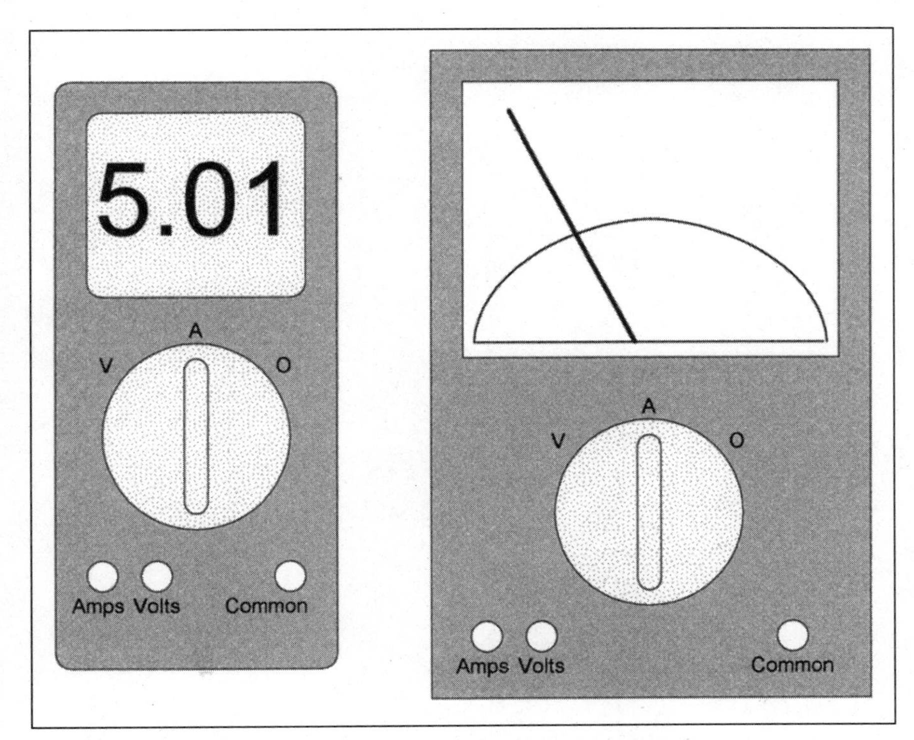

Because of the low current in most of the circuits in Chrysler electronic fuel injection systems, some voltmeters can rob enough power from the circuit being tested to cause the voltage in the circuit to drop when the meter is connected. This is the same kind of drop that you get in battery terminal voltage when the starter is engaged. This drop in voltage can cause the voltmeter to read inaccurately. To reduce this affect, make sure that the voltmeter you use for testing is a high-impedance meter. High-impedance meters are those that have an internal resistance of 10 million ohms (10 megohms or more). This specification includes most digital meters, but only a few analog meters.

Units of Measure

The basic unit of measure for current flow is the amp. To the engineer, one amp of current flow means 6.28 billion billion electrons per second passing any point in the circuit when pushed by 1 volt. In servicing automotive electronic systems there are very few times when amperage will need to be measured directly.

Units of measurement for current flow are amps and milliamps. The milliamp is 0.001 amp.

Resistance

Resistance is anything which impedes or inhibits the flow of electricity through a conductor. All conductors have some natural resistance in them. Resistance can be demonstrated with a simple high school physics experiment.

Using a number 2 pencil, draw a dark line on a piece of standard notebook paper. Place the black lead of an ohmmeter at one end of the line. Place the red lead of the ohmmeter right next to the black, but do not allow the leads to touch. Note the ohmmeter. Now slowly move the red lead along the pencil line away from

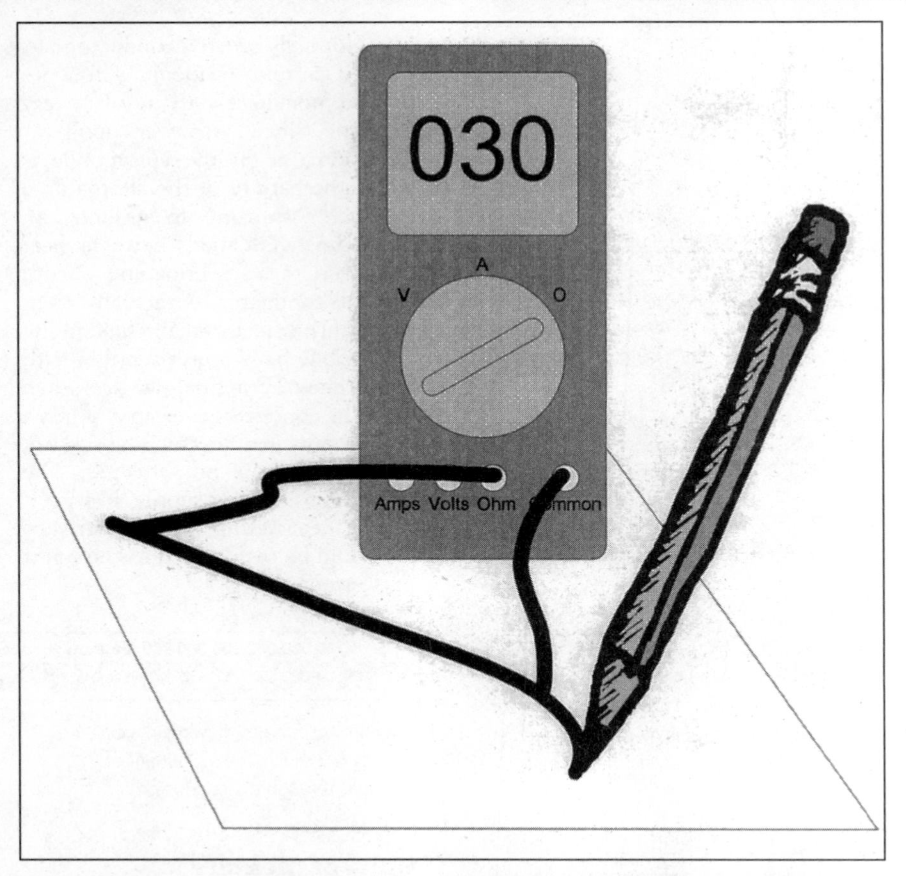

All materials that can carry a current may be considered conductors. Conductors vary in their ability to carry a current. Those that have a lesser ability are said to have a higher resistance. Conductors that have a greater ability to carry a current are said to have a lower resistance.

the black. Note that the ohmmeter reading increases. This is resistance.

There are several factors that will determine exactly how much resistance a given conductor will have.

Diameter of the Conductor

The larger the diameter of a conductor the less resistance there will be. The less resistance there is in a conductor the greater current it can carry. It is standard practice to use small wires for low current circuits (such as sensor circuits) and larger wires for higher current flow circuits (such as actuator circuits). Compare this to a water hose. A 1/2-inch hose cannot carry as large a flow, or current, as a 4-inch hose.

Length of the Conductor

The length of the conductor will affect its resistance. Resistance increases with the length of the conductor in the same way that the ohmmeter reading increases as the red lead is moved along the pencil mark. Basic wire resistors, such as those used for ballast resistors, use this principle.

Temperature of the Conductor

For most conductors, as their temperature increases so does their resistance. One notable exception is battery electrolyte.

A peculiar phenomenon occurs as a current passes through a conductor. The current generates heat. As the temperature of the conductor increases the resistance of that conductor also increases. The increasing resistance tends to impede current flow. If the temperature of the conductor rises beyond the melting point for the material forming the conductor, then the conductor will burn open.

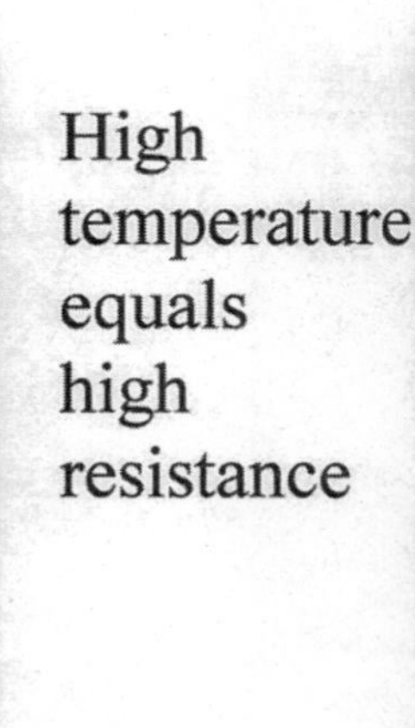
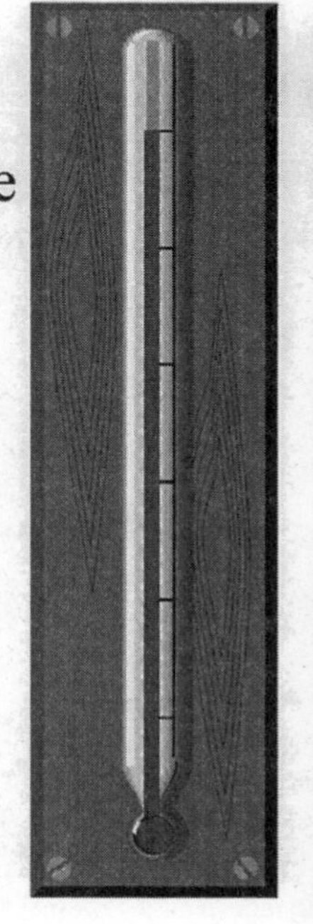

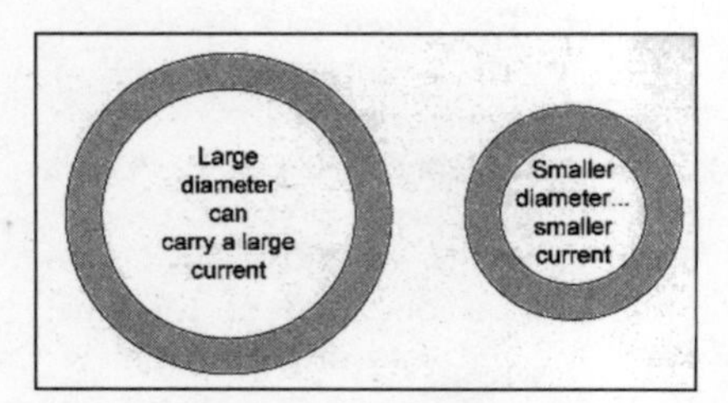

The resistance of a conductor varies directly with its diameter. The larger the diameter of the conductor, the greater its current-carrying ability.

When conductors are cooled to extremely low temperatures, -200 degrees Celsius or less, they become superconductors. This will permit extremely large currents to pass through small, lightweight conductors. Although there are currently no automotive applications for this technology, it does hold promise for emission-free electric automobiles in the future.

Material Composition of the Conductor

The material which forms the conductor may be the most critical in determining the resistance in the conductor. The best conductor is silver. One of the worst of the commonly used conductors is aluminum. Gold is used as a conductor on many critical circuits or connections. Although other materials have conductive properties superior to gold, gold does not corrode. Corrosion at connection points increases resistance in a circuit. Because of its extremely good conductive properties and relatively high resistance to corrosion, copper is the most frequently used conductor in automotive circuits.

Resistors

Opening a device with electronic circuitry will reveal that the most common electronic component is the resistor. Resistors perform many functions in a circuit. For automotive purposes, we can summarize two ways in which they are used.

First, they are used for current limiting. Any resistance in a circuit reduces current flow in that circuit. Resistors are placed in many automotive electronic circuits to limit current flow. When a wire becomes grounded, resistance in the series circuit is reduced. If reduced enough, the increased current flow from the power supply could either damage the power supply or the current-carrying wire. Putting a current-limiting resistor in the series circuit inside the power supply module will prevent this.

Second, resistors are used to create a voltage drop. When a resistance is placed in a circuit carrying a current, voltage will be reduced as the current passes through the resistance. By using resistors that change value as events occur, it becomes possible for a computer to monitor these events through the changes in voltage.

Resistance is measured in ohms. One ohm is the amount of resistance it takes to limit 1 volt applied to a conductor at 1 amp of current flow. An ohmmeter is used to measure resistance.

The ohmmeter contains an internal voltage source, such as a 1.5-volt battery, and is connected in series with the component being tested. All power supplies must be removed from the component being tested. The ohmmeter passes a known current through the component, measures the outbound voltage, measures the return voltage, and uses the voltage drop to calculate the resistance.

Be careful when testing components, such as diodes, which have a polarity. Some ohmmeters apply the positive voltage to the black lead when testing while others apply it to the red lead. Familiarity with the meter being used will prevent condemning a good component.

There is a mathematical relationship between volts, amps, and ohms. Stated in the simplest terms it takes 1 volt of electrical pressure to push 1 amp of current through 1 ohm of resistance. For the automotive technician, there are two important facts concerning Ohm's law.

First, assuming we keep the voltage the same, as the resistance in a circuit decreases the current flow in the circuit increases.

Second, assuming we keep the voltage the same, as the resistance in a circuit increases the current flow in the circuit decreases.

LEFT: The temperature of a conductor has an affect on the resistance of that conductor. The higher the temperature, the greater the resistance. This principle is used in heated oxygen sensors to control the temperature. When the current flowing through the oxygen sensor heater warms up, its resistance increases. The increasing resistance causes the current flowing through the heater to decrease. The decreasing current generates less heat. The oxygen sensor heater cools. This causes the resistance in the heater to drop. The lower resistance allows the current to increase. The temperature of the heater rises, the resistance increases, the current drops, and the temperature drops. The effect of this is self regulation of the oxygen sensor temperature.

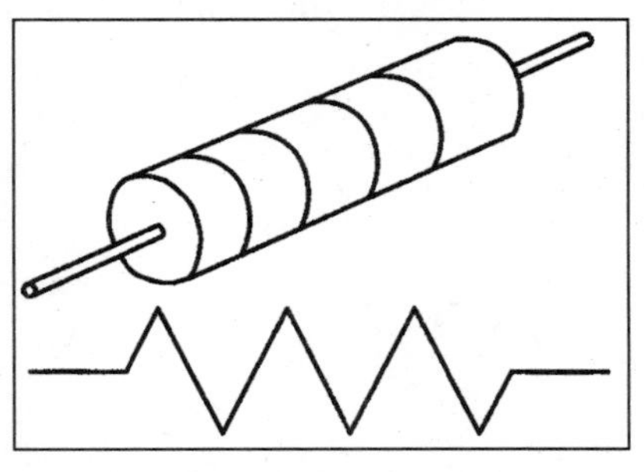

Resistors are probably the most common electronic component. They are used to control voltage and to alter voltage. In automotive electronics, there are many kinds of resistors used.

The second fact of Ohm's law is what enables the current-limiting resistors used in computer circuits to prevent overload of the power supplies and voltage regulators.

Kirchhoff's Laws For Voltage Drop

Kirchhoff's second law states that the algebraic sum of the voltage drops in a series circuit equals source voltage. This principle is self-explanatory, thus we will not address it here.

There is another electrical law built on the principles of Ohm's law. Kirchhoff's second law states that as a current passes through resistances in a series circuit, voltage will be lost. The amount of voltage lost is directly proportional to the resistance. If a circuit is powered by 5 volts, and if there are five equal resistances in the circuit, there will be 1 volt lost as the current passes through each resistance.

A slightly more scientific explanation involves the notion of voltage drop per ohm. If the source voltage in a circuit is 5 volts, and the total resistance in that circuit is 5,000 ohms, the drop in voltage as the current passes through each ohm would be 1 millivolt (.001 volt). Therefore, if the 5,000 ohms consists of a 2,000-ohm resistor, a 1,000-ohm resistor, and a second 2,000-ohm resistor, then the voltage drop across each resistor will be as follows:

- 2,000 ohms x .001 volts = 2 volts drop
- 1,000 ohms x .001 volts = 1 volt drop
- 2,000 ohms x .001 volts = 2 volts drop

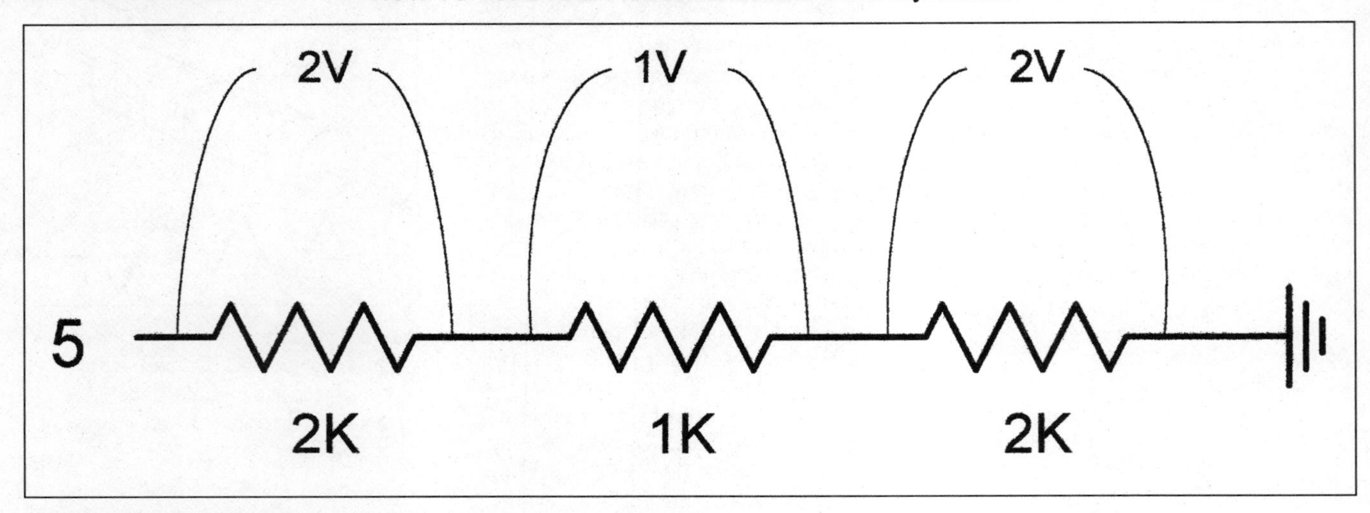

Kirchhoff's Law states that the sum of the voltage drops in a series circuit will equal the source voltage. What this means in a practical manner to automotive electronics technicians is that every time the current flows through a resistance, there is a decrease in voltage that is proportional to the ratio of that resistance to the total amount of resistance in the circuit. When a current passes through electrical connectors, there should be no measurable voltage drops. If there is, there is resistance in that connector.

The sum of the three voltage drops in the above example equals 5 volts, or source voltage. Implied in this is that if the voltage is measured after the first resistance, the result would be 3 volts. Measuring after the second resistance would yield 2 volts and after the third resistance 0 volts.

This law remains in effect regardless of the number or size of resistances in a series circuit.

1 ohm X .00000119999892 volt = .0000012 volt drop leaving 11.9999988
1 ohm X .00000119999892 volt = .0000012 volt drop leaving 11.9999976
1 ohm X .00000119999892 volt = .0000012 volt drop leaving 11.9999964
1 ohm X .00000119999892 volt = .0000012 volt drop leaving 11.9999952
1 ohm X .00000119999892 volt = .0000012 volt drop leaving 11.9999940
1 ohm X .00000119999892 volt = .0000012 volt drop leaving 11.9999938
1 ohm X .00000119999892 volt = .0000012 volt drop leaving 11.9999926
1 ohm X .00000119999892 volt = .0000012 volt drop leaving 11.9999914
1 ohm X .00000119999892 volt = .0000012 volt drop leaving 11.9999902
10,000,000 X .00000119999892 = 11.9999892 volts drop leaving 0 volts

Even when there are extremes of resistance values, the principles of Kirchhoff's Law still applies. Since the normal resistance values of the electronic sensor circuits in Chrysler fuel injection systems are quite high, it takes a great deal of unplanned resistance to have a significant affect on the operation of the system.

For Current Flow

According to Kirchhoff, and according to everyone who has studied current since Kirchhoff, the sum of the currents flowing through a parallel circuit is equal to the amount of current leaving and returning to the source. This law is very important to us in practice although of little intellectual or mathematical interest to the average technician, either professional or backyard.

In the course of my real job, I often spend several weeks every year working with journeyman technicians on Guam. One of the more popular performance modifications on the island does not relate to the engine or even the drivetrain. The most popular performance modification on Guam is to the radio. Those wimpy little factory units are rapidly removed and replaced with gargantuan amplifiers, tuners, and speakers. Often the bed of a pickup will be filled with speakers. The amplifiers can direct as much as 1,100 watts toward the speakers in momentary pulses, momentary loads. This 1,100 watts represents momentary loads on the charging system, battery, and electrical system as high as 87 amps for moments at a time. The typical automotive alternator is about 80 amps. This means that a load in excess of maximum alternator output will come and go with the baseline of the music. These intermittent loads on the electrical system can cause voltage fluctuations that can affect the stability of all the other electrical and electronic systems on the vehicle. Quite literally, the type of music you play can have an effect on the operation of your fuel injection system. Imagine this: a customer comes into a repair shop and says, "My car runs fine when I play polka music; however, when I play rap, it cuts out and misfires." Do not laugh, it happens.

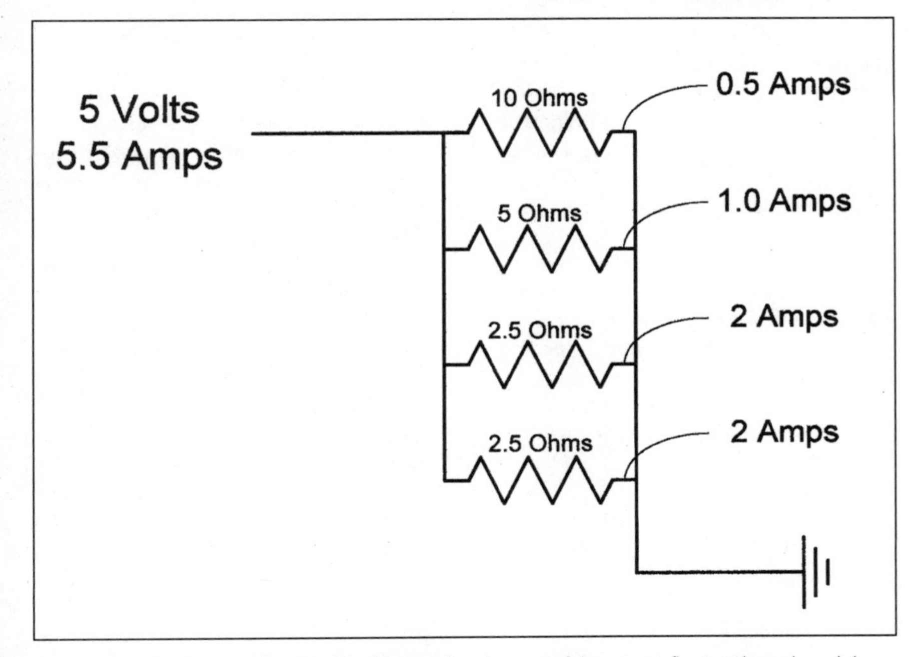

When dealing with current flow, Kirchhoff stated that the sum of the current flowing through each leg of a parallel circuit is equal to the current flowing from the source. In other words, when you connect a voltmeter to a circuit, the current path through that voltmeter adds to the electrical load on the circuit. This is why the accuracy of the readings may be questionable when a low-impedance meter is chosen.

Sensor Circuits

Sensor circuits are the electronic circuits and devices that the computer uses to gather information about what the computer is thinking, hearing, feeling, smelling, and doing. These circuits are typically very low current circuits, often carrying flows less than a milliamp (0.001 amp).

Switch to Voltage

This is one of the simplest, but rarest, of sensor circuits we will call "switch to voltage." This input circuit is used to answer a "yes or no" question about what is going on in an engine or other electronically controlled device. When the switch is open, the computer sees no volt coming from the sensor. When the monitored event occurs, the switch closes and the voltage seen by the computer rises. The computer then knows that the event has occurred and can make appropriate output decisions.

When I teach courses on electronic fuel injection, I always find it very difficult to think of examples where this type of circuit is used. On Chrysler fuel injection applications, the uses for this circuit are limited to the brake switch, the park/neutral switch, and the electric back light (rear window defogger).

Switch to Pull Low

The switch to pull low circuit, like the switch to voltage circuit, is used to detect the answer to a yes or no question. Unlike the switch to voltage circuit, however, this circuit is extremely common.

A 5-volt reference voltage inside the computer pushes a voltage through a resistor inside the computer. The voltage continues through a wire to a

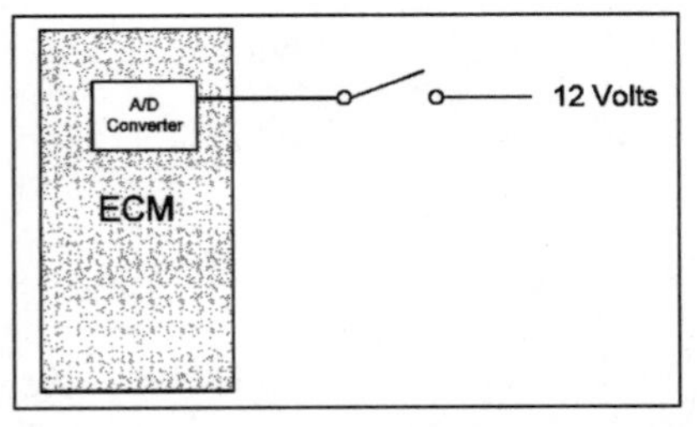

There are three major categories of components in the Chrysler electronic fuel injection systems: the control module, actuators, and sensors. Although there is much mystique about the complexity of these components, the fact is that they are actually quite simple. The simplest of these is the switch-to-voltage circuit. The circuit answers a simple yes or no question for the computer. Examples of where this circuit is used include the brake switch and park/neutral switch.

The switch-to-pull-low circuit is more common than the switch-to-voltage. Here the current is supplied from a 5-volt reference. It passes through a resistor located inside the computer. This resistor allows a voltage drop to occur. The current then proceeds to the switch installed to monitor some operation of engine management. If the switch is open, there is no current flowing and there will be no drop in voltage. A computer chip monitoring this voltage sees it as 5 volts, and the computer knows that the event being monitored has not occurred. When the switch closes, current begins to flow. Since there is only one resistance in the circuit (the internal resistor), the voltage downstream of this resistance drops to zero and the computer knows that the event being monitored has occurred.

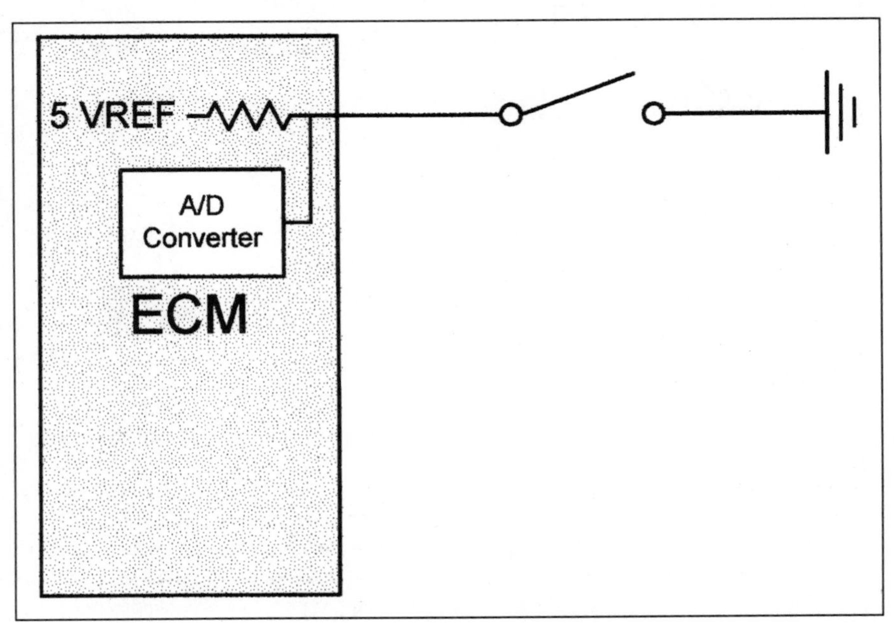

switch placed to monitor the occurrence of a specific event. The computer monitors this voltage between the resistor inside the computer and the switch before the wire leaves the computer. When the switch is open, the voltage seen by the computer is high. When the switch closes, the voltage drops to zero. When the computer sees 0 volts on the wire, it knows that the monitored event is occurring; when the computer sees 5 volts on the wire, the computer knows that the monitored event is not occurring.

Variable Resistance to Pull Low

The most common circuits that fall into this category are those that are used to measure temperature. Thermistors are used to measure temperature. There are two basic types of thermistors: the positive temperature coefficient thermistor and the negative temperature coefficient thermistor. Both types are resisters that change value when exposed to different temperature. All conductors change resistance when their temperature changes and thermistors have been designed to maximize this change. The positive temperature coefficient (PTC) thermistor behaves like most conductors; as the temperature increases, its resistance increases. The negative temperature coefficient (NTC) thermistor decreases resistance as the temperature increases. Most all thermistors used in automotive electronics are NTC.

Thermistors are used to measure coolant temperature, air temperature, intake manifold temperature, car interior temperature, and sun intensity. With all these different uses, many share the same resistance specs. Almost all Chrysler thermistors have the following specifications.

-40 degrees F= 100,700 ohms
0 degrees F = 25,000 ohms
20 degrees F = 13,500 ohms
40 degrees F = 7,500 ohms
70 degrees F = 3,400 ohms
100 degrees F= 1,600 ohms
160 degrees F= 450 ohms
212 degrees F= 185 ohms

Note that the above are examples, and although they would work for most of the applications described, be sure to check the reference manual for the vehicle being tested.

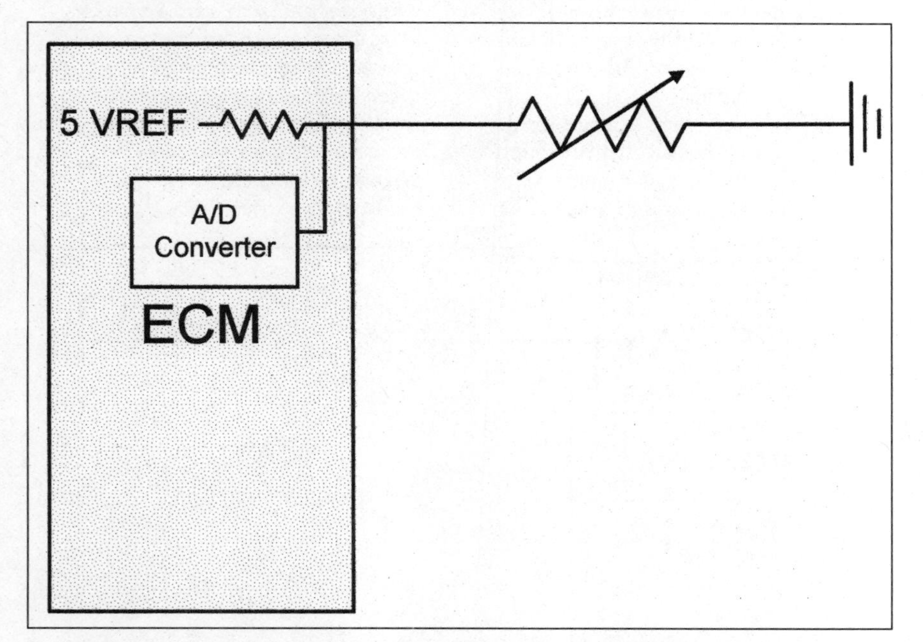

A modification of the switch-to-pull-low circuit, called variable-resistance-to-pull-low, can be used to answer a question that has no yes or no answer. For instance, if the computer asks the question, "How hot or cold is it?" the sensor must respond with a temperature. A variable resistor replaces the switch. This resistor might be designed to measure the intensity of light or the pressure in a line. However, the most common use is to measure temperature. A thermistor replaces the switch. As the temperature rises, the resistance in the thermistor decreases, and this causes the voltage sensed by the computer to decrease accordingly. The voltage is inversely proportional to the temperature.

The Thermistor Circuit

The thermistor circuit behaves a little differently than the electrical circuits with which most automotive technicians are familiar. A power supply (usually 5 volts) supplies a reference voltage to the circuit. Before leaving the computer, the 5-volt current passes through a fixed value resistor causing a voltage drop. The current then continues through the thermistor and on to ground where the voltage is zero. As the resistance of the thermistor changes, the voltage on the wire between the fixed-value resistor and the thermistor will also vary. The computer measures this voltage on the outbound side of the fixed-value resistor to determine the temperature.

When the computer sees a comparatively high voltage on the wire to the NTC thermistor, it knows that the resistance in the thermistor is high; therefore, the temperature of the thermistor is low. A low voltage on this wire means that the resistance is low and therefore the temperature must be high.

Chrysler throws a bit of a curve ball on this circuit. Applications after 1986 may have a dual-stage coolant temperature sensor circuit. When the engine is cold, about -40 degrees F, the voltage on the signal wire is close to 5 volts. As the engine warms, the voltage on the signal wire drops rapidly. When the engine temperature rises to about 120 degrees F, the voltage on the signal wire is sitting at about 1.2 volts. At about this point the voltage will suddenly rise to over 3 volts. The Chrysler fuel injection computer uses a two-stage coolant temperature circuit. This trick is performed by placing a second internal resistor in the current path to the thermistor. When the engine is cold, the computer uses the current path that contains a 10,000-ohm resistor. When the temperature of the engine reaches about 120 degrees F, the computer switches to a 909-ohm resistor. This causes the voltage on the sensor signal wire to rise to a little over 3 volts. Well that is interesting, but what is the reason for doing all this? In a standard coolant temperature sensing circuit, there is realistically 5 volts, actually a little less, in which to measure about 300 degrees of temperature change. By using a two-stage unit, the circuit has effectively 10 volts in which to measure the

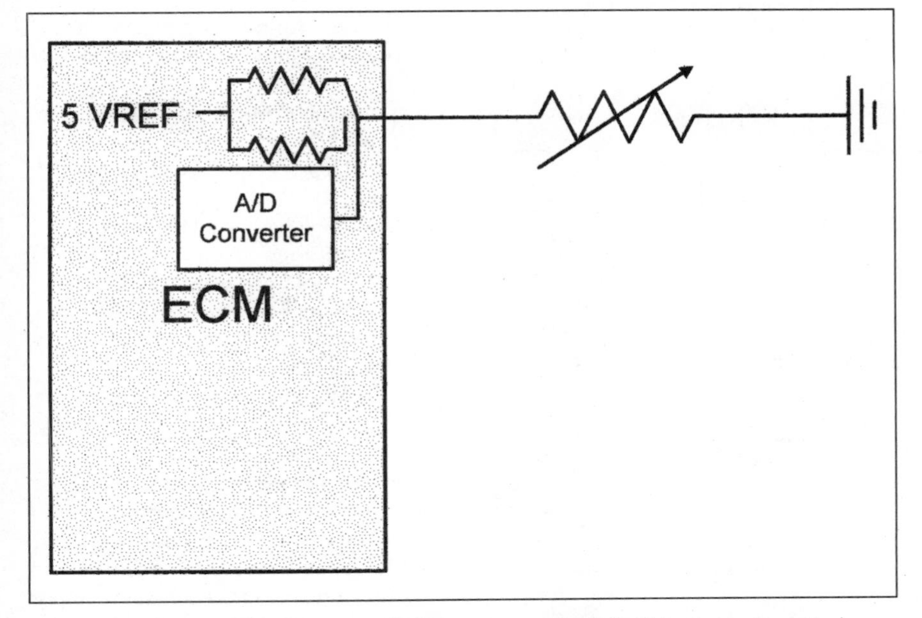

Chrysler also led the way in what is now a fairly common practice in fuel injection systems. Dual-range coolant sensor circuits permit the computer to use what is effectively 10 volts instead of 5 to measure about 300 degrees of temperature change. When the coolant temperature is below about 120 degrees, the computer uses an internal resistor with a value of about 10,000 ohms. Above a perceived 120 degrees, the computer uses a resistor with a value of 909 ohms. Therefore, if the voltage is being monitored during a warm-up, it will drop rapidly and then suddenly leap to over 3.5 volts and drop again.

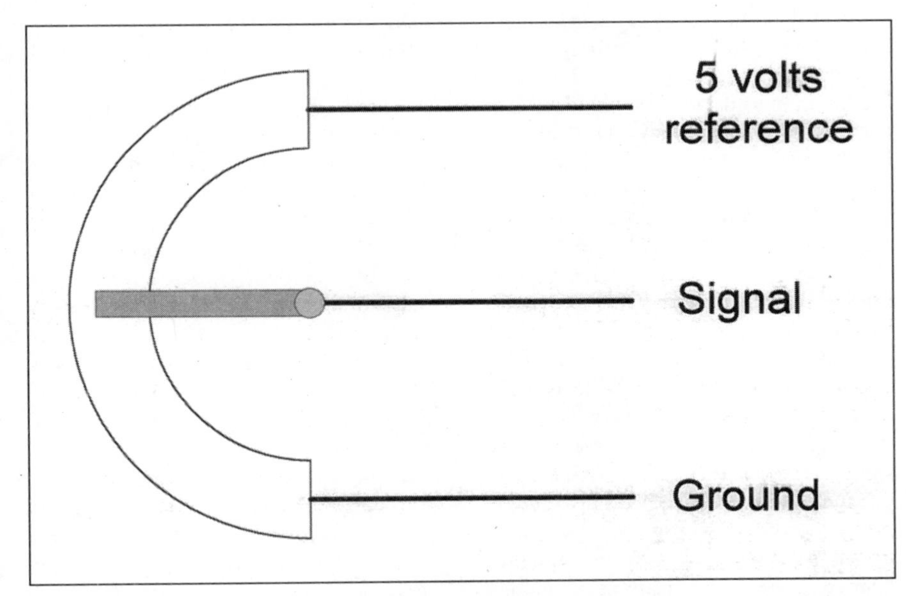

A potentiometer is used to measure varying positions in a system. The modern potentiometer is a piece of ceramic material with a carbon-based paint covering it. Five volts is connected to one end of the paint and a ground to the other. If voltage were sampled from the 5-volt end to the ground end, it would be noticed that it starts high and decreases proportionally as the voltmeter lead is moved toward ground. The potentiometer contains a wiper that does just that. As the wiper moves, it reports changes in sampled voltage to the computer.

300 degrees of change. This means that each volt of change represents about 30 degrees instead of about 60 degrees. This means more accuracy.

Circuit Fault Characteristics

There are three types of circuit faults: open, short, and ground.

An open circuit can occur at any point in a thermistor circuit (other than inside the computer itself). The voltage at all points before the open circuit will rise to reference voltage. The reason is that when a circuit has an open fault, no current is flowing; where there is no current flow there is no voltage drop. The voltage after the open will be 0 volts. When the NTC thermistor circuit goes open, the computer sees the high voltage and assumes that the temperature is low (in fact about -40 degrees F).

When the NTC thermistor circuit is shorted to voltage, the computer will believe that the temperature is low. The voltage on the wire between the computer and the thermistor would be unusually high.

Should the wire between the computer and the thermistor become grounded, the voltage seen by the computer will drop to zero. The computer will believe that the temperature of the thermistor is more than 275 degrees F.

Variable Resistance to Push Up

There are rare, and happily forgettable, applications where the PTC thermistor is used. The temperature conditions sensed by the computer would be the opposite of what is described above. Some versions of the old and infamous Chrysler "Lean Burn" system used a PTC thermistor.

The operation of this circuit is much like that of the variable resistance to pull low circuit. The result of changes in the monitored variables causes the voltage on the signal wire to rise instead of decrease.

Three-Wire Variable Voltage Potentiometers

The modern throttle position sensor is a potentiometer. A potentiometer is basically a piece of ceramic material sporting a carbon-based paint. A piece of metal rubs back and forth across the paint. There are three wires connected

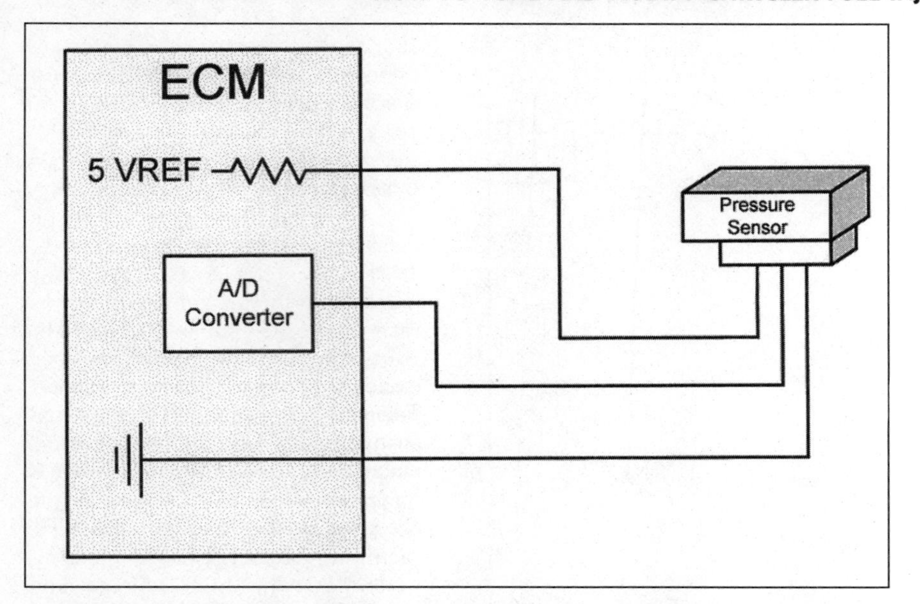

Pressure measurement is done by the computer primarily to determine how much air is flowing to the engine. The pressure sensor receives 5 volts from the computer. This is used as a reference to indicate maximum pressure. The sensor is grounded through the computer and a third wire, the center wire, carries the signal back to the computer. A high pressure is indicated by a high voltage; a low pressure is indicated by a low voltage.

to the TPS. There is a 5-volt reference, a ground (sometimes called signal return), and the signal wire. As the throttle is opened, the signal voltage will proportionately increase from a low of about 0.5 to 1.0 when the throttle is closed to a high of more than 4 volts when the throttle is wide open.

To test this sensor, turn the ignition switch to the ON position, there should be 5 volts on one wire, .5 to 1 volt on another wire, and 0 volts on the third wire. If the voltage on the signal wire is well over 1 volt with the throttle closed, inspect the ground wire for an open circuit before attempting to adjust or replace the TPS.

If the signal voltage is between .5 and 1 volt, gradually open the throttle and monitor the voltage (this is most accurate with an analog voltmeter). The voltage should gradually increase to more than 4 volts. If the increasing voltage hesitates or drops as the throttle is opened, replace the TPS.

Pressure Sensors

Today's most commonly used technology for measuring air flowing into the intake manifold is called speed/density. This method is used on almost all of the Chrysler products on the road today. Barometric pressure is compared to the pressures in the intake manifold. When these pressures are far apart, it is assumed that very little air is entering the intake manifold. When the pressures are close together, it is assumed that a large amount of air is entering the engine, causing the intake manifold pressure to rise. As early as 1968, the LVDT of the Bosch D-Jetronic system was using this method of measuring air flow.

The modern replacement for the linear variable-differential transformer (LVDT) is the Variable Voltage Manifold Absolute Pressure (MAP) Sensor. This sensor consists of an extremely thin diaphragm strung between four variable resistors that form a "Wheatstone Bridge." When the pressure on the diaphragm increases, the resistors are stretched, causing their resistance to change. The result is the output voltage of the sensor varies from reference voltage as the pressure on the diaphragm changes. For most applications using this type of sensor, full atmospheric pressure causes an output voltage of about 4.5 volts (where reference voltage is 5 volts). If the sensor is being used to measure manifold pressure, this voltage will drop to about 1.5 volts when the engine is idling. (These voltages are approximate.) This drop in voltage is proportional to the drop in manifold pressure that occurs when the engine is at an idle.

The following mathematical expressions show how this occurs:

Normal barometric pressure
= 29.92 inches of mercury
= 15 psi
= 100 kPa
Variable voltage MAP output
= 4.5 volts
Manifold pressure at an idle
= 10 inches of mercury
= 5 psi
= 35 kPa
MAP sensor output voltage
= 1.5 volts

Note: These pressures are about one-third of atmospheric pressure.

Turbocharged applications that use a variable voltage pressure sensor to measure manifold pressure simply cut the above voltage readings in half. This accommodates the accurate measurement of pressures greater than atmospheric when under boost.

Testing Pressure Sensors

The voltmeter is the best instrument for testing pressure sensors. With the harness connected, the ignition on, and the engine not running, the three wires should read 5 volts (VREF), 4.5 volts (pressure signal), and 0 volts (ground). If these readings are obtained, apply a vacuum by using a hand-held vacuum pump or by starting the engine. The signal voltage only should drop to 1.5 volts. If these readings are obtained, then the pressure sensor is good.

Open-circuit voltage on the harness can behave in a somewhat unusual manner. VREF should have 5 volts in all cases. The ground wire should have 0 volts in all cases. The signal wire will usually have 0 volts. This would be as expected since the current path is from the computer's 5-volt power supply through the sensor then back to the computer. Some applications, however, have an extremely high value resistor inside the computer between VREF and the signal wire. When there is an open circuit in either the VREF wire or the signal wire, current travel through this "pull up" resistor simulates a high pressure. This supplies a limp-in signal in the event of an open circuit. Open circuit voltage for these applications is 5.0 VREF, 4.8 volts (approximately) on the signal wire, and 0 volts on the ground wire.

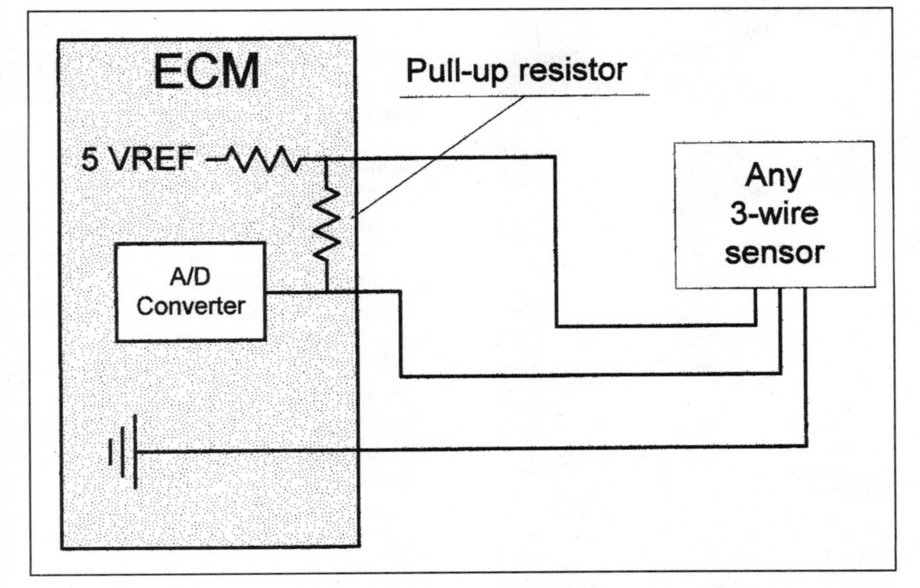

The three-wire variable voltage sensors used by Chrysler, both the potentiometers and the pressure sensor circuits, have a very high-impedance resistor located between the 5-volt reference wire and the signal wire. This resistor is not a significant part of the circuit until there is an open in either of these wires. At that point, the resistor flows a tiny current, which causes the voltage at the signal wire to rise to nearly 5 volts. This alerts the computer that there is a defect in the circuit.

There are three significant wires connected to the variable frequency sensor. One wire is a 5-volt power supply to the sensor, a second wire carries the pressure sensor signal to the computer, and the third wire is a ground.

Testing Variable Frequency Sensors

The variable frequency sensor can be tested with an oscilloscope. Square wave pulses with a 50 percent duty cycle should be observed. With the engine off there should be a little less than two pulses per millisecond. With the engine idling there should be a little more than one pulse per millisecond. As the pressure applied to the variable frequency sensor changes, it should continue to produce a steady, uninterrupted signal. If the signal breaks up or is interrupted when revving the engine or through changing pressures with a hand-held vacuum pump, the pressure sensor should be replaced.

A signal voltage that is in the midrange and changing very little with changes in pressure may indicate an open condition in the ground wire.

Common Failures

The most common type of pressure sensor is the MAP sensor. A vacuum hose connects the MAP sensor to the intake manifold. Hydrocarbons and acids can build up in this hose and damage the MAP sensor.

DC Frequency Generator Variable Frequency Pressure Sensors

Sensors that create a variable frequency are used only in the small imported Chrysler products featuring Mitsubishi drive trains. The fuel injection system of these imported Chryslers are not under the purview of this book. Chrysler's competitor, Ford Motor Company, chose to use this technology to measure manifold and barometric pressures. The Ford pressure sensor produces a frequency that varies from 92 to 150 hz. As the pressure being measured increases, the output frequency increases. At sea level, the output frequency is about 160 hz; at an idle the output frequency is about 110 to 120 hz. For short periods on deceleration the frequency may drop as low as 92.

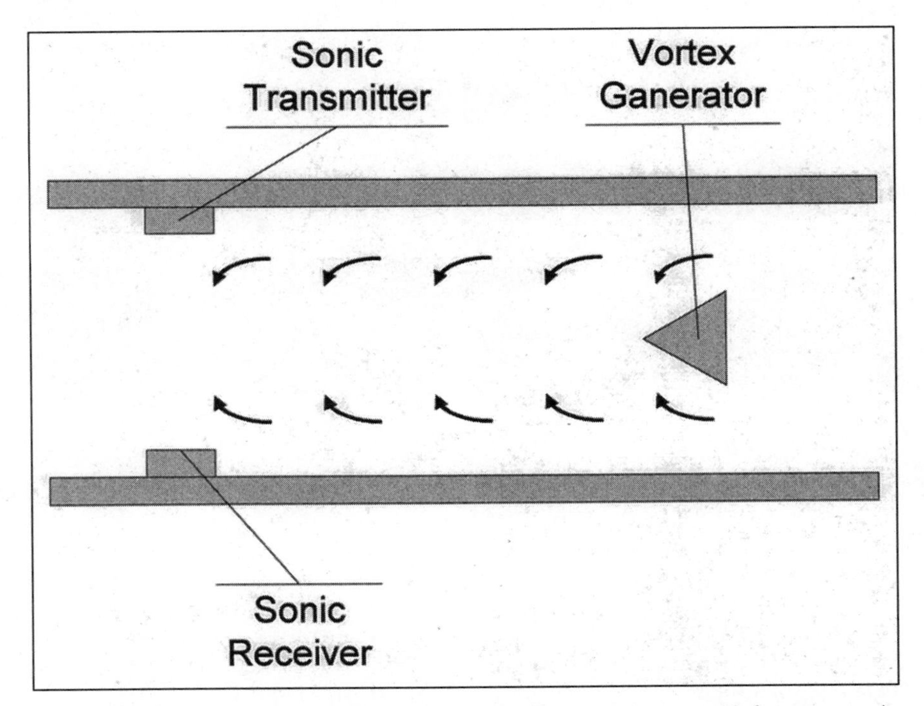

The Karman Vortex sensor used in some Chrysler import applications creates a variable frequency signal. The Karman Vortex sensor creates a pulse. The frequency of this pulse varies as the air flow measured into the engine varies. This sensor is interesting because it hears the mass of air entering the engine. This sensor has a vortex generator across the path of air flow, an ultrasonic generator, and an ultrasonic receiver. As air flows through the sensor on its way to the intake manifold, it contacts the vortex generator. Vortices are set up. The frequency of these vortices is directly proportional to air flow. The ultrasonic generator emits waves perpendicular to the path of air flow. As these waves meet with the vortices, they are accelerated, increasing their frequency in a manner similar to the Doppler effect. The ultrasonic receiver interprets these frequencies as volumetric air flow and sends a variable frequency square wave to the computer.

An oscilloscope can be a very handy tool for testing electronic components. This is particularly true when the output of the sensor is a pulsed voltage. Although "automotive oscilloscopes" can be quite expensive, I have used a scope like this quite effectively for years. It is not pretty. It is not highly portable. However, it does the job.

A voltmeter may be used to test a variable frequency sensor. However, a voltmeter's capability will be limited to determining if the sensor is producing a signal; it cannot be used to determine if that signal is accurate. At any pressure, the sensor output (the center of the three wires) will read 2.5 volts. As the pressures changes, anything other than a very slight variation from 2.5 volts would indicate a defective sensor.

A dwell meter can be used to test the variable frequency pressure sensor like the voltmeter. At any pressure the dwell meter should read about 45 degrees on the four-cylinder scale. This dwell reading should not fluctuate as the pressure is changed.

A second best to the oscilloscope for testing the variable voltage pressure sensor is a tachometer. On the four-cylinder scale, with the ignition on and the engine not running, the tachometer should read about 4,800 to 5,000 rpm. When the engine is started, the tach reading should drop to about 3,500 rpm. If a hand-held vacuum pump is connected, and the pressure to the sensor is slowly changed, the tach reading should slowly change. This change should correspond to the pressure change.

Common Failures

Variable frequency pressure sensors usually fail completely when they fail, putting out no signal at all. Occasionally, however, there will be interruptions in the signal that occur only at specific pressures.

Rotational AC Pulse Generators

Variable Reluctance Transducer

The variable reluctance transducer (VRT) is more commonly known as the pickup coil. The typical automotive technician will think of the electronic ignition distributor when this component is mentioned. In reality, there are many uses for this device. The VRT is used to measure crankshaft rotational speed on some distributorless ignition systems, wheel speed for four-wheel anti-lock braking systems, differential speed on rear-wheel anti-lock braking systems, and in some vehicle speed sensors. The VRT produces an AC sine wave with a frequency directly proportional to the speed of the rotation. The primary advantage of the VRT over other rotational speed sensors is its simplicity. Consisting of a coil of wire, a permanent magnet, and a rotating reluctor, there is very little that can go wrong with it. Its main disadvantage is its inability to accurately detect low-speed rotation. At low rotational speeds the VRT is unable to produce a signal.

How the VRT Works

A coil of wire sits in a magnetic field created by a permanent magnet. A metal wheel with protruding reluctor teeth rotates through the magnetic field. As it rotates, and one of the teeth approaches the magnetic field, the magnetic field is bent toward the approaching tooth. As it is bent it passes across the coil of wire inducing a voltage. Continuing to rotate the reluctor tooth drags the magnet field across the coil of wire, eventually bending it in the opposite direction. The result is an AC signal.

Testing the VRT

Most books on troubleshooting electronic ignition systems will describe

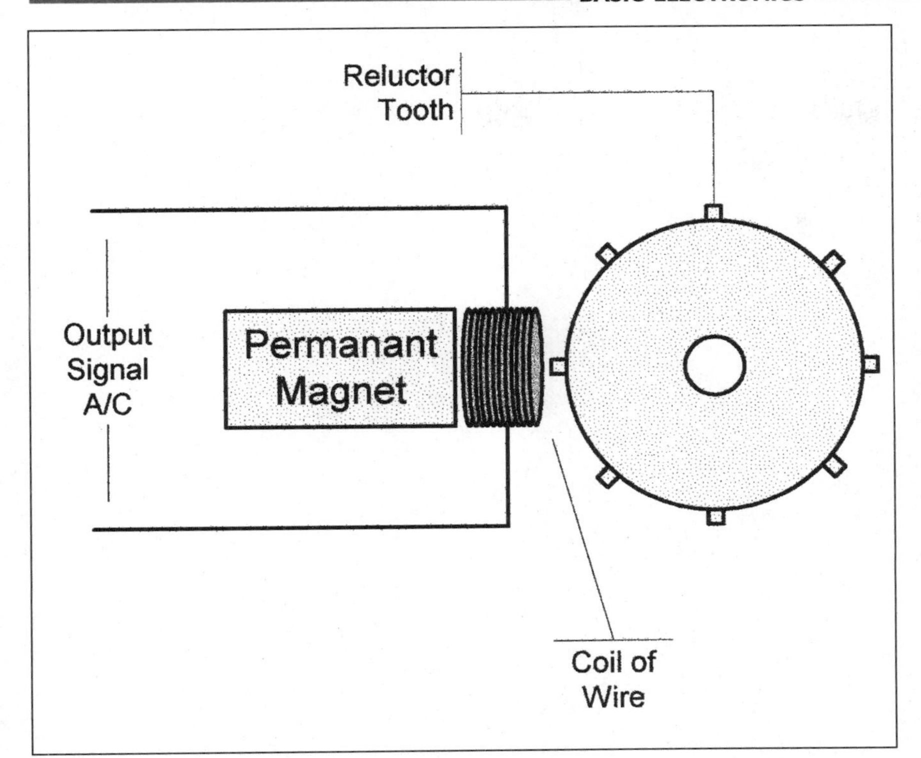

Early Chrysler electronic ignition systems used a reluctance pickup to determine the speed of the engine. Although this sensor is no longer used on domestic Chryslers, it is used extensively on the imports. The sensor uses a permanent magnet to cast a magnetic field across a coil of wire. When a rotating reluctor tooth approaches the magnetic field, it distorts the field across the coil of wire. This movement induces a voltage. As the tooth continues to rotate, it eventually aligns with the center of the magnetic field. At that moment, the field is no longer moving and therefore no voltage is induced. Soon the tooth rotates off in the opposite direction, dragging the magnetic field in the opposite direction, and inducing a voltage in the opposite direction. End result: alternating current or AC voltage.

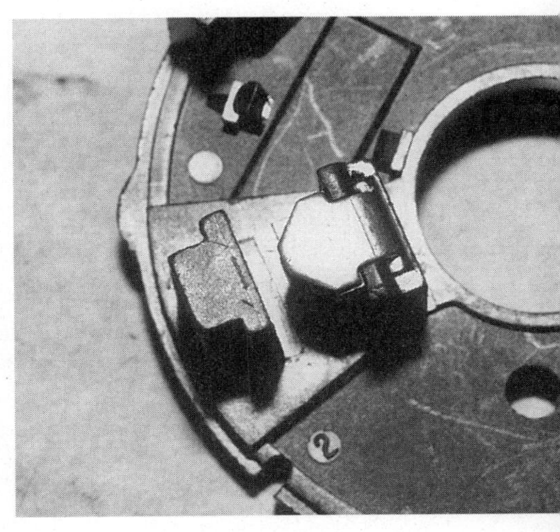

There are two types of sensors used by Chrysler to produce a pulsed signal proportional to crank or cam speed. Both produce a DC pulse. The first is the Hall Effects. In this picture, the base plate of the sensor supports a protruding plastic pillar. Inside that pillar is the Hall Effects sensor chip. Opposite the pillar is a permanent magnet.

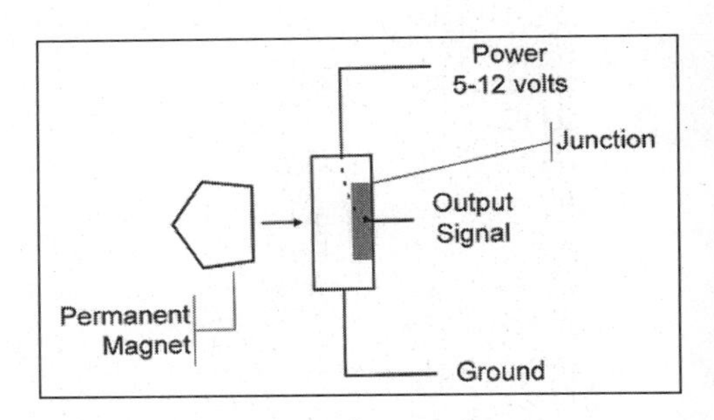

In the Hall Effects sensor there is a chip, a piece of semiconductor material. Voltage and ground are connected to that chip so that current can pass through. Embedded in this chip is a second piece of semiconductor material. The point at which these two pieces of material are joined together is called a junction. The resistance through the main portion of the chip is high; the resistance across the junction is very high. The current will take the path of least resistance. Introducing a magnetic field perpendicular to the direction of current flow is like throwing up a road block. The presence of the magnetic field makes the effective resistance through the main portion of the chip very, very high. The current will now take the path of least resistance, across the junction, and the output signal wire voltage will go high.

using an ohmmeter to test a VRT. Although this is not a totally invalid test, it only tests the coil of wire. Reluctor air gap, condition of the permanent magnet, and adequate rotational speed are not tested with this method.

Disconnect the pickup from the ignition module or vehicle wiring harness leads. Connect the ohmmeter to the pickup coil leads and measure the resistance. Typical ohmmeter readings for a good reluctance pickup coil would be between 500 and 1,500 ohms.

A much better way to test a VRT is with an oscilloscope. Connect the scope to the pickup coil leads and rotate the reluctor (crank the engine, rotate the wheel). A series of ripples should appear on the scope. If the line remains flat, there is an open condition in the coil of wire, the permanent magnet has been damaged, or the air gap between the pickup and the reluctor teeth is too large.

The AC voltmeter is a practical and effective alternative to the oscilloscope. Connect the AC voltmeter in the same manner as described for connecting the oscilloscope. Rotating the reluctor at minimum speed (cranking the engine, rotating a wheel at 1 rpm) would yield between .5 and 1.5 volts. If the AC voltmeter does not produce a voltage, there is an open condition in the coil of wire, the permanent magnet has been damaged, or the air gap between the pickup and the reluctor teeth is too large.

Notes on Testing

Since the VRT is primarily a coil of wire and a permanent magnet, it is prone to intermittent failures with changes in temperature and vibration. If the failure being diagnosed is intermittent, the sensor should be heated and tapped while testing.

Rotational DC Pulse Generators

Hall Effects Sensor

The Hall Effects sensor is often used as an alternative to the VRT. Many ignition systems, both distributorless and distributor type, use a Hall Effects device. Its primary advantage over the VRT is its ability to detect position and rotational speed from 0 rpm to tens of thousands of rpm. Its primary disadvantage is that it is not as rugged as the VRT and is more sensitive to errant magnetic fields. An intense magnetic field can shut down the proper operation of a Hall Effects sensor.

How Does It Work?

A Hall Effects pickup is a semiconductor carrying a current flow. When a magnetic field falls perpendicular to the direction of that current flow, part of that current is redirected perpendicular to the main current path. The semiconductor is placed near a permanent magnet. A set of metal blades, or armature, attached to a rotating shaft or other device passes between the Hall Effects semiconductor and the permanent magnet. As the armature rotates, the magnet field is alternately applied to the Hall Effects and interrupted. The result is a pulsing current perpendicular to the main current path. This frequency is directly proportional to the speed of armature rotation. Since the output is only dependent on the presence of the magnetic field, the Hall Effects unit is capable of detecting armature position even when there is no rotational speed.

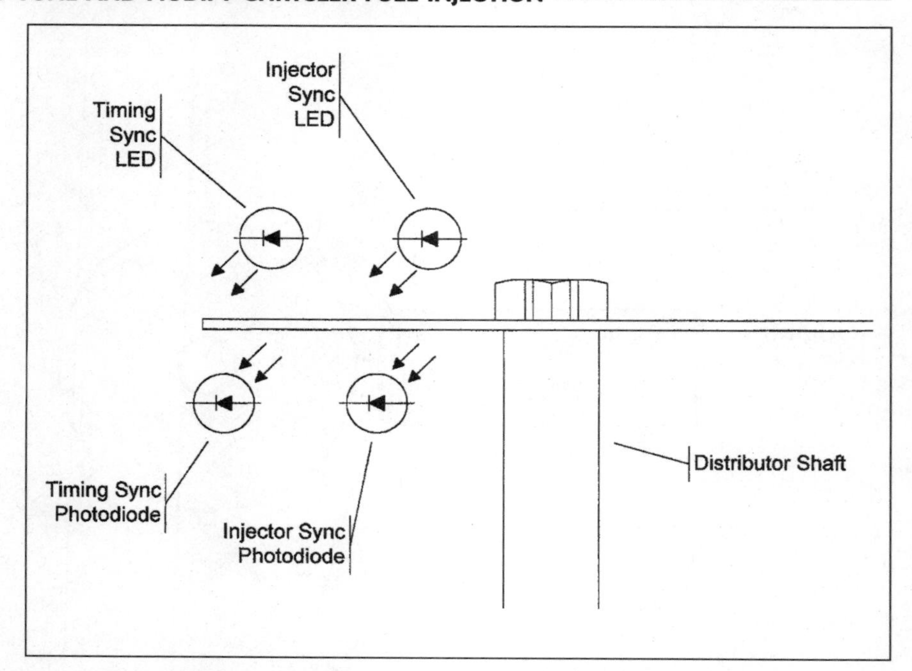

The Chrysler 3.0-liter engine, which is manufactured by Mitsubishi, uses an optical sensor in the distributor. This sensor actually consists of two elements. The outer element reports the exact position of the camshaft to assist the computer in accurately controlling ignition timing. The inner element triggers the fuel injectors. Each element consists of an infrared light emitting diode (LED) and an infrared photo diode.

Testing the Hall Effects Sensor

A Hall Effects sensor is best tested with an oscilloscope. Connect the oscilloscope to the Hall Effects signal lead. Rotate the armature. Depending on the number of blades and the rotational speed of the armature, the scope pattern could appear either as a square wave or a flat line that rises and falls with rotation.

A voltmeter can also be used for testing. Connect a voltmeter to the Hall Effects output lead. The voltmeter should display either a digital high (4 volts or more) or a digital low (around 0 volts). Slowly rotate the armature while observing the voltmeter. If the voltmeter had read low it should now read high; if the voltmeter had read high it should now read low. If the voltage fluctuates in this manner as the armature is rotated, then the Hall Effects is good.

Another method of testing involves the use of a dwell meter. Since the signal generated by the Hall Effects is a square wave, the dwell meter becomes a natural for testing. Connect the dwell meter between the Hall Effects output and ground. Rotate the armature as fast as possible (for instance, crank the engine); the dwell

If a set of metal blades is introduced between the magnet and the Hall Effects unit, the magnetic field will be deflected, and the current will once again flow through the main portion of the chip. Attach the blades to a rotating shaft and the sensor will produce a signal, the frequency of which will be directly proportional to the rotational speed of the shaft.

meter should read something other than zero and full scale. If it does, the Hall Effects is good.

An alternate method of testing requires a tachometer. As with the dwell meter, the tachometer is also a good tool for detecting a square wave. Connect the tachometer between the Hall Effects output and ground. With the armature rotating as described in the paragraph on the dwell meter, the tachometer should read something other than zero if the Hall Effects is good.

Optical Sensor

The 3.0-liter Mitsubishi engine uses an optical rotational speed and position sensor. The signal produced by the optical sensor is identical to the one produced by the Hall Effects sensor. The signal, however, is produced by an armature interrupting light. A light emitting diode (usually infra-red, invisible light) sits opposite an optical receiving device such as a photodiode or phototransistor. An armature is rotated between the LED and the receiver. Unlike the Hall Effects sensor, the armature in the optical sensor can be metal, plastic, or any translucent material. As the armature rotates, light alternately falls on and is kept from falling on the receiver. As this occurs, the current flowing through the receiver is turned on and off, creating a square wave with a frequency directly proportional to armature rotation. In some cases, such as many GM vehicle speed sensors, the light is reflected off of rotating blades rather than interrupted.

The main advantage that the optical rotational sensor has over the VRT and the Hall Effects sensor is its ability to produce extremely high frequencies. The 3.0-liter Chrysler distributor produces a frequency of 17,250 hz (.54 megahertz!) at just 3,000 rpm. The primary disadvantage is its sensitivity to dirt, oil, and grease, creating erroneous signals.

Testing the Optical Sensor

An optical sensor is best tested with an oscilloscope. Connect the oscilloscope to the optical sensor signal lead. Rotate the armature. Depending on the number of blades and the rotational speed of the armature, the scope pattern could appear either as a square wave or a flat line that rises and falls with rotation.

A voltmeter can also be used for testing. Connect a voltmeter to the optical sensor output lead. The voltmeter should display either a digital high (4 volts or more) or a digital low (around 0 volts). Slowly rotate the armature while observing the voltmeter. If the voltmeter had read low, it should now read high; if the voltmeter had read high, it should now read low. If the voltage fluctuates in this manner as the armature is rotated, then the optical sensor is good.

The windows used to produce the high-frequency timing control pulse of the optical distributor are seen here. The optical sensor is the black box. All of these components are located under a dust-resistant metal cover.

Another method of testing involves the use of a dwell meter. Since the signal generated by the optical sensor is a square wave, the dwell meter becomes a natural for testing. Connect the dwell meter between the optical sensor output and ground. Rotate the armature as fast as possible (for instance, crank the engine); the dwell meter should read something besides zero and full scale. If it does, the optical sensor is good.

An alternate method of testing requires a tachometer. As with the dwell meter, the tachometer is also a good tool for detecting a square wave. Connect the tachometer between the optical sensor output and ground. With the armature rotating as described in the paragraph on the dwell meter, the tachometer should read something other than zero if the optical sensor is good.

15° gap updates computer on position of TDC number 1.
345 slots
6 slots

The disk that separates the LEDs from the photo diodes contains 351 slots. There are 345 slots located around the outer edge of the disk. These are the slots that inform the computer about the position of the camshaft to control ignition timing. Note that there is a gap of 15 degrees where there are no slots. This gap is used to update the computer on the position of top dead center of cylinder number one during its compression stoke. The inner slots are the six that are use to trigger the injectors.

Reed Switch

The reed switch consists of a membrane that is moved to close a switch and complete a circuit in the presence of a magnetic field. A permanent magnet is attached to a rotating

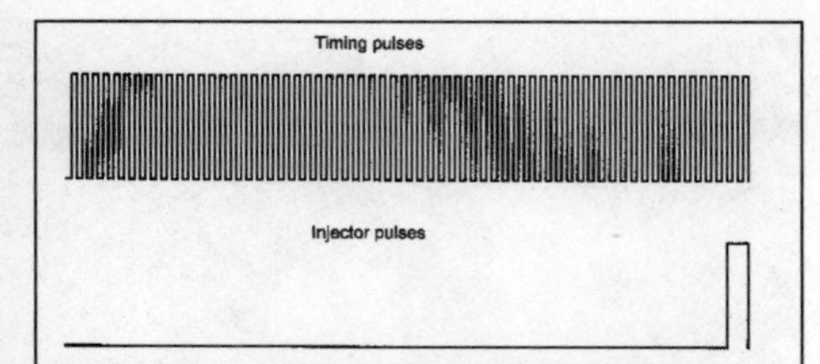

On an oscilloscope, the technician can see the closely packed high-frequency pulses of the optical sensor responsible for reporting camshaft position to the computer. This signal is always accurate to within 1 degree of camshaft rotation. The computer uses this information to control the precise timing of the ignition system. The lower pulse on this drawing represents the signal used to trigger the injectors. As you can see, there are many ignition pulses for every injector pulse.

The oxygen sensor is used by the fuel injection computer to correct errors made in the selection of injector on-time. The O_2 sensor produces a voltage that ranges from 100 to 900 millivolts during normal engine operation. When the oxygen sensor detects a large amount of oxygen in the exhaust gases, its output voltage decreases. This tells the computer that the engine is running lean. The computer responds by enriching the mixture. Conversely, when the oxygen sensor detects that there is little or no oxygen in the exhaust gases, its output voltage goes up. This tells the computer that the engine is running "not lean." The computer responds by leaning out the mixture.

shaft driven by a cable. As the permanent magnet approaches the membrane, the switch is closed. As the permanent magnet moves away from the membrane, the switch closes. The shaft carrying the permanent magnet is driven by the speedometer cable. The end result of this process produces a pulse, the frequency of which is directly proportional to the rotational speed of the speedometer cable, and therefore, of the vehicle speed.

Voltage Generators

Oxygen Sensor

The oxygen sensor, sometimes called a Lambda or EGO sensor, consists of a zirconium oxide ceramic which becomes conductive for oxygen ions at 300 degrees Celsius. One part of the ceramic is located in the exhaust stream while the other is exposed to ambient oxygen. The surface of the ceramic is covered with a thin, gas-permeable layer of platinum. When the percentage of oxygen contacting the two ends of the ceramic is equal, there is a balance of oxygen ions and no voltage is generated. When there is an imbalance, oxygen ions are attracted from the surplus side and a voltage is generated. The oxygen sensor output voltage varies from about 100 to 900 millivolts during normal engine operation.

When the engine is running rich, there is a low oxygen content in the exhaust and a high voltage is produced. When the engine is running lean, there is a high percentage of oxygen in the exhaust and a low voltage is

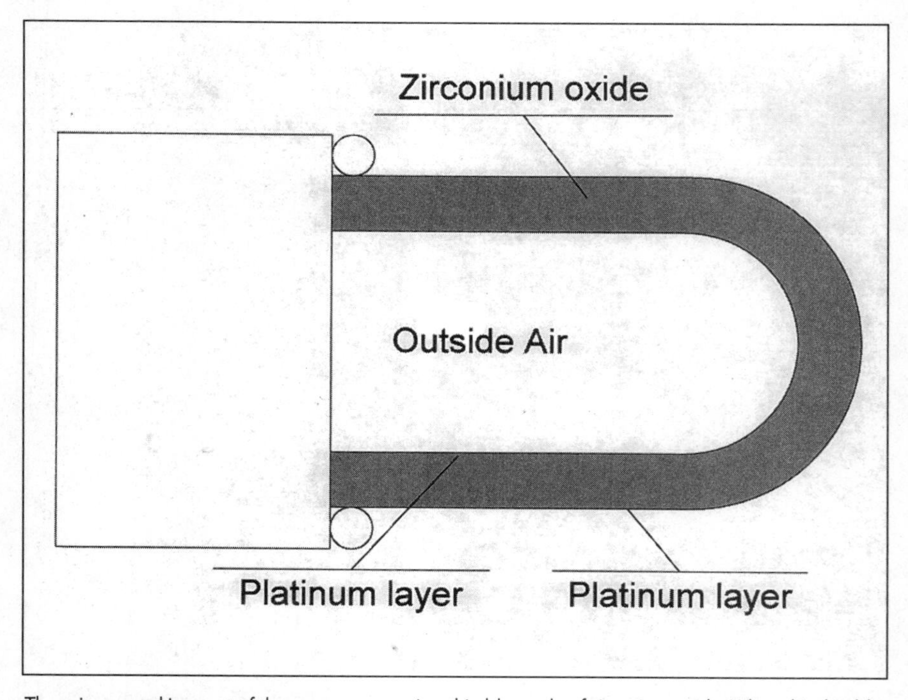

The primary working part of the oxygen sensor is a thimble made of zirconium oxide. When this thimble is heated to about 600 degrees F, it becomes conductive for oxygen ions. There is a passageway through the center of the sensor that allows fresh air containing the normal atmospheric quantity of 21 percent oxygen to come in contact with the interior of the thimble. When the oxygen content of the exhaust gases is also equal to 21 percent, the voltage output of this sensor is 0 volts. When the oxygen content of the exhaust begins to drop, the voltage output of the sensor begins to climb.

generated. At the point of perfect combustion, known as the stoichiometric point, the oxygen sensor produces 450 millivolts. This is known as the crossover point. Many applications have a very high impedance circuit which will replace a missing oxygen sensor signal with a default voltage of 450 millivolts. This voltage can be detected with a high-impedance voltmeter whenever the oxygen sensor is cold or disconnected.

The oxygen sensor will very seldom develop a defect. It is, however, very susceptible to contamination. Common contaminates include tetraethyl lead, RTV silicone, and soot.

Testing the Oxygen Sensor

Connect a voltmeter to the oxygen sensor output lead. Warm the engine to operating temperature, and raise the engine speed to 2,000 rpm for two minutes. Allow the engine to return to an idle and watch the voltmeter. If the output voltage is continuously fluctuating between 100 and 900 millivolts, the oxygen sensor and its computer circuitry are good. If the voltage fluctuations are sluggish, is stuck high or stuck low, the sensor may be contaminated and further testing may be required.

Further Testing of the Oxygen Sensor

Disconnect the oxygen sensor and connect a high impedance voltmeter to it. Connect the lead going to the computer to a good ground. This lies to the computer, telling it that the engine is running lean. The computer should respond by enriching the mixture. The oxygen sensor voltage should increase. If it does not, enrich the mixture with propane. If the oxygen sensor voltage increases, then the sensor is good and the computer, actuator circuit, or input circuit may be defective. Now touch the positive battery terminal with a finger and touch the computer's oxygen sensor input wire with another finger. This delivers a small voltage to the computer, telling it that the engine is running rich. The oxygen sensor voltage should drop. If it does not, lean out the mixture by creating a vacuum leak. If the oxygen sensor voltage decreases, then the sensor is good and the computer, actuator circuit, or input circuit may be defective.

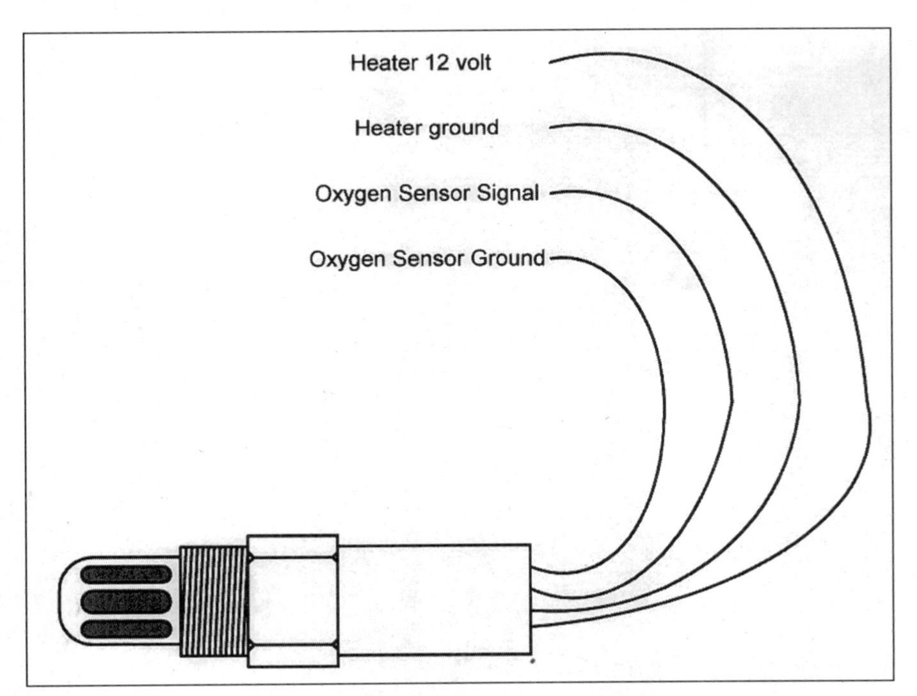

Chrysler oxygen sensors can have up to four wires. If there is a single wire, this wire is the output wire to the computer. If the sensor has three wires, one is the signal wire to the computer and the other two are power and ground for the oxygen sensor heater. The purpose of the heater is to accelerate the sensor warm-up cycle and to keep the sensor warm when the engine idles. The fourth wire, when present, is a ground for the sensor signal.

Knock Sensor

The knock sensor used by Chrysler is a piezo electric device. The term piezo electric describes the creation of an electric current by applying a pressure to a crystal. The knock sensor is used only on the turbocharged Chrysler applications. A knock, or detonation, occurs when the fuel in the combustion chamber is ignited prematurely. This "pre-ignition" can be caused by a lean running condition, overly advanced ignition timing, or the high combustion chamber temperatures that result from turbo boost.

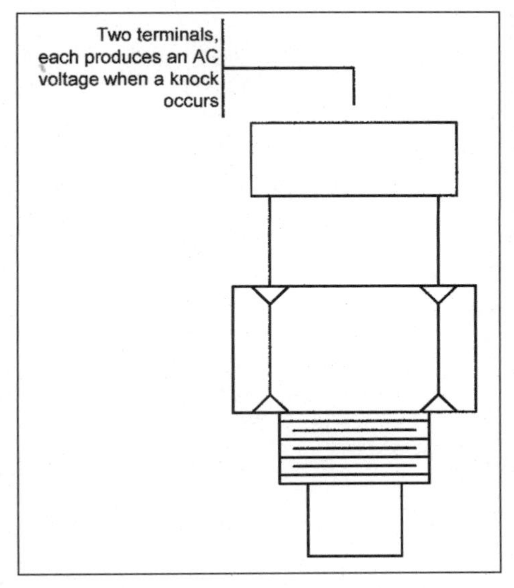

The knock sensor is a piezo-electric device that creates a small AC voltage when a detonation occurs. The computer responds by retarding the ignition timing. Only a few Chrysler applications use this sensor, primarily the turbocharged applications.

When detonation occurs, a pressure wave is sent through the metal on the engine. This pressure wave arrives at the sensor, squeezing it. This squeeze applies pressure to the crystal, causing it to produce a voltage. When the computer sees a voltage from the knock sensor, it responds by retarding the timing in an effort to decrease combustion temperatures to limit or stop the knocking.

Testing the Knock Sensor

Connect an AC voltmeter, set on the lowest possible scale, between the knock sensor output connector and a good ground. With the ignition switch in the off position, rap on the intake manifold near the knock sensor

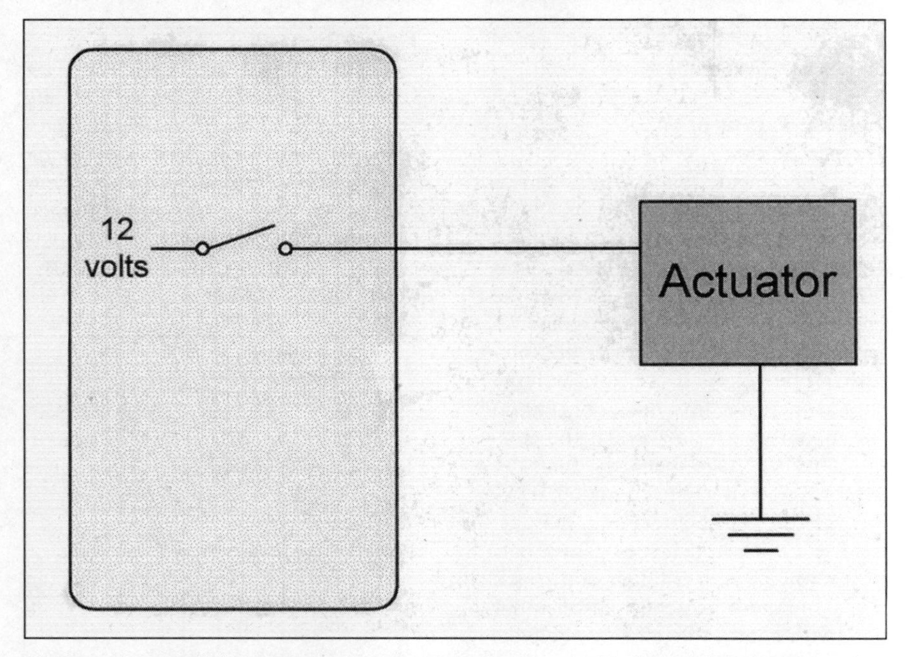

The normally grounded actuator is connected to chassis or engine block ground all the time, and the computer switches the voltage to it on and off.

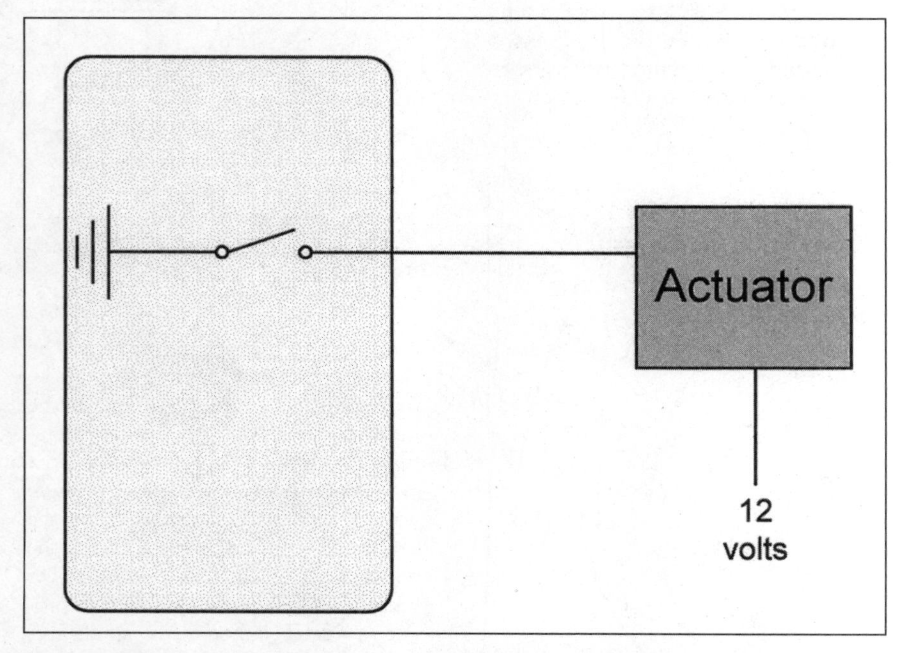

The normally powered actuator circuit is more common. In this circuit, the actuator is connected to power all the time, and the computer switches the ground on and off.

with a small hammer. A small, but noticeable, change in voltage should be observed.

Actuator Circuits

There are two types of actuator circuits: normally grounded and normally powered.

Normally Grounded

In a normally grounded actuator circuit, the controlled device is connected to ground all the time. When the computer desires to activate the device, it switches on a voltage, which pushes a current through the device, which in turn energizes the actuator. Although this is a common way for many electrical devices to be operated, it is a rare way of controlling the computer outputs.

Normally Powered

A normally powered actuator circuit applies power to the actuator device all the time, usually through a fuse. The computer provides the ground for the circuit in order to activate the device. This is the most common method used by the computer to control actuators.

The Actuators

Solenoid-Operated Valves

One of the most common devices used to control the electronic functions of a fuel-injected engine is the solenoid-operated fluid control valve. These devices are used to control the flow of fuel, vacuum, and air.

A solenoid consists of a coil of wire which can carry a current to create a magnetic field. A soft iron core sits in the center of this coil. The iron core is spring loaded in a position out of the center of the magnetic field produced by the coil. When a current is passed through the coil, the iron core seeks the center of the magnetic field. This causes the core to move against spring tension. If the iron core incorporates a valve in its design, the valve will be either opened or closed by this action. Chrysler uses solenoid-operated valves to control fuel flow (the injectors), vacuum to the manifold pressure sensor on some models, vacuum to the evaporative canister, and many other functions on the engine.

DC Motors

DC motors are the motors we typically think of when we talk about a motor on an automobile. These motors generally have a permanent magnet supplying field magnetism and an electromagnet acting as the armature. When a current is passed through the electromagnetic armature, the motor begins to spin against the electromagnetic energy of the permanent magnet energy of the motor field. Reversing the direction of armature rotation is a simple matter of reversing the current flow through the armature.

Stepper Motors

The stepper motor is a considerably different device than the standard DC motor. The armature of the stepper motor is a permanent magnet. The field

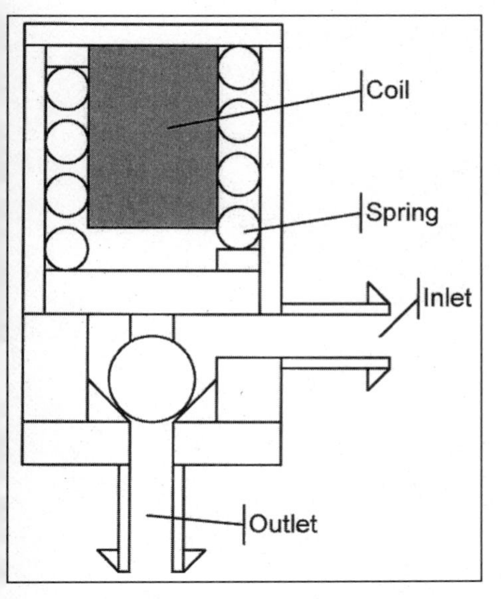

When the computer needs to control the flow of a fluid, such as air, fuel, or vacuum, it will do so with a solenoid-operated valve. When the windings are energized by the computer, the valve is pulled back against spring tension, and the fluid can flow. When the windings are de-energized, the spring pushes the valve closed. Injectors are solenoid-operated valves.

of a stepper motor is an electromagnet; there are usually two sets of field windings. When a set of field windings is energized, the stepper motor armature will rotate 1/4 to 1/2 turn. At this point, the magnetic poles of the armature and the field neutralize and the motor stops rotating. The second set of field windings must then be energized to continue the rotation of the armature. Directional control of the armature is obtained be reversing the current flow through the field windings. The advantage of this motor design is that there are no brushes, and the amount of armature rotation is precisely predictable.

The automatic idle speed control motor is a stepper motor. When a set of field windings is energized, the triangular-shaped plunger moves either in or out one step. When the current is reversed, the direction is reversed.

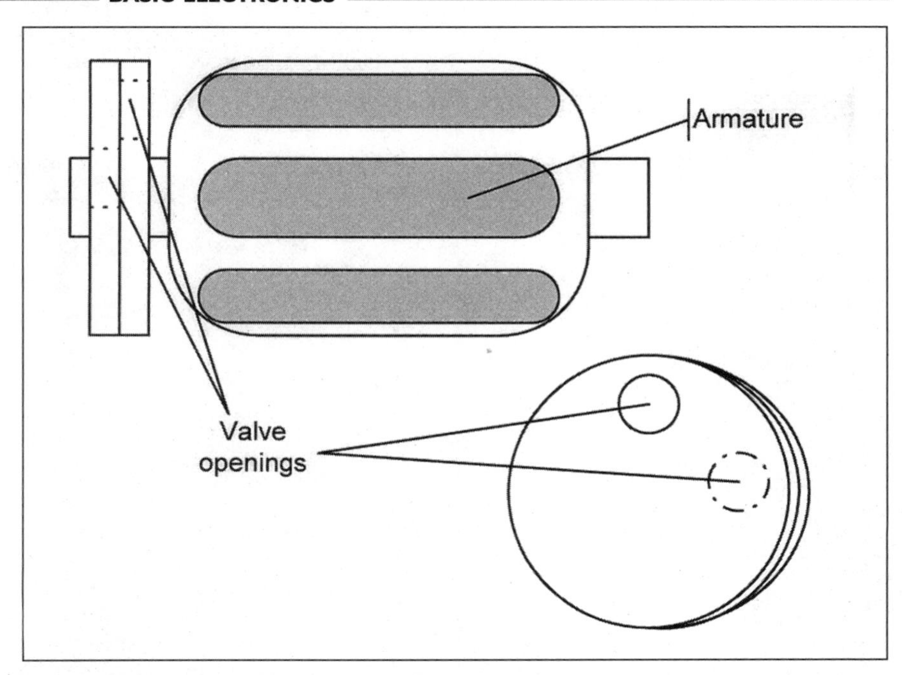

Motors are used to do a variety of jobs in the Chrysler electronic fuel injection system. In early applications, a motor is used to control or operate a valve that controls air flow. The fuel pump is also driven by an electric motor.

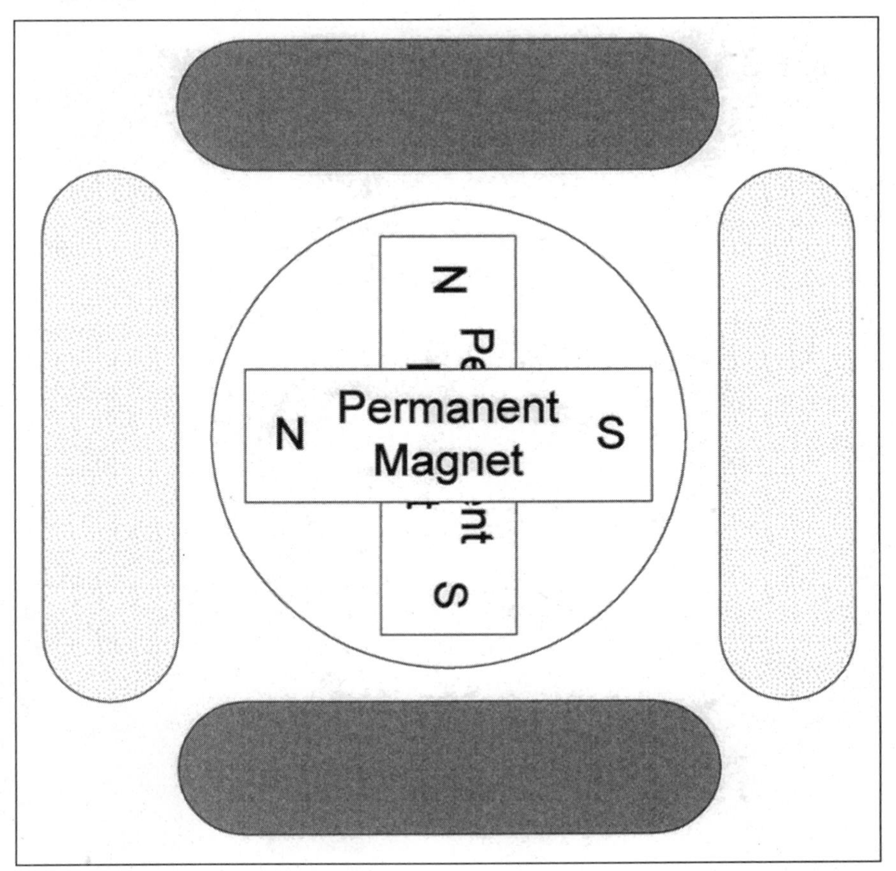

The third major category of the actuator is the stepper motor. This is basically a DC motor with one major difference. In a standard direct current motor, the armature is an electromagnet and the field is a permanent magnet. With the stepper motor, the armature is a permanent magnet and the field is an electromagnet. This allows the armature to turn a predictable amount every time a set of fields is energized.

CONTENTS

EMISSION CONTROL DEVICES

The Combustion Process

All gasoline engines, regardless of manufacturer, share some common characteristics as far as how combustion happens and the elements necessary to make quality combustion happen. The process begins when an intake valve opens and a volume of air is drawn into the combustion chamber. The piston begins moving down the cylinder just after the intake valve opens. The movement of the piston creates a low pressure area in the cylinder. Atmospheric pressure pushes air into the cylinder. With the air comes the fuel that was delivered into the induction air stream by either a carburetor or by a fuel injector. Soon the piston swings past bottom dead center and begins its journey up the cylinder. As the piston moves up the cylinder, it compresses the air/fuel charge.

The air/fuel charge in a gasoline, spark-ignition engine is a mixture of several chemicals and compounds. First there is the air. The air on planet earth consists of approximately 78 percent nitrogen and about 21 percent oxygen. The unaccounted-for 1 percent is a mixture of carbon dioxide, xenon, argon, and other inert gases. The nitrogen should exit the combustion chamber in the same condition chemically as which it entered. The oxygen should be completely consumed by the combustion process. The fuel consists of a relatively complex hydrocarbon compound. When the compound is in a ratio of approximately 14.7:1 and pressurized, or compressed, the mixture becomes ignitable.

The ignition system will fire the spark plug when the rotation of the crankshaft is a few degrees before the piston reaches top dead center. The fuel ignites. As the piston crests over top dead center, the hot gases that result from combustion begin to force the piston down the cylinder. If everything happens just right, the gases leaving the combustion chamber on the exhaust stroke will include only carbon dioxide (amounting to about 15 percent of the exhaust gases), nitrogen (amounting to about 78 percent of the exhaust gases), and water (amounting to about 5 percent of the exhaust gases). The remaining 1 percent exhaust gases consist of the inert gases mentioned earlier.

All that has been discussed so far describes what happens to exhaust gases under ideal conditions in an ideal engine with an ideal mixture and an ideal ignition system. On this planet, we are not privileged to live in an ideal world. There are many things that can go wrong with the combustion process. In the next few pages, we will look at the undesirable exhaust gases, what causes them, what the manufacturers have done to limit them, and what engine operational defects can cause them to be excessively high.

The Emission Problem

Carbon Monoxide

Carbon monoxide results from having an imbalance in the air/fuel charge. When the air and fuel are combined in the right ratios, the oxygen of the air combines with the carbon of the fuel to form carbon dioxide. If there is

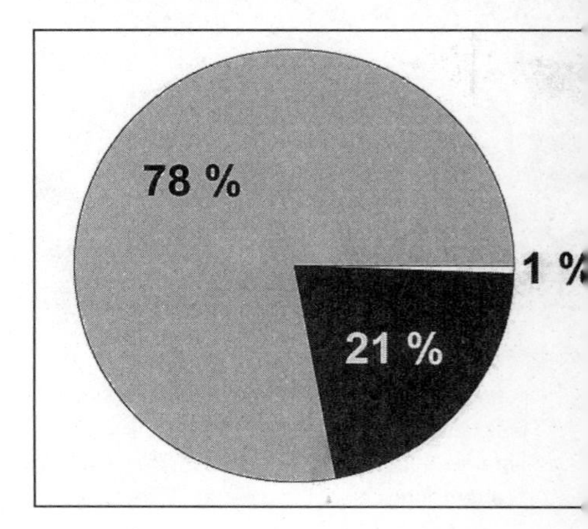

The atmosphere of planet earth consists of 78 percent nitrogen, 21 percent oxygen, and 1 percent miscellaneous gases. During combustion, the oxygen should combine with the fuel to yield carbon dioxide and water. If the temperature of the combustion process exceeds 2,500 degrees F, the nitrogen will combine with the oxygen and produce NOx.

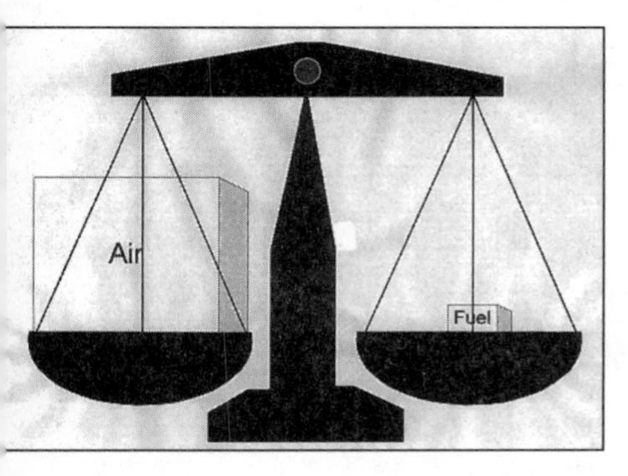

Gasoline is a very temperamental fuel. Proper combustion can only occur when the ratio of air to fuel is very close to 14.7 weight units of air to one weight unit of fuel.

too much fuel or not enough air, then the oxygen gets consumed very early in the combustion process. During the latter part of the combustion process, there is not enough oxygen left in the combustion chamber to complete the process that creates the CO_2 and the process stops short at CO.

Carbon monoxide is an odorless, tasteless, colorless, and potentially lethal gas. Inhaling as little as a 1 percent concentration of CO for as little as 30 minutes can be deadly. The gas displaces oxygen in the blood stream. In fact, the red blood cells, whose job it is to distribute oxygen to the other cells of the body, have a 240 times greater affinity for carbon monoxide than for oxygen. This means that the red blood cells consider carbon monoxide to be Dom Perignon while it considers oxygen to be well water. The simple fact is that given the choice, the blood cells will take the CO every time.

Carbon monoxide is a product of the combustion process simply because there was not enough oxygen available to complete the combustion process. The engine was running rich. Earliest efforts to reduce CO emissions in the automobile engine amounted to leaning out the mixture of the engine. This was very successful at reducing CO, however it meant that engine was always running on a borderline misfire. Anything additional that kept the burn in the cylinder from being perfect would cause the cylinder to misfire and cause a massive increase in hydrocarbon emissions. A lean mixture also causes high combustion temperatures. When the mixture got lean enough, the combustion temperature would exceed 2,500 degrees F. At that point oxygen begins to combine with nitrogen and oxides of nitrogen are produced.

The next thing that came along as an effort to reduce CO emissions was the air pump. Commonly referred to in the early days as a "smog pump," the pump would inject air into the exhaust manifold. This air would combine with the CO, converting it to CO_2. Although this device had a reputation for robbing the engine of power, the reality of the smog pump is that it required about 0.7 horsepower to drive it. Only the smallest engines, and only at idle on these small engines, were power affected.

In 1975, the catalytic converter was introduced on the federally delivered automobiles. In those days, the catalytic converter acted on two gases, one of them was CO. The catalytic converter uses oxygen in the exhaust, or oxygen supplied to the exhaust by the air pump, to combine with CO to make CO_2.

Hydrocarbons

Hydrocarbon emissions are basically raw, unburned fuel. These emissions went into the combustion chamber as a mist being carried by the air destined for the combustion chamber and left the engine. The only change in the compound during combustion was that it changed from a mist into a gaseous state.

Many hydrocarbon compounds are known to be carcinogenic. Whereas carbon monoxide is odorless, colorless, and tasteless, hydrocarbon fumes can be very noxious, burning the eyes, throat, and other soft tissues. Many of these chemicals, in the presence of nitrogen oxide and sunlight, produce photochemical smog. This is the element of smog that burns the eyes, nose, and irritate the mucous membranes. While to the environmentally conscious hydrocarbon emissions amount to a veritable soup of foul substances, we will treat them as a single substance.

While allowable CO levels are relatively high, one to two parts per 100 (percent) is allowable even for late model cars in most jurisdictions, the allowable concentration of various hydrocarbon emissions is extremely low, only about 100 to 300 parts per million. To compare, 1 percent equals 10,000 parts per million or one part per million equals .00001 percent. In percent, therefore, an allowable level of hydrocarbons is only about .002 to .003 percent.

Excessive hydrocarbon content in the exhaust can be caused by poor secondary ignition quality, vacuum leaks, engine running too rich, lean misfire (engine running too lean), and low compression. Efforts to reduce hydrocarbon (HC) emissions began with the air pump in the late 1960s. Not only would the air pump allow additional oxygen to combine with the carbon monoxide to yield carbon dioxide, it would also provide oxygen to be combined with HC to yield CO_2 and water.

Later the manufacturers added the evaporative canister to the emission control devices. The job of the evaporative canister is to capture and store HC fumes from the gas tank and the carburetor fuel bowl. The canister contains activated charcoal which store the hydrocarbon vapors until the engine can accept the release of the extra vapors. The catalytic converter also adds oxygen to the hydrocarbons coming through the exhaust system.

Nitrogen Monoxide and Nitrogen Dioxide

Oxides of nitrogen include both nitrogen monoxide and nitrogen dioxide. Usually these two chemicals are grouped together and referred to as NOx. Nitrogen monoxide is an odorless, colorless, tasteless gas which is basically harmless to the environment. However, created along with the NO is nitrogen dioxide. Nitrogen dioxide is a reddish-brown poisonous gas which destroys lung tissue. To further complicate matters, nitrogen monoxide combines with atmospheric oxygen to become nitrogen dioxide.

NOx is created when combustion temperatures exceed 2,500 degrees F. At these temperatures, the nitrogen combines chemically with oxygen (burns). Although NOx is generally not measured for diagnostic purposes, it is the principle emission responsible for oxygen sensor-controlled fuel injection systems. The allowable level of NOx since 1982 for new cars delivered in the U.S. has been 1.0 gram per mile.

Since there is currently no economical way to measure NOx, it is of no diagnostic value. NOx levels will increase, however, when the combustion temperatures increase as a result of incorrect initial timing, a lean air/fuel ratio, high compression (such as caused by carbon build-up on the pistons), or a vacuum leak. In some states, as this is being written, it is required that NOx be tested during the annual or bi-annual emission test.

Attempts to reduce oxides of nitrogen include reducing the compression ratio, using an exhaust gas recirculation valve, and using a reduction catalyst in the converter. The reduction catalytic converter was introduced in 1978. This converter adds a second bed ahead of the oxidizing section. This second section is charged with the task of reducing the NOx into its base elements of oxygen and nitrogen.

Carbon Dioxide

Currently, the carbon dioxide emission is of no concern to the automobile industry. Listed as one of the results of complete combustion, the goal of most of the pollution control devices we have on the spark ignition gasoline engine today is to increase the output of carbon dioxide. Yet carbon dioxide is not totally harmless. Increasing levels of atmospheric carbon dioxide has been linked to the "Greenhouse Effect." It is theorized that the burning of fossil fuels as well as the depletion of worldwide vegetation are bringing levels to the point where solar heat will be trapped in the earth's atmosphere, increasing temperatures worldwide. However, the automotive industry has not yet been mandated to lower carbon dioxide emissions.

Carbon dioxide levels at the tailpipe is a valuable diagnostic tool. A reading between 10 percent and 15 percent indicates that the quality of combustion is good and there are no leaks in the exhaust system. The closer the reading is to 15 percent, the better the overall quality of combustion.

Low carbon dioxide levels can result from poor ignition quality (bad plugs, cap, rotor, plug wires), low compression (head gasket, valves, rings), exhaust leaks (diluted sample), incorrect air/fuel ratio (too rich or too lean), and vacuum leaks.

Water

A byproduct of complete combustion is water. I have lived in the Seattle, Washington, and Fairbanks, Alaska, areas during the past two decades. In both places the mornings are often quite cool. Cool mornings make the water vapor produced during combustion visible. Automobiles have been producing large amounts of water during combustion for over two decades. To this day I am still asked if the large amount of water vapor created during initial startup is normal. The answer is yes. However, if the vapor persists for several minutes after startup, it could be indicative of a serious problem, such as a blown head gasket.

Oxygen

Oxygen is both a desirable and undesirable byproduct of combustion. While there is certainly nothing wrong with returning oxygen to the environment, it also is an indication that the air/fuel ratio is lean. A lean air/fuel ratio can result in high combustion temperatures, which can increase the production of NOx. So while emitting oxygen is in itself not bad, it can be indicative of other, harmful, emissions.

Sulfur and Sulfuric Acid

All fossil fuels contain some sulfur. If a large amount of gasoline is put into the combustion chamber, such as when the engine is running rich, then a significant amount of the sulfur will be oxidized. The odor of oxidized sulfur is often compared with that of rotten eggs. Some of the sulfur will also be converted into sulfuric acid, or H_2SO_4.

Pre-Combustion Methods of Limiting Harmful Emissions

Engine Modifications

There are two pre-combustion methods of limiting harmful emissions: lowering the compression and overlapping the valves.

By dropping the compression ratios in the mid 1970s, the manufacturers were able to lower combustion temperatures and thereby lower the production of NOx. This caused fuel economy to plummet and totally destroyed the potential for performance.

Increased valve overlap allows for some of the exhaust gases to be drawn back in during the beginning of the intake stroke. These exhaust gases are

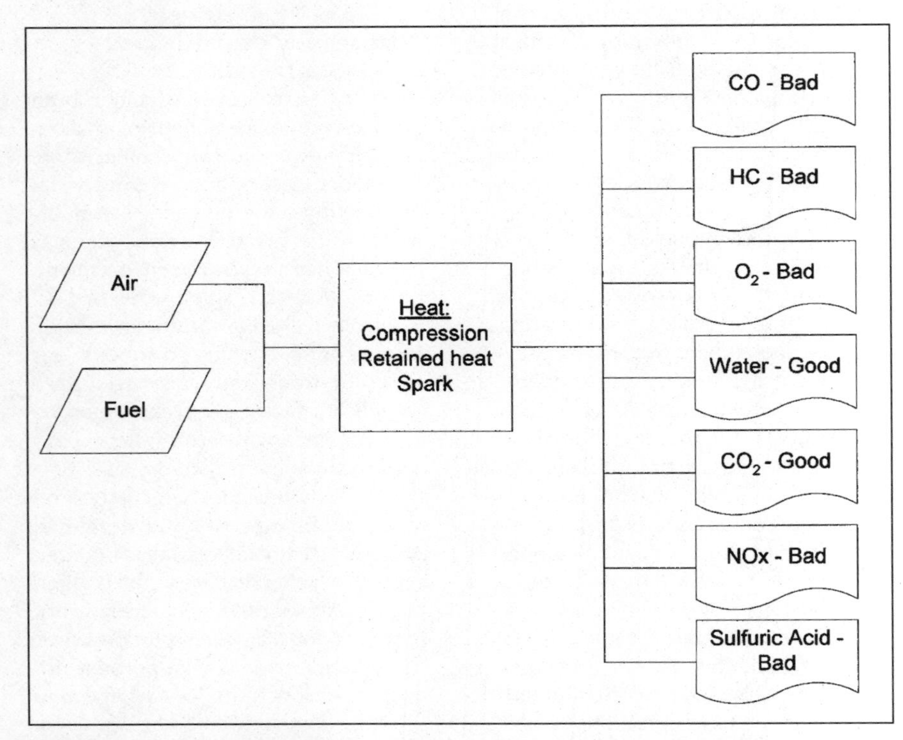

When combustion is less than perfect, the resulting exhaust gases contain a virtual plethora of noxious substances. These gases range from the fatal to the carcinogenic to the corrosive to the harmless.

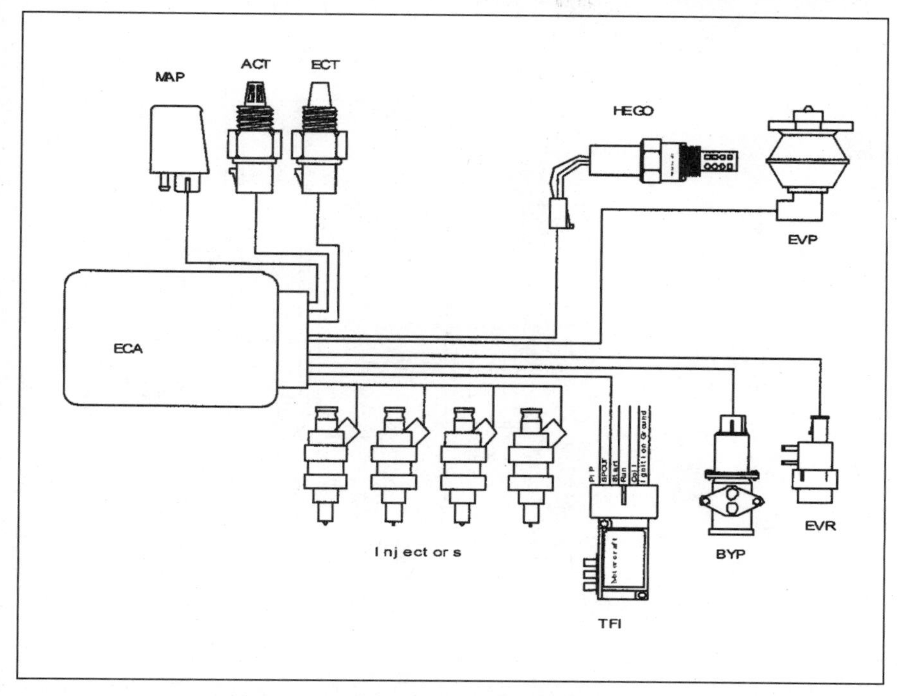

Although there are many advantages to fuel injection over carburetors, it took emission control regulations by the federal government to inspire the auto manufacturers to use fuel injection. Fuel injections systems are themselves one of the most effective emission control devices in the manufacturers' emission control arsenal.

inert; the fuel and the oxygen were burned out of them during the previous combustion cycle. As the next combustion cycle begins, these inert gases act as a heat sink to lower the combustion temperature and decrease the production of NOx.

The Injection System

Electronic fuel injection has been used as a method of controlling automotive emissions since they became an issue in the mid 1970s. The domestic manufacturers did not hold it to be a credible solution to the problem in those days. The D-Jetronic system was used on a variety of European applications, including SAAB, Volvo, and Mercedes through the early 1970s until 1976. In spite of a lack of proper understanding of how the system worked by those confronted with servicing it, the system persisted and introduced the service and diagnostic procedures of electronic fuel injection to the American mechanic. And yet, this system was largely seen by the auto repair industry as a fluke, a one-shot deal.

Cadillac introduced the first mass-produced domestic electronic fuel injection system in September 1975 as standard equipment on the 1976 model Cadillac Seville. This system was developed with cooperation from Bendix and Bosch. It bore an amazing resemblance to the Bosch D-Jetronic system. By this time, the manufacturers had begun to address the need for a systematized troubleshooting system to aid in the servicing and repair of fuel injection. These procedures have evolved into the flow charts which are the industry standard today.

The Cadillac/Bendix system was used until the introduction of the next technological improvement of fuel injection, the digital computer. Cadillac introduced its "Digital Fuel Injection" system in 1980. The "DFI" system had been conceived as a multipoint system with one injector per cylinder. Simplicity and economy won out over technology and DFI was introduced as a two-injector throttle body injection system.

For Bendix, the idea for digital control of fuel injection dated back to patents it filed for in 1970, 1971, and 1973. Benefits which could be found with the digital computer included the more accurate control of the injectors plus the ability of the computer to control a wide variety of engine support systems. With the use of digital computer ignition timing, air pump operations, torque converter clutch functions, as well as a wide variety of emission-related items, could all be controlled by a single compact control module. Additionally, the stored memory potential of a digital computer meant that it would be possible for the sensors to "reprogram" the computer for changes in the overall condition of the engine. This meant that the fuel injection system could detect and compensate for old, tired engines. Service intervals could be greatly increased, because the injection system could compensate for the deterioration of ignition components. A digital computer could also detect and store memories concerning circuit failures in the system. These memories could later be called on by a technician to assist in troubleshooting.

Pontiac was the first GM division to bring fuel injection to the masses. In 1982, it mounted a single injector version of the throttle body injector system (first used by Cadillac) on its 2.5-liter "Iron Duke." That year also saw Chevrolet installing the TBI Iron Duke in its X-body Citation and F-body Camaro.

Chevrolet introduced the 5.0-liter Camaro and 5.7-liter Corvette with what was dubbed "Crossfire Injection" in 1982. This system featured two throttle body single injector units mounted on a common intake manifold. The throttle bodies supplied fuel to opposite sides of the intake manifold. They mounted the throttle bodies on a common manifold with the left unit feeding the right side of the engine and the right unit feeding the left side of the engine. This crossfire concept allowed for increased air velocity and good fuel atomization. The Crossfire Injection system provided a 20-horsepower gain over carburetion on the 5.0-liter Camaro.

This Rochester throttle body injection system was also fitted to the Brazilian-built 1.8-liter engine in 1982. This overhead cam engine was used in the J-body Cavalier. In 1983, its Pontiac cousin, the J-2000, was the first GM car to break the EPA 50-mpg barrier.

Although other GM divisions were using multipoint injection systems as early as 1984, Chevrolet held out until 1985 when it introduced the port fuel injection system on the 2.8-liter engine used in the Camaro, Cavalier, Celebrity, and Citation. Chevrolet's ultimate production fuel package was also introduced in 1985 on the Camaro 5.0-liter engine and the Corvette 5.7-liter mill; Chevy calls this the "Tuned Port Fuel Injection" system. Tuned Port Injection boasts an independently documented 434 cfm. To the old carburetor guys out there this may sound like a trifle when compared to an 850 double pumper, but a little math would show that the cfm potential of the 5.7-liter at 5,000 rpm is only (and here we assume an impossible 100 percent volumetric efficiency) 506 cfm (a more realistic cfm rating would be 405).

There is a lot of research and activity still going on with fuel injection systems. Several aftermarket manufacturers are building and marketing throttle body and Tuned Port clones for older carbureted engines. The original equipment manufacturer (OEM) suppliers are looking again at the pluses and minuses of direct injection. As happens with many technologies, we have reached a level of development and sophistication where simplicity and refinement are becoming the order of the day. Gone are the days of scrambling to simply produce a workable fuel injection system, much of the inspiration for which came from government-mandated emission and fuel economy standards, and arrived are the days of performance and a little fun.

In 1965, Hillborn fuel injection was fitted to the Ford four-cam V-8 engine developed for the Indy cars. A four-cylinder, 16-valve Lotus engine equipped with Lucas fuel injection was used in a few European Ford Escorts during the 1970 model year. It was not until 1983 that any Ford division decided to use fuel injection in a serious manner. That year European Ford began to use the Bosch K-Jetronic system. This fuel injection system had been widely used by northern European manufacturers since the early 1970s.

Meanwhile, beginning in 1978, North American Ford went through three generations of electronically controlled carburetors. The EEC I, EEC II, and EEC III systems were intended to meet the ever-tightening emission standards of the late 1970s

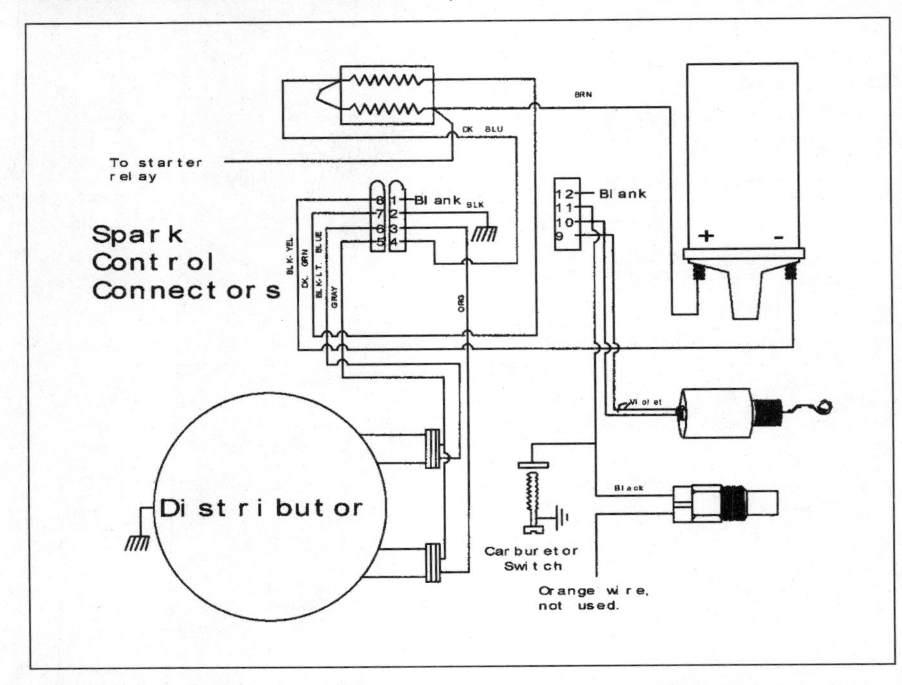

Ignition timing control systems preceded electronic fuel injection as a method of reducing exhaust emissions. One of these systems, the lean burn system of the mid-1970s, caused such frustrating driveability anomalies that many staunch Chrysler owners moved to other makes.

and early 1980s. From an outsider's perspective, either Ford, along with her North American competitors, had a fear of marketing fuel-injected cars or they were holding off in order to perfect their systems.

In 1980, Ford introduced its high-pressure CFI on the EEC III-equipped 5.0-liter Versailles. In 1981, the usage was expanded to the LTD and Marquis. The 1983 model year saw the introduction of multipoint injection on the 1.6-liter applications. With the introduction of the EEC IV system during the 1984 model year, carburetion became the exception rather than the rule for Ford. As we entered the 1990s, the only Fords still equipped with carburetors were special equipment packages, such as police applications and towing packages.

Chrysler first ventured into electronic fuel injection with the hydraulic plate fuel injection system used on the Chrysler Imperial in the early 1980s. In 1984, Chrysler introduced its high-pressure throttle body injection system. This system was designed by Bosch for Chrysler. Later in 1984, Chrysler introduced its multipoint injection system on its Turbo models. In 1987, the first non-Turbo application featuring multipoint injection made its debut. This injection system was used on the 3.0-liter Mitsubishi engine found in the Caravan and Voyager. By 1988, the entire Chrysler product line was fuel injected.

Ignition Timing Control System

Electronic timing control offers a number of advantages over standard mechanical ignition timing control. First of all, advancing the timing at an idle can reduce emissions at an idle. However, advancing the timing with standard mechanical systems causes detonation upon sudden acceleration. This is because the mechanical timing control system is not capable of retarding the timing fast enough to prevent detonation. An electronic ignition timing system can effectively retard the timing for acceleration in as little as 1/15 of a second. In a few models, this timing change can occur in as little as 1/5400 of a second.

The ignition timing must change as the engine is running to adjust to different engine speeds and loads. When the engine is running at an idle, the spark must begin at a point in crankshaft rotation which will allow for the spark to extinguish when the crankshaft is about 10 degrees after top dead center. Since the length of time the spark is jumping the gap is a relative constant, about 2.5 milliseconds, the spark must start sooner as the engine speed increases.

As an example, on a hypothetical engine, the spark occurs at 10 degrees before top dead center when the engine is running at 1,000 rpm. At this speed, the spark extinguishes at about 10 degrees after top dead center. This means that the crankshaft has rotated 20 degrees since the initiation of spark. As the engine speed increases, the crankshaft rotates more degrees in the 2.5 milliseconds that the spark is jumping the gap. At 2,000 rpm, the crankshaft will rotate twice as much. If the timing at 1,000 rpm should be 10 degrees before top dead center (TDC), then the timing at 2,000 rpm should be about 30 degrees before TDC. As the engine speed continues to increase, the timing will need to continue to advance. The amount of total advance, the upper limit of the advance, will vary depending on the design of the engine.

In the old days, the change of timing in response to rpm was accomplished through a set of spring-loaded weights. As the speed of the engine increased, the weights swung out against spring tension. The cam that opened and closed the points, although it was mounted on the distributor shaft, was not part of the distributor shaft. The swinging weights caused the cam to rotate with respect to the distributor shaft. This advanced the timing.

Vacuum advance actually retards the timing when the engine is under a load. In most applications, the vacuum advance is connected to ported vacuum. The advance unit receives no vacuum at an idle, but when the throttle is opened, the vacuum advances the timing. As the load on the engine increases, the vacuum drops. As the vacuum drops, the timing is not advanced as much; it retards. Retarding the timing lowers the combustion temperature, and therefore prevents detonation and decreases the potential of damage to the engine. On the Chrysler fuel injection system, the vacuum advance is replaced by the manifold absolute pressure sensor.

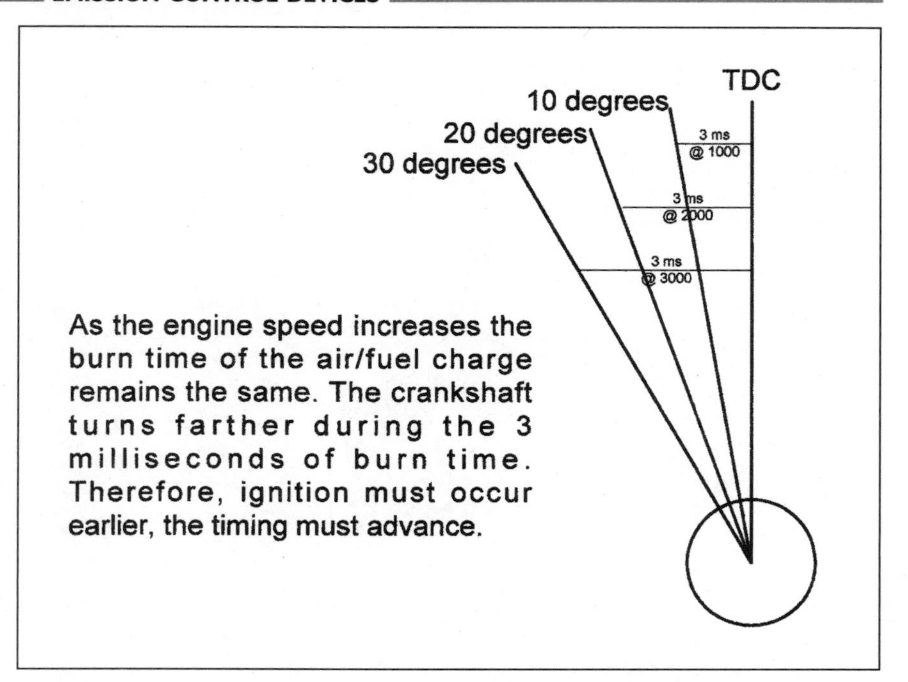

Ignition timing must advance as the engine speed increases to provide enough time for the burn in the combustion chamber to be completed.

Evaporative Canister

Since the 1970s, cars have been equipped with an evaporative control system. This system consists of a canister filled with activated charcoal. The canister is then connected to the fuel tank (and in the case of carbureted cars to the carb fuel bowl) by hoses. As gasoline (HC) evaporates from the fuel tank, the fumes are stored in the activated charcoal. A device known as the canister purge valve opens when the engine can accept the extra fuel, and manifold vacuum sucks the fuel into the engine to be burned.

Exhaust Gas Recirculation

The exhaust gas recirculation (EGR) valve is used to reduce the production of oxides of nitrogen. As the engine rpm rises above idle, either vacuum applied to a diaphragm or a solenoid lifts the pintle of the EGR valve, allowing exhaust gases into the combustion chamber. Approximately 7 percent of the combustion chamber volume will lower the burn temperature by about 500 degrees F. This lower temperature reduces the production of NOx.

Post-Combustion Air Pump

The air pump pumps air into the exhaust manifold while the engine is cold. Since a rich mixture is present in the combustion chamber at this time, the resulting exhaust gases will be laden with CO and HC. When the oxygen in the air comes in contact with the CO and HC, the temperature increases dramatically and "afterburning" takes place. The afterburn consumes much of the residual CO and HC.

As a side benefit from this afterburning, a great deal of heat is generated. The heat is used to assist in bringing the catalytic converter and the oxygen sensor up to their proper operating temperature.

Before the computer goes into its closed loop mode, it is necessary for the air being pumped by the air pump to be diverted either to the atmosphere or downstream of the oxygen sensor. On some Chrysler applications, such as the 5.2-liter and 5.9-liter light truck applications, the air will be upstream, even after entering closed loop, whenever the engine idles.

Catalytic Converter

When the catalytic converter was introduced in 1975, it spelled the beginning of the end for leaded gasoline. This

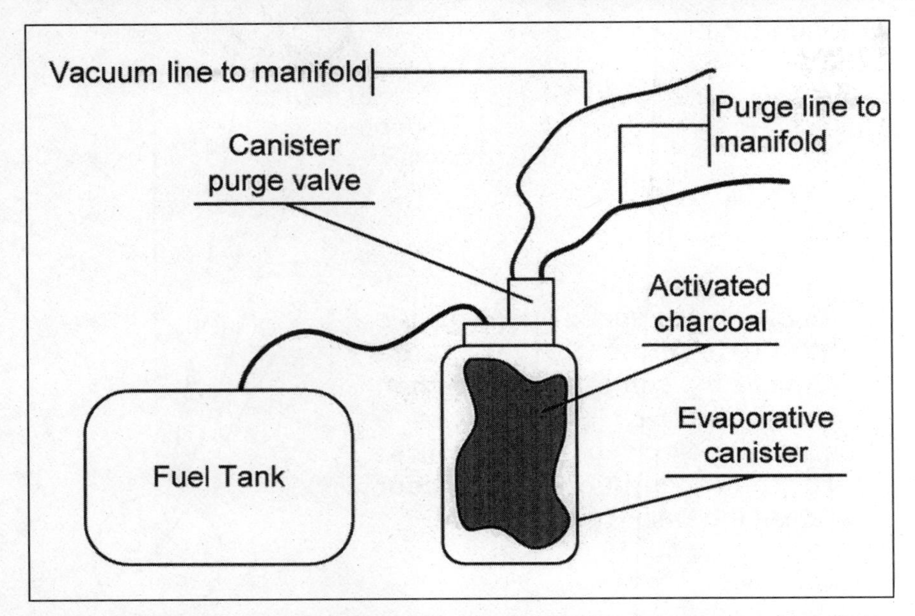

Gasoline slowly evaporates in the fuel tank. This increases the hydrocarbon emissions of the vehicle. For over two decades, the manufacturers have used a container filled with active charcoal to store fuel tank evaporates. When the engine is running under medium rpm moderate load conditions, the evaporative canister control valve opens and allows HC vapors to travel to the manifold and be burned by the engine.

"oxidizing" catalytic converter consisted of a platinum/paladium coating over an aluminum oxide substrate. The substrate over the years has been in the form of pellets, beads, and a honeycomb monolith. The platinum was not compatible with tetraethyl lead, requiring it to not be present in the fuel used in catalyst-equipped vehicles. The job of the catalyst is to provide an environment where enough heat can be generated to allow further combustion of the HC and CO to occur. The converter is heated by a chemical reaction between the platinum and the exhaust gases. The minimum operating, or lightoff, temperature of the converter is 600 degrees F with an optimum operating temperature of about 1,200 to 1,400 degrees. At a temperature of approximately 1,800 degrees, the substrate will begin to melt. These excessive temperatures can be reached when the engine runs too rich or is misfiring. I once attended a meeting of the Society of Automotive Engineers where a representative of a company that built the converter substrates stated that a 25 percent misfire (one cylinder on a four-cylinder engine) for 15 minutes was enough to begin an irreversible self-destruction process in the converter. This type of catalytic converter requires plenty of oxygen to do its job; this means that the exhaust gases passing through it must be the result of an air/fuel ratio of 14.7:1 or leaner.

The years between 1978 and 1982 saw the gradual introduction of the "dual bed" converter. This converter adds a second "rhodium" catalyst, known as the reducing section, ahead of the oxidizing section. The rhodium coats an aluminum oxide substrate and reacts with the NOx passing through it. When heated to in

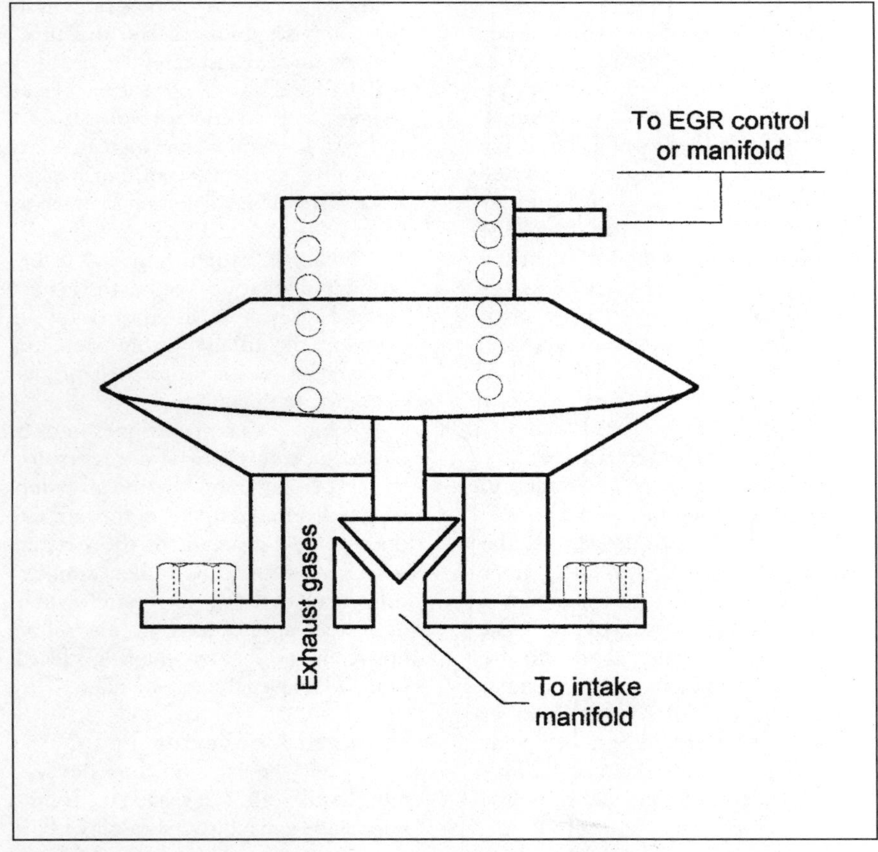

The EGR valve is one of the earliest emission control devices used to control oxides of nitrogen. When the engine is under a load, combustion temperatures rise. When they rise above 2,500 degrees F, nitrogen combines with oxygen. The EGR allows exhaust gases to pass from the exhaust into the intake. These exhaust gases contain very little oxygen or fuel; therefore, they are inert in the combustion process. Their presence in the combustion chamber, however, robs heat from the combustion. The result is lower combustion temperatures and lower emissions of NOx.

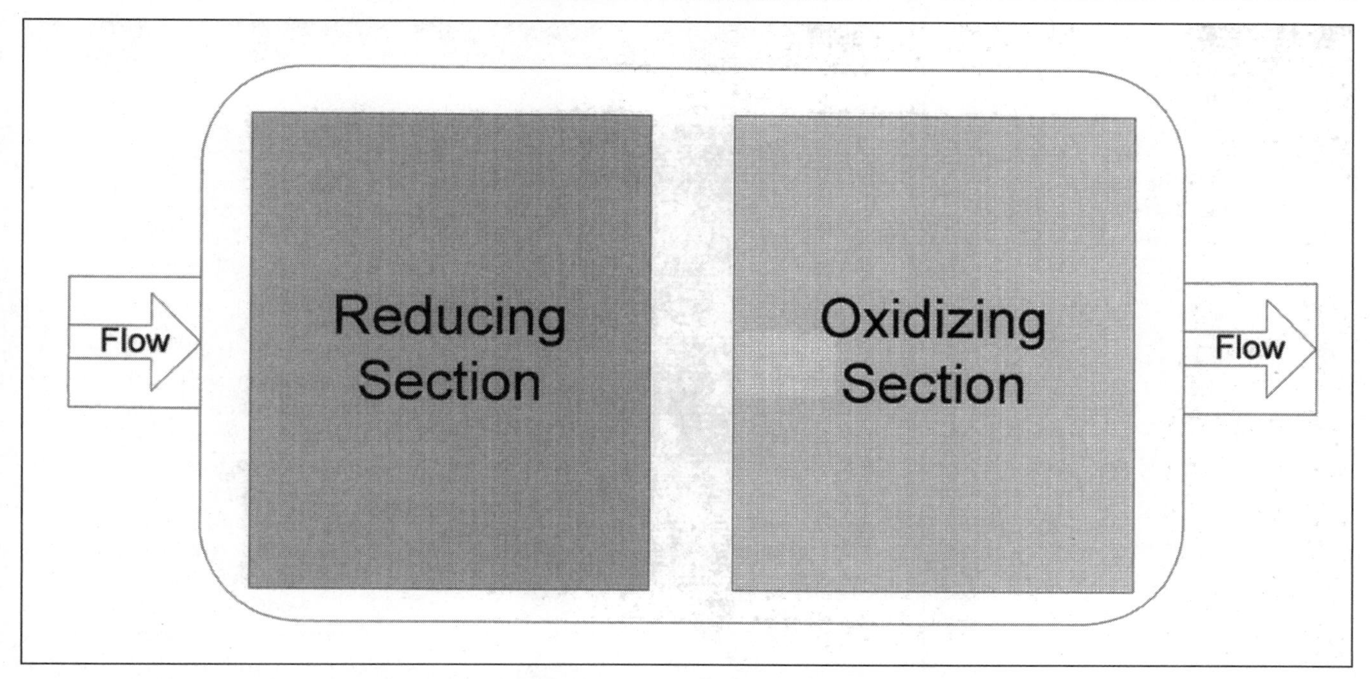

The catalytic converter has two sections. The first section that the exhaust gases pass through is known as the reducing section. Here the NOx is separated into oxygen and nitrogen. The second section is the oxidizing section. Here the CO and HC is combined with oxygen to yield CO_2 and H_2O.

excess of 600 degrees F, the nitrogen and oxygen element of the NOx passing through it will be stripped apart. Although only about 70 to 80 percent efficient when coupled with the EGR, it does a dramatic job of reducing NOx. Since the job of the reducing catalyst is to strip oxygen away from nitrogen, it works best when the exhaust gases passing though it are the result of an oxygen-poor air/fuel ratio of 14.7:1 or richer.

The only air/fuel ratio that will permit both sections of the converter to operate efficiently is an air/fuel ratio of 14.7:1. The job of the oxygen feedback fuel injection systems being used today is to precisely control the air/fuel ratio at 14.7:1 as often as possible.

On many applications, the air from the air pump, after completing its role of preheating the catalytic converter, will be directed between the front reducing section and the rear oxidizing section of the converter to supply extra oxygen to improve the efficiency of the oxidizer.

Contents

CHAPTER THREE

TOOLS

Tools for the Backyard Mechanic

Timing Light

The timing light is a tune-up instrument, not an engine rebuild necessity. The timing light is used to synchronize the primary ignition system to the position of the crankshaft.

Tachometer and Dwell Meter

Still handy for checking engine speeds, a digital tachometer also has other uses. One of the new diagnostic instruments needed on fuel injected cars is a frequency counter. Since many professional technicians and most enthusiasts do not own a frequency counter, let me suggest the use of a digital tachometer.

The tachometer measures the number of primary ignition pulses per minute and mathematically converts the number of pulses into the number of crankshaft rotations. It sends that calculation to the display as crankshaft revolutions per minute. With some very simple math, we can reverse this mathematics to get a frequency reading in cycles per second, or hertz.

Switch the tachometer to the four-cylinder scale; it does not matter whether the engine you are working on is a four-cylinder, six-cylinder, or eight-cylinder (remember, we are counting pulses per second, not crankshaft rpm); always use the four-cylinder scale. This technique will work equally well on the six- or eight-cylinder scale, but the math is more complex. Connect your tachometer to the wire (usually ground side of an actuator circuit or the output of an appropriate sensor) you want to test and a good ground. Observe the reading on the tachometer and divide by 30. That will be the reading in hertz.

Since the advent of electronic ignition, the dwell meter has been gathering dust. No longer do we make adjustments on points during a tune-up. However, as the tachometer was actually measuring the frequency of the primary ignition system, the dwell

Ignition timing remains an important factor in good engine performance. The rumors of a decade ago that the computer will automatically set the ignition timing where it wants are only true when the initial timing is correct. The timing light even serves a function on the distributorless engines where adjustment is not possible. The timing light can be used to prove that the timing actually does change.

One of the handiest troubleshooting tests involves shutting down the spark to each of the cylinders and measuring the drop in rpm. At one time, a tool to do this test was several thousand dollars. The used engine analyzer pictured here cost its owner only a few hundred dollars. Smaller, even more portable machines can be purchased new for about the same price today.

meter was measuring its "duty cycle," therefore, when duty cycle needs to be measured, the dwell meter becomes again a handy tool.

Remember that the duty cycle is a measurement of the relationship of the on time and the off time. Place the dwell meter on the four-cylinder scale (again, do not be concerned about how many cylinders the engine has that you are working on). Connect your dwell meter to the wire you want to read the duty cycle on and observe the reading. Multiply the reading by 1.1. In reality, there will almost never be a time in automotive troubleshooting where that degree of precision is required; simply using the observed reading will suffice.

Cylinder Contribution Tester

Several tool companies produce a "cylinder shorting" tach/dwell meter. These devices electronically disable one cylinder at a time while displaying rpm. Engine speed drops can be noted. Unfortunately, these testers can cost $500 or more and most will not do a cylinder balance on a distributorless engine.

Another valid method does the old test light technique one better. Cut a piece of 1/8-inch vacuum hose into four, six, or eight sections, each about an inch long. With the engine shut off and one at a time, so as not to confuse the firing order, remove a plug wire from the distributor cap, insert a segment into the plug wire tower of the cap and set the plug wire back on top of the hose. When you have installed all the segments, start the engine. Touching the vacuum hose "conductors" with a grounded test light will kill the cylinder so that you can note rpm drop. Again, the cylinder with the smallest drop in rpm is the weakest cylinder.

Whichever methods you use, follow this procedure for the best results:

1. Adjust the engine speed to 1,200-1,400 rpm by blocking the throttle open. Do not attempt to hold the throttle by hand, you just will not be steady enough.
2. Electrically disconnect the IAC motor to prevent its affecting the idle speed.
3. Disconnect the oxygen sensor to prevent it from altering the air/fuel ratio to compensate for the dead cylinder.
4. Perform the cylinder kill test. The rpm drop should be fairly equal between cylinders. Any cylinder that has a considerably lower rpm drop than the rest is weak. Proceed to step 5.
5. Introduce a little propane into the intake, just enough to provide the highest rpm. Repeat the cylinder kill test. If the rpm drop from the weak cylinder tends to equalize with the rest, then you have a vacuum leak to track down. If there is no significant change in the power output from the weak cylinder, then proceed to step 6.
6. Open the throttle until the engine speed is about 1,800-2,000 rpm. Repeat the cylinder kill test adding propane. If the rpm drops are now equal, then the most likely problem is that the EGR valve is allowing too much exhaust gas to enter the intake at low engine speeds. Remove the EGR valve and inspect for excessive carbon build-up and proper seating. If the rpm drop on the cylinder in question remains low, then the problem is most likely engine mechanical. Proceed to step 7.
7. Perform both a wet and dry compression test. If you have low dry compression and low wet compression, then the problem is a bad valve or valve seat. If the dry compression is bad, but the wet compression is good, then the problem is the piston rings. If the

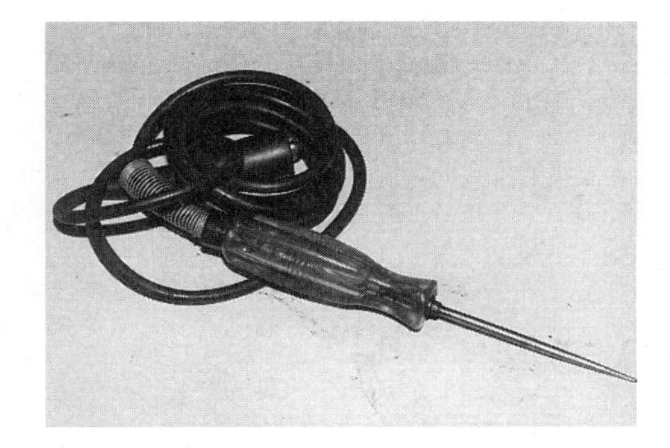

For those with a Scottish ancestry, like myself, there is a cheaper way to detect a weak cylinder. A set of sacrificial spark plug wires and a sharp test light. With the key off, disable the automatic idle speed (AIS) motor. Start the engine. Connect the alligator clip of the test light to a good ground. Probe the distributor cap end (or coil end, if the application is distributorless) of the plug wires. This will kill the affected cylinder. Record the rpm drop on each cylinder.

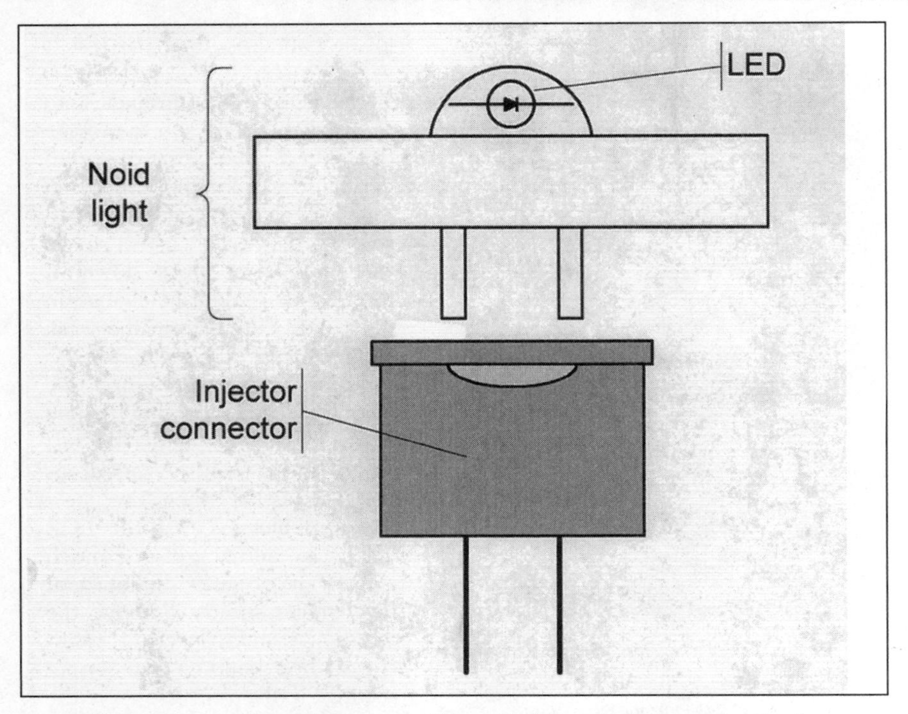

The noid light is designed to tell the technician if there is a firing pulse being received by the injectors. Place the noid light in the harness connector to an injector. Crank the engine. If there is a signal being sent to the injector, the noid light will flash.

compression is good both dry and wet, then the problem is in the valvetrain, such as the camshaft, lifters, or pushrods.

8. After testing is completed, reconnect anything that was disconnected or removed for testing.

If you were to buy a tester capable of doing a cylinder balance on an engine equipped with a distributorless ignition system, you would need to plan on spending many thousands of dollars. However, the vacuum hose trick described above will work nicely.

Be sure to disconnect the IAC, the oxygen sensor, and stabilize the rpm at 1,200-1,400 for the first balance test. The rest of the testing procedure is exactly the same as for the distributor type ignition.

Noid Light

The noid light is a small bulb designed specifically to be installed in the fuel injector harness in place of a fuel injector. The engine is then cranked. While the starter is engaged, or even after the engine has started, pulses intended to fire the injector will cause the noid light to blink. This tool is used in diagnosing no-start conditions primarily, but can also be used to identify problems in the fuel injection wiring harness.

Digital Volt/Ohm Meter

The most important tool in working with and troubleshooting modern Chrysler fuel injection systems can be one of the cheapest and easiest to acquire. Digital voltmeters that are adequate to do the job range in price from a low of about $25 to a high of well over $500. There is an advantage in *some* of the more expensive meters in that they combine many of the functions that we will later describe using tachometers and dwell meters for into one package.

The digital voltmeter boasts a high impedance input (10 million ohms or more); this allows the voltmeter to be connected to very small current flow circuits without affecting the voltage reading. Voltmeters with a low input impedance tend to rob power from the circuit being tested; this causes the voltage readings to be lower than they really are. For this reason, the digital meter is what should be used any time precise voltage readings are required.

There is a downside to the use of a digital voltmeter. It is a fact that since it is digital, it merely samples voltage and displays it. There are major gaps between these samples. Transient fluctuations are completely missed; there may be a device, such as the throttle position sensor potentiometer, which is supposed to create a steadily increasing voltage as the throttle is opened. As the TPS wears, there may be places where the wiper no longer makes contact with the carbon film strip; this would result in a sudden drop in voltage. If the digital voltmeter's sampling did not happen to coordinate with the drop in voltage, the fluctuation, which could be the cause of a major driveability problem, would be missed. For this reason there is a better tool for measuring variations in voltage.

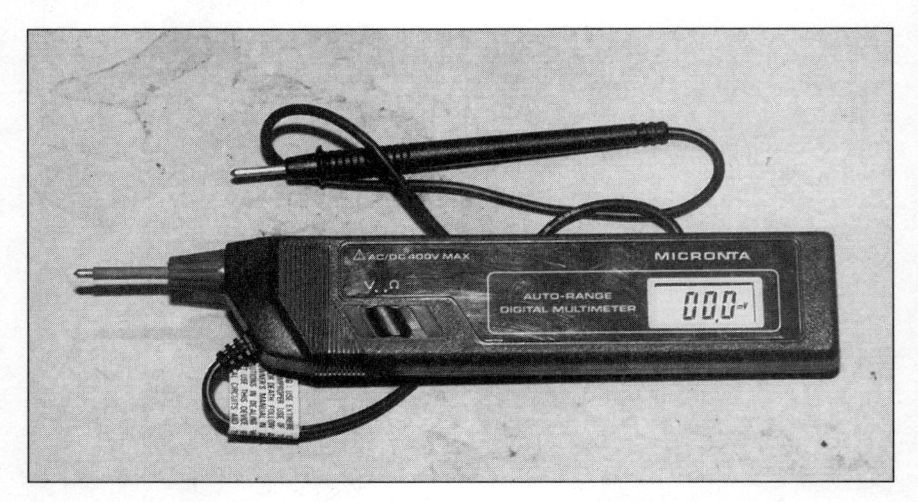

Digital voltmeters come in prices that range from about 30 dollars, like the one shown here, to several thousand dollars. For the typical consumer, or even the typical technician, the cheaper digital voltmeter will serve quite well.

Like the digital meter, the analog meter comes in a wide variety of price ranges. Many analog meters, like this one, rob power from the circuit being tested. This has led to stories about computers being damaged by these meters. For safety and accuracy, a digital meter should be used for most testing.

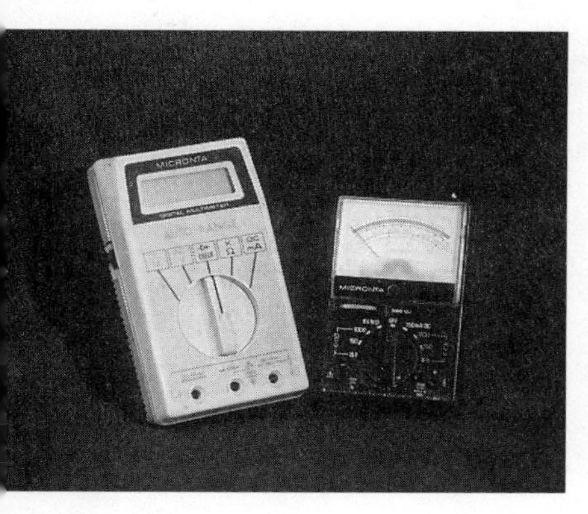

When accuracy and precision are necessary, the digital voltmeter is the best choice. When changes in voltage, especially smooth changes in voltage, are to be monitored, the best choice is the analog meter. The meter shown in this picture is quite adequate and cost less than $15.

Analog Volt/Ohm Meter

Where the digital voltmeter displays its reading as digits, the analog voltmeter uses a needle moving across a scale to display its readings. The benefits of the analog voltmeter have suffered much abuse since the introduction of the electronic engine control systems at the end of the 1970s. This is because the majority of inexpensive analog meters have a relatively low input impedance. As previously mentioned, a low impedance meter can distort readings. Rumors about technicians ruining computers and other components by using analog meters to take measurements are largely exaggerated.

The analog meter will detect fluctuations in voltage much better than the digital voltmeter. When a transient voltage change occurs, it will show up in the analog meter as a fluctuation in the needle.

Use the analog meter when you are looking for fluctuations in voltage and the digital when you are looking for precise readings.

Because of the extremely low current output of the oxygen sensor, most analog voltmeters will ground out the O_2 reading; the meter will display 0 volts continuously. Always use a digital voltmeter or a 10 megohm input impedance analog when taking O_2 readings.

The AC voltmeter has a place in the troubleshooting of electronic ignition and fuel injection systems just as it does in the testing of a home appliance. The single electronic device that it tests effectively is the reluctance type pickup. When the reluctor is being rotated at approximately one revolution per second, the AC voltage reading should be between .2 and 1.5 volts, depending on the number of teeth.

Tools for the Professional Technician

Engine Analyzers

During the 1980s, the high-tech, high-buck engine analyzer became popular as a method of troubleshooting driveability problems. To an extent, these gargantuan pieces of equipment tended to be pricey beyond their practical capability. However, these machines served as a great instructional tool.

The modern engine analyzer is a single tool capable of testing the starting system, the charging system, the primary ignition system, the ignition timing and timing control system, the secondary ignition system, and the exhaust emissions. When the skilled technician has completed the analyzer's comprehensive test, he or she has a complete picture of the operating characteristic of the engine.

Fuel Pressure Gauge

If you do not have one of these, you might as well forget about troubleshooting fuel injection systems. All good troubleshooting begins with a fuel pressure test. The fuel pressure gauge that is combined with your vacuum gauge will no longer fit the bill, however, you will need a gauge capable of accurate readings of up to 75 psi. A variety of fittings and adapters will also be necessary. Shop around a little, the prices on these gauges with fittings can run anywhere from $100 to $1,000. A little ingenuity and a trip to a local hydraulics or air conditioning supply store could probably yield a more-than-adequate gauge and a savings of dozens of dollars.

Diagnostic Scanner

Probably one of the most essential tools for the reader that is going to be involved in troubleshooting or fine tuning on a regular basis is the diagnostic scanner. A scanner connects to the diagnostic connector located under the hood, and translates computer code from the engine controller into digital information about what the computer is seeing, thinking, and doing.

Scanners come in all sizes, price ranges, and user friendliness. Scanners are incorporated into engine analyzers costing tens of thousands of dollars and in hand-held units, such as those marketed by Snap-On, OTC, and others, at between $600 and $2,000

Although a scanner is not a necessary tool for the average enthusiast, it is definitely a very valuable tool. Its real value comes from its use in conjunction with flow charts for diagnosing problems related to trouble codes.

Lab Oscilloscope

This valuable tool is new to the arsenal of automotive diagnostic weaponry. The "engine analyzer" scope has been used for decades to troubleshoot primary and secondary ignition systems. The manufacturers of these analyzers early in the 1980s saw the need to test patterns and wave forms at much lower voltages than what you would experience in the ignition system. They began to incorporate low-voltage functions in their professional-quality engine analyzer scopes.

However, for considerably less than the thousands of dollars a shop has to invest in one of these testers, a low-voltage oscilloscope can be found. When a used one is found, I can personally testify to the fact that these scopes can cost as little as $100. Even new, through an electronics hobby store, they cost as little as $500.

Keep in mind that this type of scope cannot be used to analyze either primary or secondary ignition.

Engine analyzers come in all sizes, price ranges, and levels of antiquity. Without exception, any engine analyzer is only as good as the technician using it. Ironically, the technician who has sufficient skill to use an engine analyzer is also a technician that has the ability to troubleshoot most problems without it.

AIS Tester

The automatic idle speed motor (AIS) is a stepper motor controlled valve which the computer moves in order to control the speed of the engine at idle. The AIS can be moved to any one of 256 positions by the computer to ensure the correct idle speed regardless of changes in engine load due to the transmission, power steering, alternator, air conditioning compressor, or anything else. At an idle, the AIS will be at about position 20 with no loads on the engine. As engine loads increase, the rpm will tend to drop. As the rpm

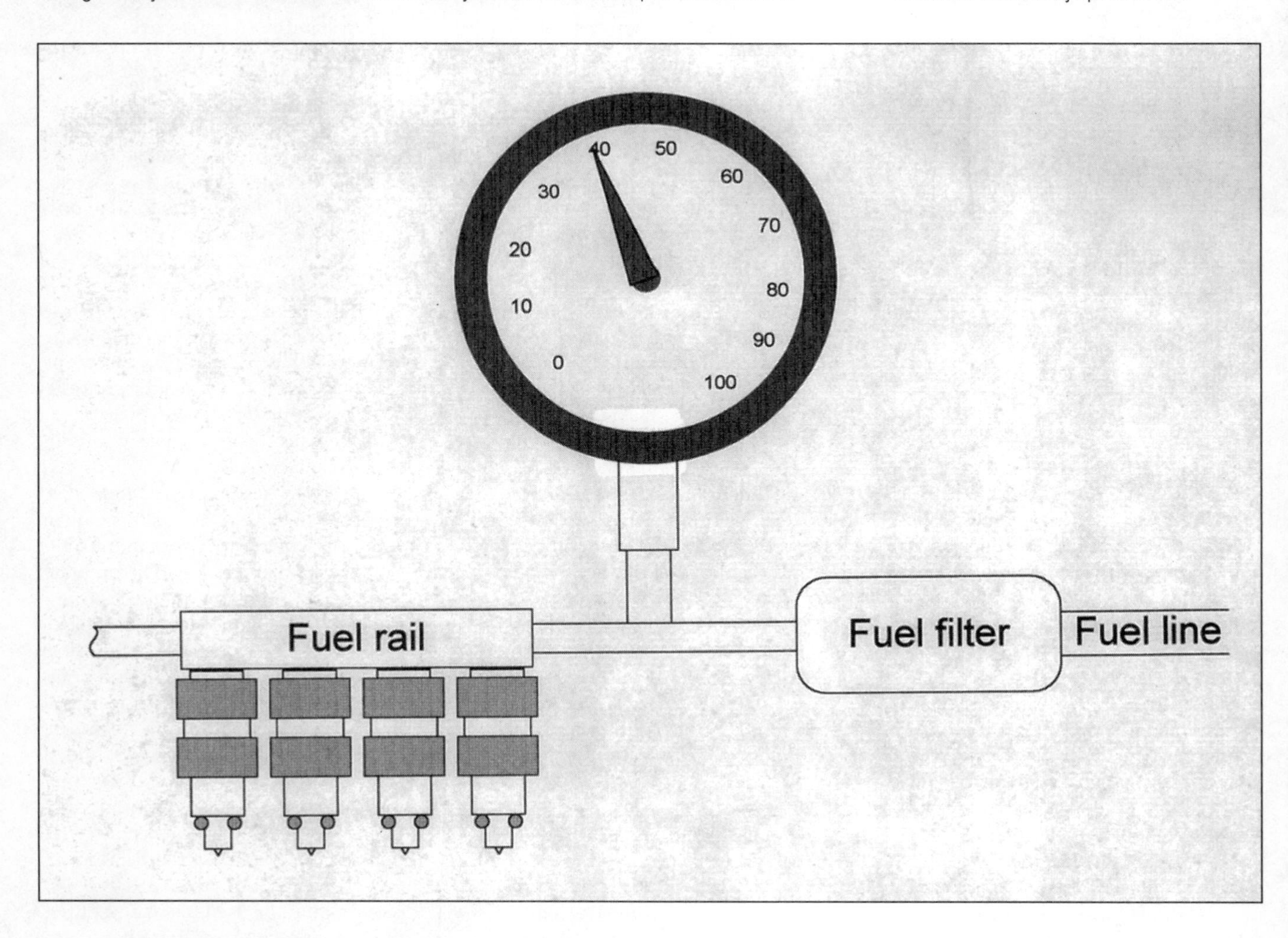

Perhaps the single most important tool in the arsenal of fuel injection tools is the fuel pressure gauge. The gauge should be connected between the fuel filter and the fuel injectors. Since many driveability problems are the result of problems from reduced volume, the fuel pressure should always be tested under the conditions where the symptom occurs.

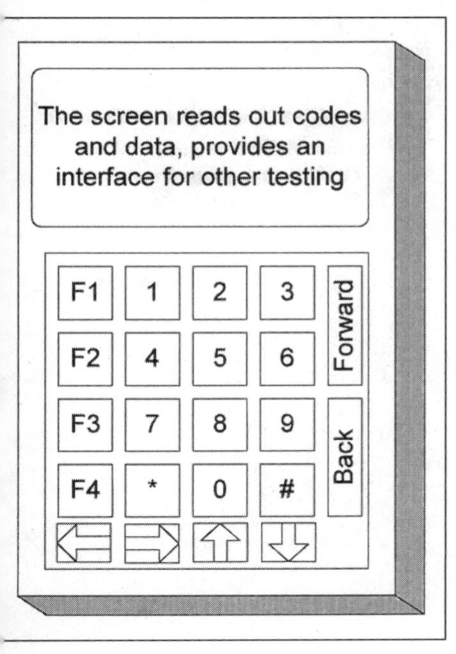

The scanner is thought of by many in the auto service profession as a magic tool. Those that think of it as a magic tool, however, are not those that use it as a tool. Basically, a scanner is only an interface box which allows the technician to extract codes and observe what the computer is seeing, hearing, feeling, smelling, thinking, and doing. There are many brands on the market; one has little real advantage over another. Every scanner has four critical buttons. One that moves forward through the menus, one that moves back through the menus, and two that scroll between choices. The scanner manufactured by Snap-On uses a large knob to scroll between choices.

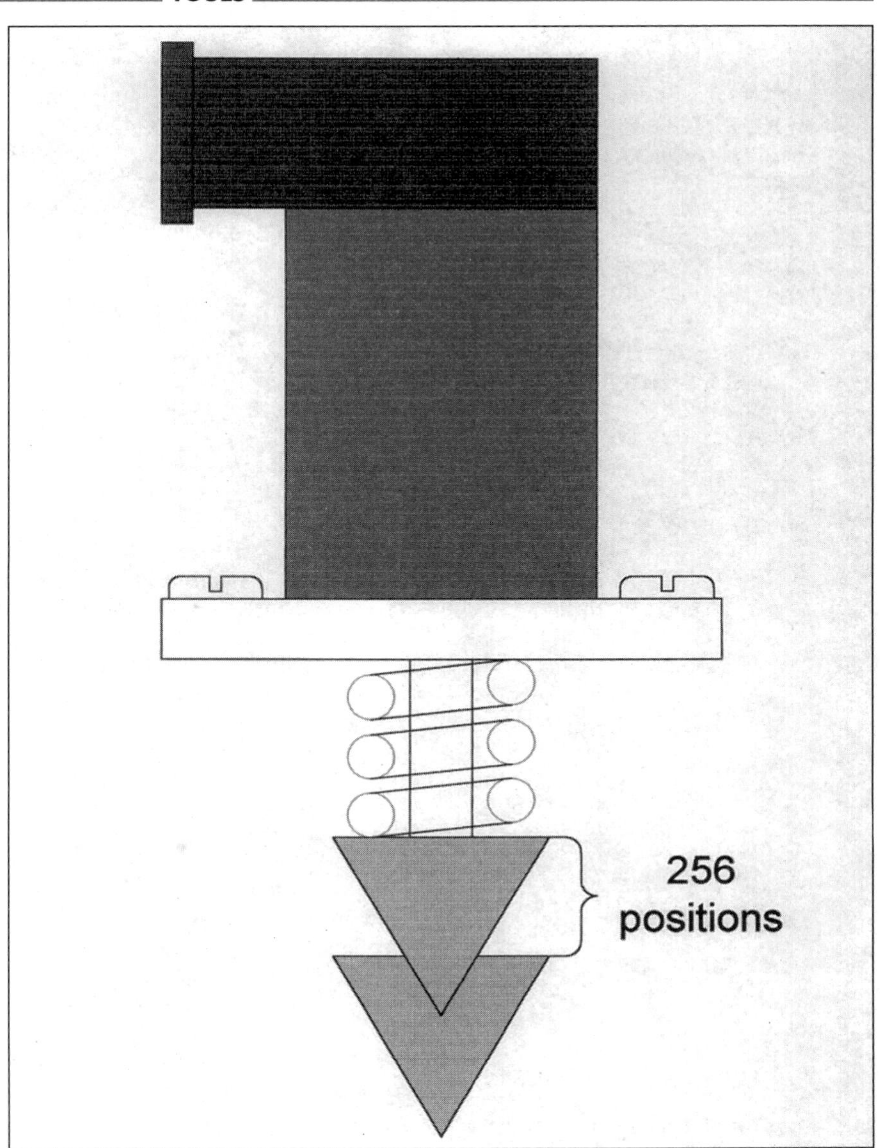

The stepper motor type AIS motor can be difficult to test. Thexton builds a tool that connects to the battery and to the AIS motor. When energized, the tool can drive the stepper motor through its 256 positions.

drops, the computer steps the AIS to a more open position (higher number); as the rpm decreases, the AIS is stepped in. AIS position is displayed through serial data and is an important piece of troubleshooting information.

The AIS motor can be tested using a special tester. This tool sends extend and retract pulses to the AIS motor, thereby testing the motor's ability to control idle. If the idle control system of the engine does not appear to be functioning, this tool can be used to confirm the operation of the AIS motor. If idle speed seems to be incorrect, then the problem must be in the wiring or the computer.

Injector Cleaner

During the latter half of the 1980s, car manufacturers, General Motors, in particular, had problems with fuel injectors becoming restricted. This was not a noted problem during the 1970s when only European and Japanese cars were equipped with multipoint fuel injection. Ever faithful to the demands of the marketplace, several of the tool and automotive supply companies came to the rescue with injector cleaning systems.

These systems use a highly caustic solvent. In most cases, the engine is fueled by the cleaner as the cleaner does its job. The price of the systems runs from as low as $100 to well over $1,000.

Intake System Cleaner

A new tool to be offered to the professional fuel injection technician is the intake system cleaner. These systems clean the entire air intake system from the throttle plate to the intake valve. These systems usually use a strong caustic substance to clean. One system uses crushed walnut shells. The shells are blasted into the intake system with a tool resembling a sandblaster. After the cleaning is accomplished, the bulk of the remaining crushed walnut shells are removed with a suction gun. Any residual shell fragments are simply burned in the combustion chamber when the engine is started.

THE COMPUTERS

CONTENTS

Dual Module

Chrysler introduced its electronic fuel injection system in 1984. This early version of Chrysler's fuel injection system used a two-module computer. Under the hood was a unit directly responsible for controlling the actuators related to engine operation. This module was known as the power module. In the passenger's kick panel, the second module could be found. Known as the logic module, it was charged with gathering information from the various sensors, making a decision based on that information, then ordering the power module to carry out those orders.

Compared to its domestic competitors, Chrysler chose a different approach in configuring its underhood electronics. When I teach journeymen technicians about troubleshooting Chrysler fuel injection, I like to kid them a little about this configuration. "If you ever need to replace the ignition module, it is going to be real expensive...but you get a free computer. If you ever need to replace the voltage regulator, it will be real expensive...but you get a free computer and a free ignition module. If you ever need to replace the cruise control module, it will be real expensive...but you get a free computer, a free ignition module, and a free voltage regulator." Basically everything under the hood is controlled by the power module in response to commands from the logic module. This philosophy of electronic condensation has a good and bad side. The good side of condensation allows for low costs on the assembly line. Basically, the engine can be pre-assembled and pre-wired—computer, injectors, everything. The car comes down the assembly line, and the engine is rolled under the car and shoved in from underneath. The computer is then bolted to the inner fender. Engine and engine wiring harness installation is complete. The down side of the condensation has to do with the expense of replacing what are ordinarily inexpensive components.

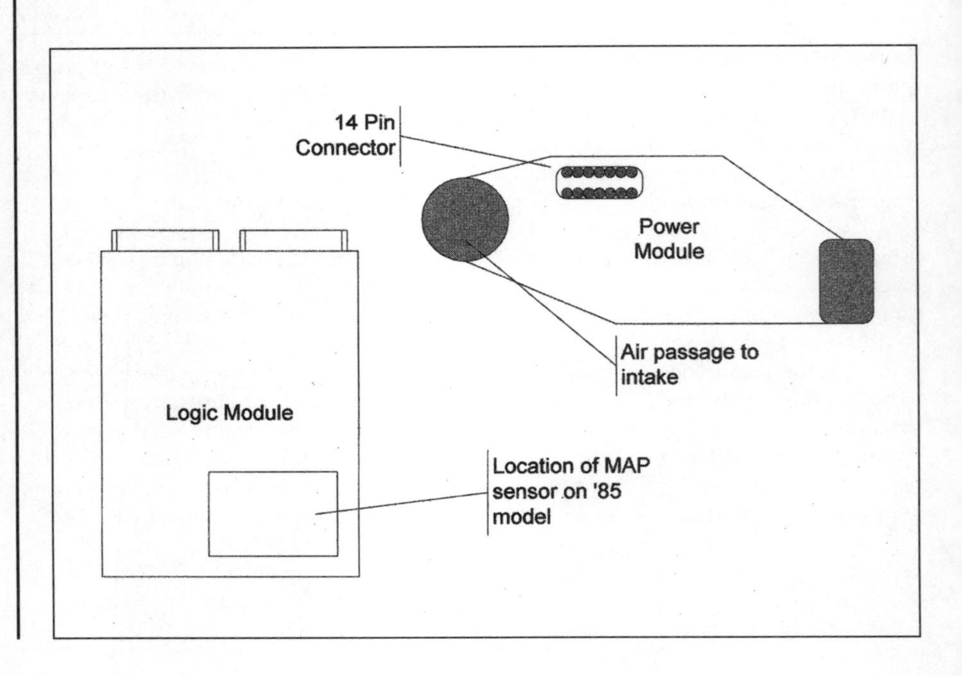

The first Chrysler electronic fuel injection systems used a dual-module configuration. The logic module gathered data from the various engine sensors and processed that information. Upon reaching a decision, it sends its commands to the power module. The power module handles all of the high-current actuators. It is the ignition module, the injector driver module, the cruise control module, and the charging system voltage regulator all built into one box.

The logic module is located in the passenger's kick panel. All the primary engine sensors are connected to this component. The logic module receives signals from these sensors, interprets the signals, and makes decisions about fuel delivery, ignition timing, A/C compressor clutch control, emission component control, and most other engine control-related items. After the decisions are made the logic module directs the power module to carry out its decisions.

Dual Module Self Diagnostics (Without a Scan Tool)

The self-diagnostic ability of the dual module configuration was, for its day, very sophisticated. There were three major test procedures that could be done. The first was the fault code delivery procedure. To obtain fault codes, the technician could either use the Chrysler Digital Readout Box (DRB) or the ignition switch. To use either, it was necessary for the technician to indicate to the logic module that he or she wished to receive the codes. This indication was delivered by cycling the ignition switch on and off three times within 5 seconds, leaving the ignition switch in the on position. On the instrument panel was a light labeled "Power Loss" or "Check Engine." This light would begin to flash codes once the diagnostic mode was entered with the ignition switch. These codes were delivered in a two-digit format. Generally, these codes related to perceived failures in the sensor circuits.

The logic module was ready for a second test procedure as soon as the codes were delivered. This procedure is called the switch test. The technician watches the Power Loss or Check Engine light while operating the switches that are driver inputs to the logic module. These switches include the air conditioning switch, the automatic transmission shifter switch, the brake switch, and the electric back light (EBL) switch.

After the technician reads the trouble codes through the flashing light, he or she can perform the switch test simply by using them. For instance, when the A/C switch is pushed to the on position, the Check Engine or Power Loss light will come on; when pushed again, it will go out. When the brake pedal is depressed, the Check Engine or Power Loss light will come on and go out

This is the logic module from a 1985 model. This was a unique year because the MAP sensor was mounted on the logic module.

when the pedal is released. This procedure can be repeated for each of the driver input switches.

Dual Module Self Diagnostics (With a Scan Tool)

After the switch test, there is nothing more that can be done without the DRB or other scan tool. The scanner can read and deliver fault codes in a two-digit numerical format delivered on the screen of the scan tool. After completing the fault code test, the technician can perform the switch test. When the logic module detects a change in state on one of the driver-activated switches, the scanner will respond by changing what it displays.

After the switch test, the scanner is also able to perform an additional test. This test is often referred to as the engine running test. Depending on

The logic module contains several microprocessors. These components are sensitive to static discharge. Never disassemble any powertrain control module. While the unit is an assembly, there is plenty of metal to absorb the spikes of static voltage that might come from your body through handling. When disassembled, the chips are open to damage from the static in your body.

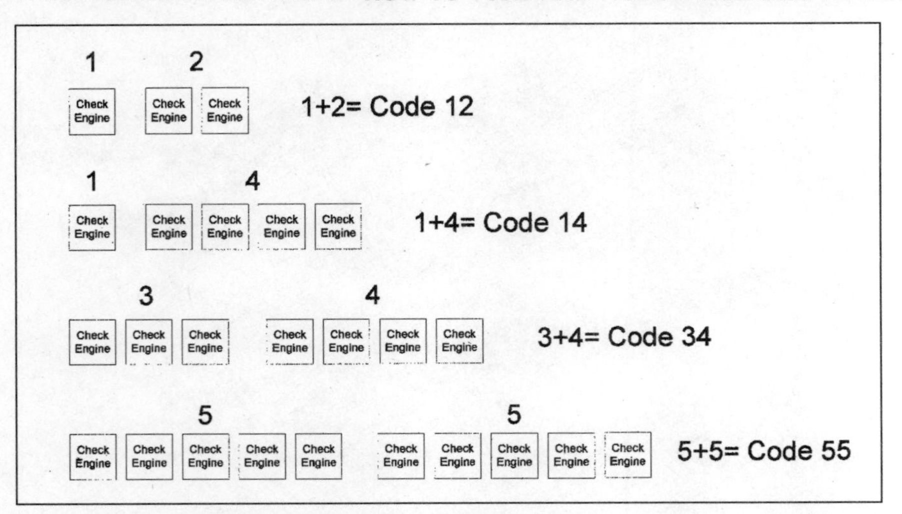

The computer can communicate with the technician whenever he desires. Cycle the key on and off three times leaving the key in the on position. The computer will begin to deliver codes through the Check Engine light. The term "Check Engine" is actually just an industry vernacular. The light will usually say Power Loss or Service Engine Soon. One flash followed by a pause and then two flashes is a code 12. Three flashes, a short pause, then four flashes is a code 34.

the year and model, it may only give the technician an indication of rich and lean O_2 sensor reading, or it may also give readings from the other sensors. More detailed information about diagnostics and scanner data will be discussed later in this book.

The two-module configuration began to disappear in 1987 with the introduction of the Single Module Engine Controller (SMEC) on the 3.0-liter version of the Dodge Caravan and Plymouth Voyager.

Single Module Engine Controller (SMEC)

Until 1987, the minivans were notoriously underpowered. During the second half of the production year, Chrysler introduced the 3.0-liter Mitsubishi engine. The introduction of this engine was accompanied by the introduction of the Single Module Engine Controller. This control module was a major leap forward in control and diagnostics. This configuration placed the logic module's printed circuit inside the power module. These modules are easy to recognize; they have two connectors. One is a 14-terminal connector with relatively large wires. This roughly equates to the power module connections. The second connector is a 60-terminal connector that roughly equates to the connectors of the logic module.

An interesting feature of the SMEC is its self-recovery feature. With the dual module configuration, a fault detected by self diagnostics would place the logic module's programming into a limp-home mode, and the Power Loss (Check Engine) light would come on. Even if the fault corrected itself while the vehicle was being driven, the logic module would remain in the limp-home mode. The SMEC, however, would detect that the problem has been corrected and exit the limp-home mode and the Power Loss/Check Engine light would go out.

With the introduction of the SMEC came a whole new generation of capabilities. These capabilities required the microprocessors to the SMEC to work very hard, which in turn caused them to generate a great deal of heat. Both GM and Ford place their fuel injection computers inside the passenger compartment of the vehicle to keep them cool. Chrysler places the SMEC under the hood to keep it cool. At first this might seem a bit silly, however, there is a great deal of cool air traveling at high velocity available under the hood to cool the computer. The SMEC is therefore usually located in the air induction systems to keep it cool.

Single Module Self Diagnostics (Without a Scan Tool)

The diagnostic capabilities of the SMEC without a scan tool are the same as the dual module configuration. The technician can obtain trouble codes and perform the switch test. All other tests require the use of a scan tool. A scan tool that was introduced about the same time as the SMEC is the DRB II (Digital Readout Box Series 2). At the Dodge dealership that I used to work at, the popular name for this unit was the "dribble."

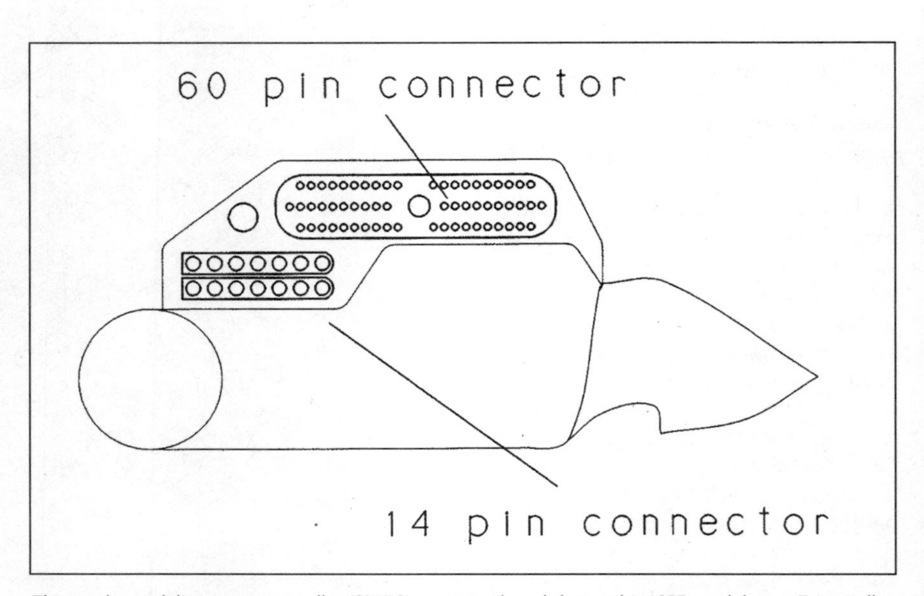

The single module engine controller (SMEC) was introduced during the 1987 model year. Essentially, the board from the logic module has been added to the power module. There are two connectors on the SMEC. There is a 14-pin connector with relatively large wires. This is roughly the equivalent of the power module. The 60-pin connector is the equivalent of the logic module.

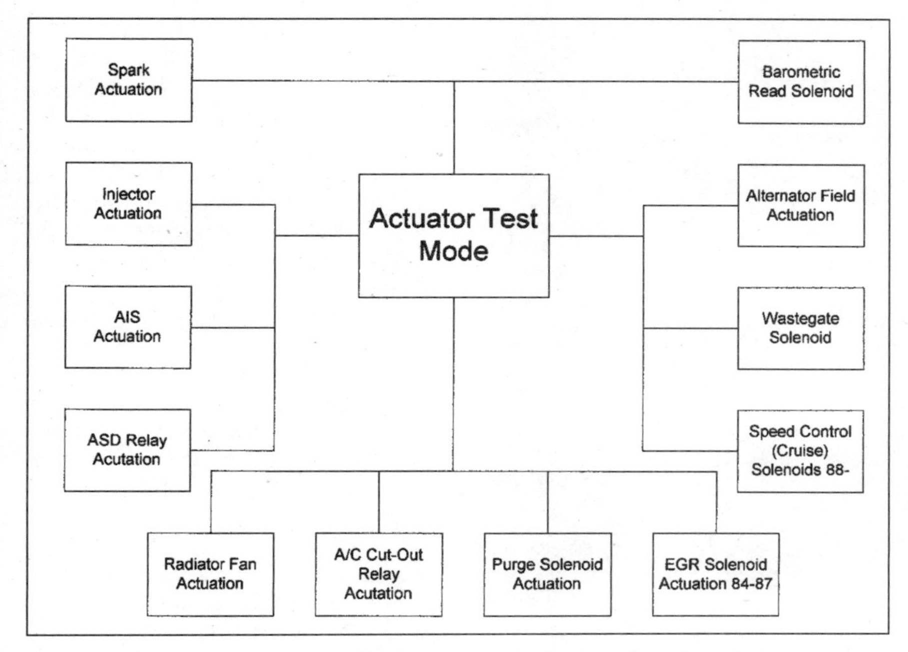

The actuator test mode can be initiated with a scanner. This function allows the technician to test the computer's ability to control its actuators. When these modes are activated, the computer will attempt to operate the selected actuator every two seconds.

Single Module Self Diagnostics (With a Scan Tool)

The DRB II and similar products have opened up a great deal of diagnostic opportunities in the SMEC. A scanner can extract trouble codes, as you might expect, and perform the switch test, as you might expect, and can also perform an actuator test and a data stream test.

Actuator Test Mode (ATM)

The actuator test is used to verify the computer's ability to control the devices which the SMEC controls. When the scan tool is placed in the ATM mode, it allows the technician to choose a device and circuit to test. For instance, the scanner can be instructed to tell the computer to fire the ignition coil every two seconds. With this test mode, the technician is able to ask the SMEC to activate the computer-controlled actuators. This test is unique in the industry. The on-board diagnostics with other applications makes it difficult to test the computer's control of actuators. When the actuator mode for a specific actuator is activated, the technician can test the controlling circuit while the SMEC is attempting to control the device. With a test light or voltmeter, the technician will be able to determine the root cause of a circuit malfunction.

Data Stream Test

The data stream test is now standard throughout the industry. Chrysler and General Motors led the charge in making this technology standard. The serial data stream carries coded information about what the computer is seeing, hearing, feeling, smelling, thinking, and doing. To read this information, the technician will need a scan tool such as the DRB or one of the aftermarket products. There are two types of data supplied by the SMEC: values and voltages. A very important fact of Chrysler troubleshooting is that the voltage readings on the scanner are almost always accurate representations of what is really happening in the system, while the value reading may be inaccurate. In many instances when the SMEC detects readings from a sensor that indicate a faulty circuit, the SMEC will substitute replacement values.

An example of value replacement relates to the MAP sensor circuit. A technician starts the engine and allows the engine to warm up and go into closed loop. While watching the MAP value data field, measured in inches of mercury, he observes the reading is 18 inches hg. The technician then snaps the throttle, the MAP

With the introduction of the single board engine controller (SBEC) in 1990, the computer has only one connector. All of the sensor inputs, power supplies, grounds, and actuator outputs travel in and out though a single 60-pin connector.

If you should ever need to replace the computer on any vehicle, be sure to record the part number and the serial number. The parts people will need this information recorded accurately to get the correct replacement.

reading drops to 5 inches hg, then as the throttle is closed, it climbs to 22 inches hg then levels off at 18 inches hg. The technician now disconnects the vacuum hose that runs from the intake manifold to the MAP sensor. Both when the vacuum hose is disconnected and when the electrical connector is disconnected, the SMEC notices that the MAP input does not match engine operating conditions being indicated by the other sensor inputs. The SMEC then substitutes values for the MAP sensor that are in alignment with the values indicated by these other sensors and sends these substituted values to the scanner. The end result of all this is that the SMEC delivers values to the scanner that are indicative of the reading the MAP sensor should be delivering instead of readings that are indicative of what the MAP is actually delivering.

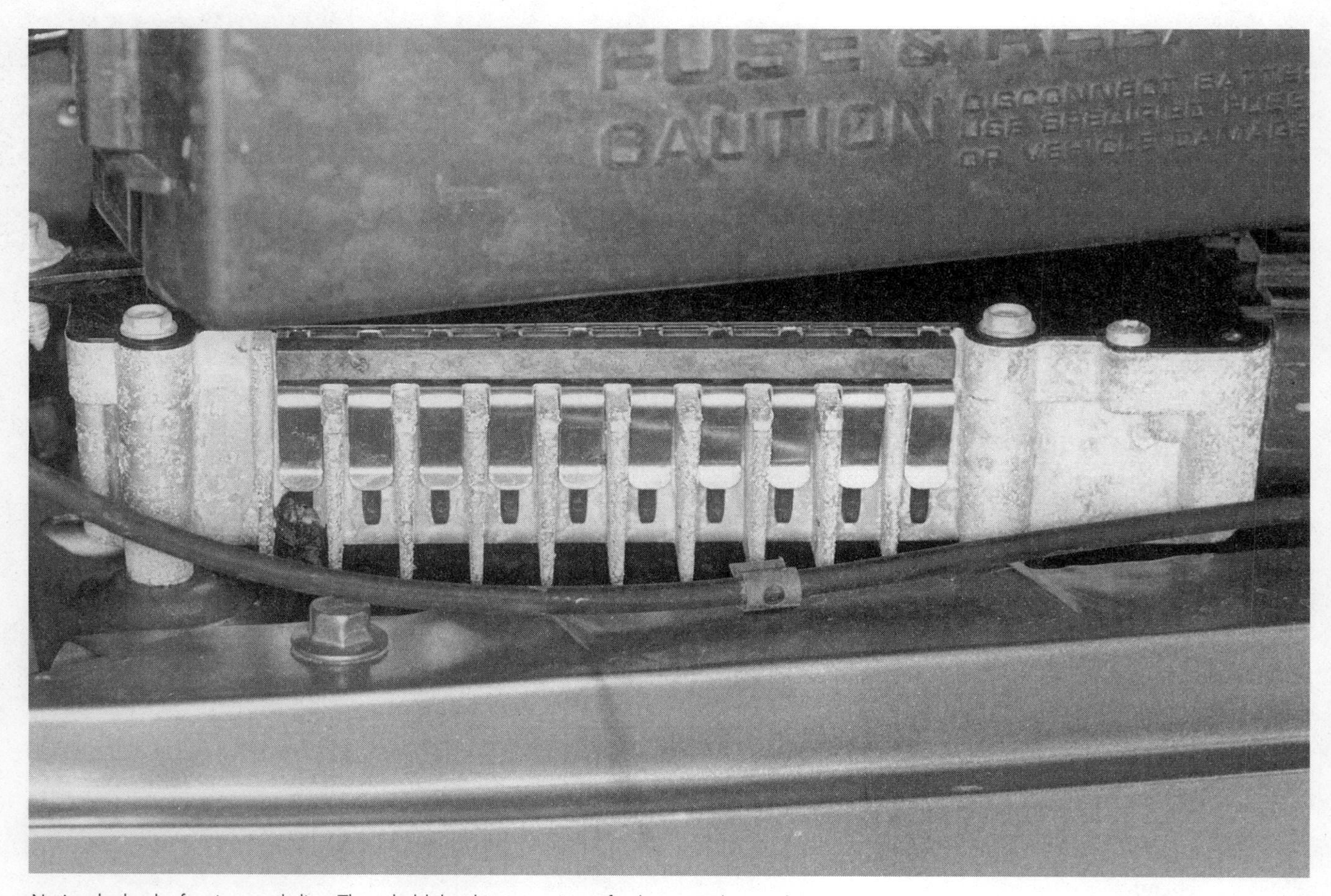

Notice the bank of spring steel clips. These hold the driver transistors firmly against heat sinks to dissipate heat.

The electronics in the SBEC easily rival that of the most sophisticated laptop computer. In addition, the SBEC features a bank of transistors to send current through the actuators.

Therefore, a sensor that is malfunctioning will appear to be functioning when observed on the scan tool. To avoid being tricked, use the voltage data fields. They indicate what is really happening in the electronic engine control system.

Single Board Engine Controller (SBEC)

During the 1990 model year, Chrysler introduced the Single Board Engine Controller, or SBEC. Although very similar in form and function to the SMEC, it eliminates the 14-pin connector. The on-board diagnostics remain pretty much the same from the perspective of the technician, yet are much more thorough.

Interface with the A604 Electronically Controlled Transaxle

A communication link exists between the SMEC or SBEC and the control module for the A604 Electronically Controlled Transaxle. If you remember back to about 1990 or 1991, several of the TV news magazine shows did features on a troublesome Chrysler transmission. The transmission was the A604. This book does not intend to evaluate the problems with this transmission, problems that within the industry are still clouded somewhat in mystery. However, the electronic control module for the A604 uses information about engine operating conditions and parameters that are supplied by the SMEC or SBEC. This information originates with the fuel injection sensors.

THE FUEL SUPPLY SYSTEM

CONTENTS

Fuel Tank

The fuel tank on an electronic fuel-injected car is not significantly different from its carbureted counterpart. The outbound line from the tank is larger to accommodate the increased volume of fuel required by the injection system. There will also be a large return line. At times 90+ percent of the fuel passing through the fuel pump will travel completely through the fuel system and return to the tank.

In some applications, the fuel tank is made of a polymer (that is a polite way of saying plastic). Be careful with the type of fuels used in these applications. Although federal and local regulations are relatively stiff concerning the use of fuel additives and alcohol blends, the simple fact is that gas station pump labeling is not always as accurate as it should be. Methanol in particular has been shown to cause rapid deterioration on many plastics. Choose a gas station and stick with it. When traveling stay with one of the "name brand" fuels.

Fuel Pump and In-Tank Filter

There are two types of fuel pumps used by Chrysler on its fuel-injected cars. One is a low-pressure pump which is used on the post-1985 throttle body applications. The other is a high-pressure pump that is used on the pre-1986 port fuel injection applications. It is actually inaccurate to refer to these pumps as

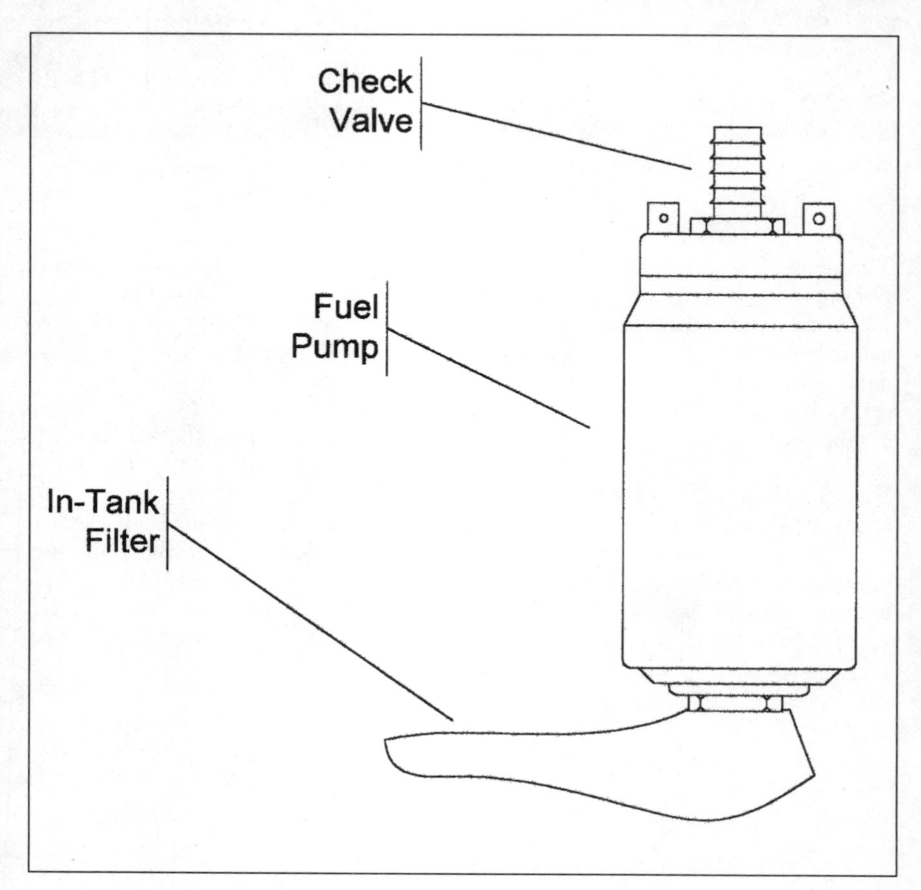

Most applications of Chrysler fuel injection use a single fuel pump located in the tank. Submerged in the fuel, the pump has no trouble pulling fuel, thus reducing the chance of vapor lock to an absolute minimum. Note that there is a filter on the inlet of the pump. This is the first layer of protection against hard particle contamination of the injectors. In the real world, this filter is seldom serviced.

The end opposite the electrical connector is the inlet for the fuel pump. The fuel is drawn in by the roller cell and pushed through the electric motor. As the fuel passes, it cools and lubricates the electric motor. Finally, it passes through the threaded fitting. This fitting also contains a check valve.

high-pressure and low-pressure because pumps do not really create pressure; they merely supply a volume of fuel. It is the fuel pressure regulator restricting the volume of fuel returning to the tank that creates the pressure. Both the low-pressure throttle body injection and high-pressure port fuel injection pumps are located in the fuel tank. The throttle body injection system uses a high-volume DC motor vane type pump that rotates at about 3,500 rpm. The port fuel injection systems use a roller vane pump. Both of these pumps are extremely efficient at pushing fuel through the injection system but are very poor at pulling the fuel from the tank. For this reason, their location inside the tank is ideal.

The design of the pump is such that fuel pulled in through the inlet passes through the electric motor portion of the pump. This fuel passing through the motor acts as a coolant and lubricant for the pump. There is a "sock" filter on the inlet side of the pump that prevents large, hard particle contamination from getting into the pump. The pump can also be damaged by foreign material that can easily pass through this filter. Such things as water and alcohol can do severe damage. Be careful where you buy your fuel and you and your fuel pump will have a long and happy life together.

The fuel pump is equipped with a check valve on the outlet side. This valve prevents fuel in the system from draining back into the tank. If the check valve should become defective, it will cause an "extended start" symptom. After the car sits for a while, the engine will have to be cranked for several seconds before it will start. The cure for this problem is replacing the pump. Keep in mind that a loss of residual fuel volume is not the only possible cause of an extended start symptom.

Fuel Filter

The fuel filter is the main line of defense against hard particle contamination in the injectors. These filters consist of a fine paper mesh filter in a metal can capable of filtering out particles as small as 10 microns (.0004 inches).

Over the years, I have seen many inappropriate installations of parts. A fuel injection fuel filter is one of the most potentially dangerous of these mis-applications. I remember a Volkswagen van where the owner had installed a "universal" fuel filter normally used on VW carbureted engines. This filter was designed to operate in a system where the pressure is only about 3 or 4 psi. This van was fuel injected with a system pressure of over 30 psi. The filter had "ballooned" to about 25 percent larger than its original size with white stress lines running its entire length. This was a potential rolling coffin of fire. Do not attempt to adapt an incorrect filter to your car; maximum pressures, flow rates, and differences in fittings make it either impractical or dangerous.

Some filter manufacturers stamp an arrow on the side of the filter to identify the direction of flow; others do not. When you install a new filter, ensure that it is installed in the proper direction.

The fuel filter is one maintenance part that cannot be changed too often. It is much cheaper than replacing contaminated injectors and pressure regulators. Replace the fuel filter at every major tune-up regardless of the published service interval. Also, replace the fuel filter anytime other components in the fuel system, such as a pump, injectors, or pressure regulator, are replaced.

One fleet organization I worked with a few years ago had a fuel filter service interval of about two weeks. This police fleet had underground fuel storage tanks that had been installed during the 1930s. These storage tanks were contaminated with 50 years of crud. If the technicians allowed the police cars to run more than two weeks, the fuel filters would become so restricted that the pressure of the pump would begin to force contaminants through the filter and into the rest of the fuel system.

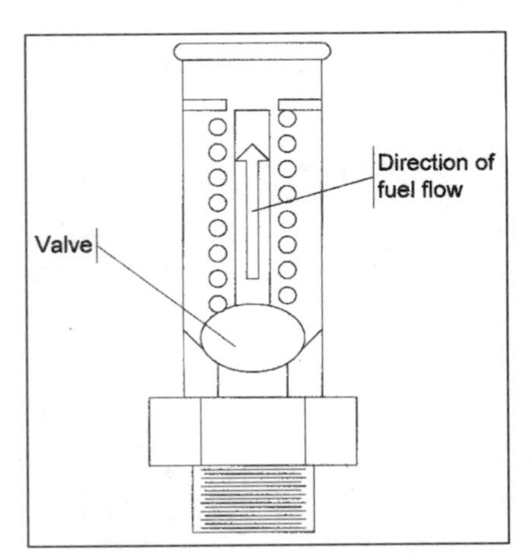

The check valve is responsible for keeping the fuel system filled while the engine is shut off. If the check valve leaks, the fuel in the system will drain back into the tank. While this will not cause any driveability problems, it can cause a problem called extended-start. In an extended-start condition, the engine has to crank several seconds before the engine starts.

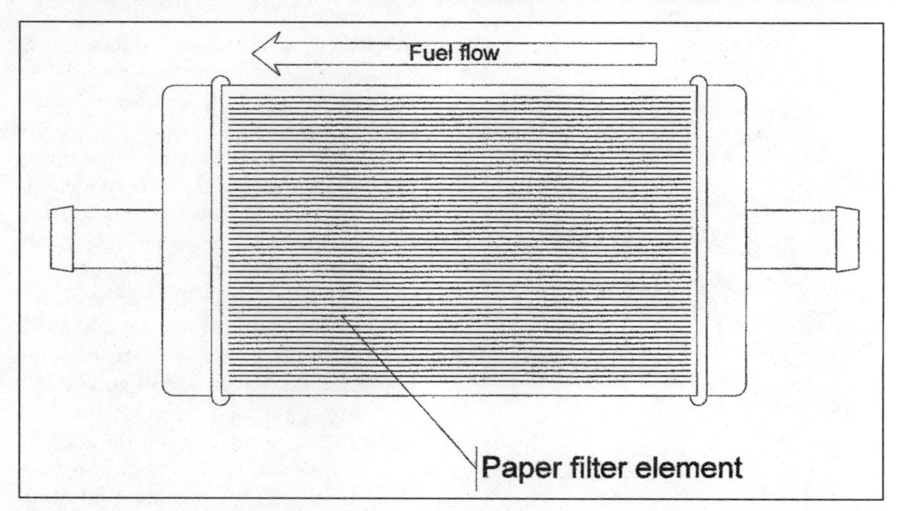

When I teach courses on fuel injection, I tell them that the fuel filter cannot be changed too often. Injectors are far more expensive than an occasional fuel filter. When the filter becomes restricted, it will reduce the fuel flow rate. This will result in a pressure drop downstream of the filter and a pressure increase ahead of the filter. The pump is capable of creating enough pressure ahead of the filter to force crud through, thereby contaminating the fuel injectors.

Fuel Rail

A line from the fuel filter runs to the engine and attaches to the "fuel rail." Chrysler's fuel rail is usually cast aluminum or tubular steel. All of the fuel system components on the engine are attached to it. Although it is less evident, the throttle body injection systems also have a fuel rail. It consists of the chambers within the throttle body assembly that connect the inbound fuel line, the pressure regulator, and the injector(s) together.

Fuel Pressure Regulator

Back in our look at the fuel pump, it was mentioned that pumps do not create pressure but rather a volume; it is the fuel pressure regulator that is the primary restriction in the system to create pressure. Proper fuel pressure is critical to maintaining the correct air/fuel ratio. If the fuel pressure is incorrect, then the electronic controls and sensors will find it difficult or impossible to meter the correct amount of fuel to provide the proper mixture.

Most port fuel injection systems run about 35 psi at an idle. There is a vacuum line connected to the top of the pressure regulator. The vacuum line is connected to manifold vacuum. When the throttle is opened by the driver, manifold vacuum drops, which causes the fuel pressure to increase.

This concept might be a little easier to appreciate when you realize that scientifically speaking, there is no such thing as manifold vacuum. In reality, what we have in the intake manifold with the engine running is a pressure. This pressure is known as manifold pressure and is lower than atmospheric pressure. Since when we measure it we are standing in atmospheric pressure, it feels like a vacuum. This pressure increases as the volume of air entering the manifold increases. Therefore, when the throttle valve is opened, manifold pressure increases, which means the pressure traveling through the vacuum hose to the fuel pressure regulator increases. As the pressure in this line increases, it causes the fuel pressure to increase.

The increase in fuel pressure as the throttle is opened averages 5 to 10 psi and is intended to act as an accelerator pump enriching the mixture during acceleration. Increasing the fuel pressure is also necessary to maintain the proper pressure differential across the tip of the injector as the intake manifold pressure increases. It should be noted that anytime a given throttle setting is maintained, manifold pressure will decrease a bit, therefore the fuel pressure will begin to drop toward the idle pressure.

Throttle body applications control fuel pressure between 15 and 20 psi. Unlike port fuel applications, the throttle body fuel pressure regulator is rebuildable. Rebuild kits are available both through the dealer and aftermarket. Caution should be used when disassembling the fuel pressure regulator as it contains a long spring under sufficient pressure to be a formidable projectile if released carelessly.

Fuel pressure does not change as the throttle is opened on most throttle body applications. Additional pressure is not needed since the fuel is being sprayed ahead of the throttle plates into the relatively constant pressure of the atmosphere.

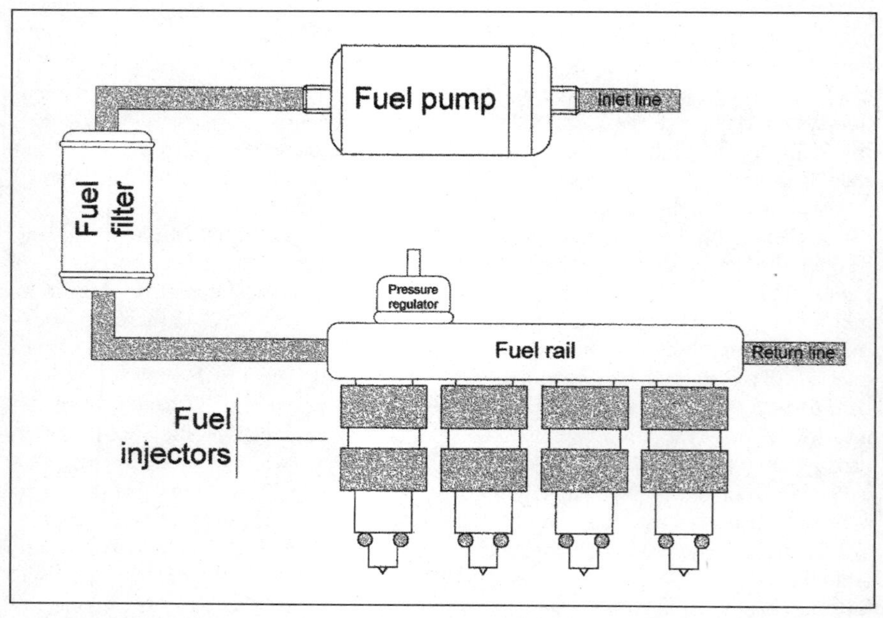

After the fuel passes through the fuel filter, it then goes to the fuel rail. The fuel rail connects the injectors and fuel pressure regulator together.

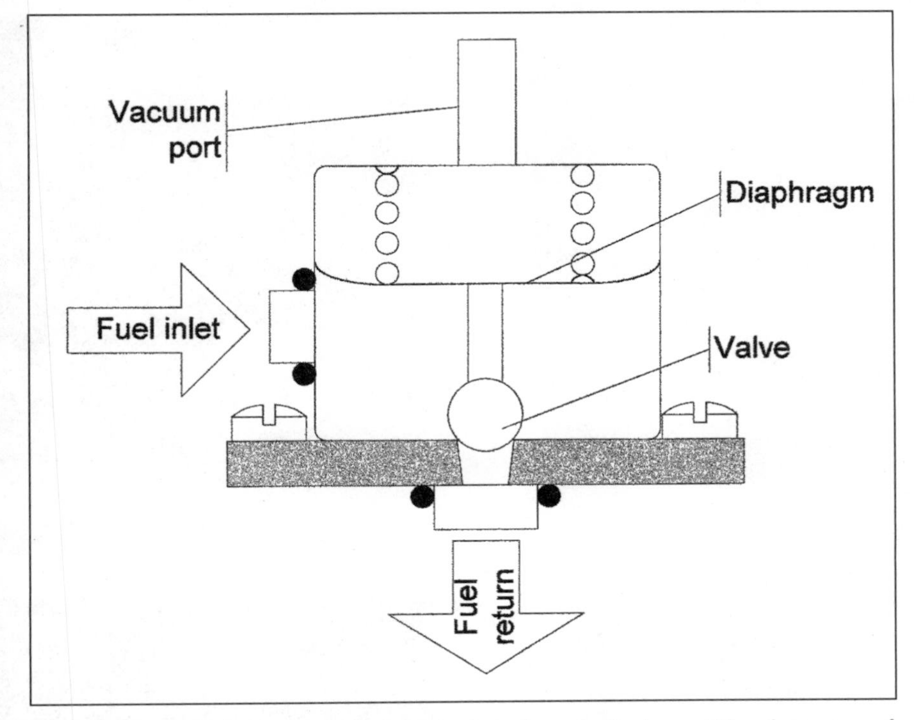

The fuel pressure regulator consists of a spring-loaded valve and a diaphragm. When the pressure of the fuel flowing into the regulator exceeds the set value of the spring, the diaphragm will push against the spring and the valve will open. Excess fuel will return to the tank and the pressure is regulated. The port at the top of the pressure regulator allows intake manifold pressure into the dry chamber above the diaphragm. As the intake manifold pressure rises, the pressure in the chamber rises. This effectively increases the spring tension against the diaphragm. The end result is the fuel pressure rises when the throttle is opened.

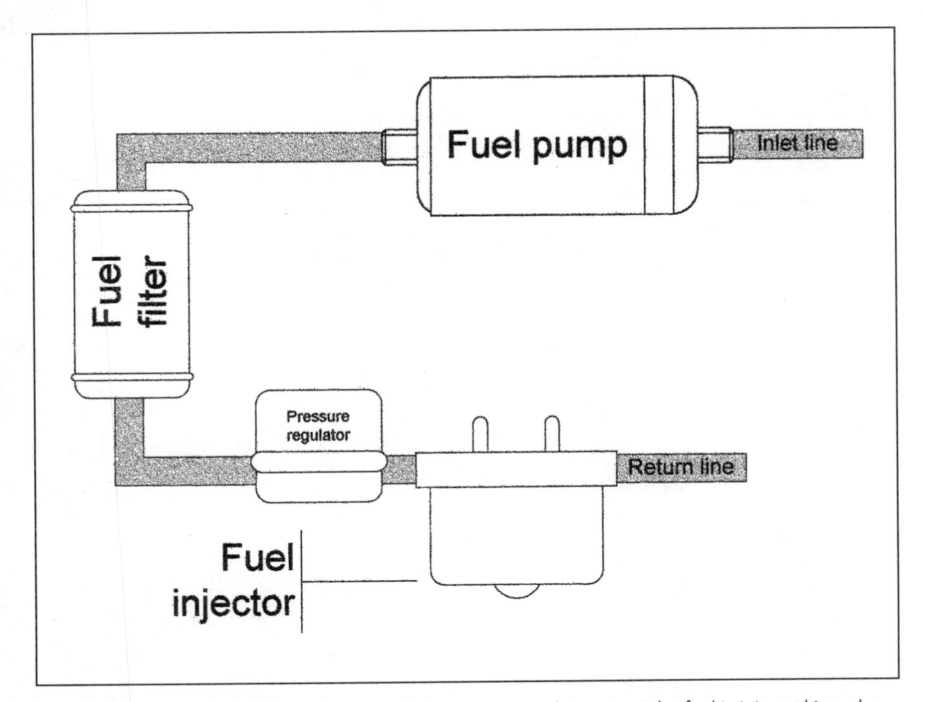

Throttle body injection systems also use a fuel pressure regulator, since the fuel is injected into the air flow stream above the throttle plates. Note there is no manifold pressure port on the top. Since the pressure at that point is relatively constant regardless of throttle position, there is no need to increase the fuel pressure when the throttle is opened.

Fuel Injectors

The modern electronic fuel injector is a normally closed solenoid-operated valve. On Chrysler applications, the injector is connected to a 12-volt power source, and the SMEC connects it to ground in order to energize and open the injector. There are two major categories of injectors used by Chrysler. The first is the top feed or high-pressure injector. This injector is used exclusively on the port fuel injection applications. They are connected to the fuel rail and sealed by an O-ring. The other end of the injector rests in the intake manifold where vacuum leaks are also sealed by an O-ring. Fuel pressures on these applications average 35 psi at an idle.

The other type of injector is known as a side or bottom feed injector. These are used exclusively on the throttle body injection applications at a typical fuel pressure of between 15 and 20 psi. Only one of these injectors is used to fuel an entire four-cylinder engine and two are used on the throttle body injection-equipped V-6s and V-8s.

The injectors on the throttle body injection applications, like those of port fuel injection applications, are connected to switched ignition 12 volts and are grounded by the SMEC to open them.

The SMEC controls the energizing of the injectors with transistors. On the port fuel injection applications, the injectors are synchronized to primary ignition and open simultaneously once every crankshaft revolution. At first this may sound a little inefficient or wasteful, but remember that the fuel sprayed on top of each of the intake valves is the correct amount to fuel that cylinder and will merely sit on top of the valve until the valve opens, allowing the air rushing into the cylinder to carry the fuel. On throttle body injection and CFI, an injector opens every time there is a primary ignition pulse. This means that on a one-injector throttle body injection, the injector opens every time a spark plug fires. On the two-injector applications, the injectors will alternate so that each injector opens every other time a spark plug fires.

Low-Pressure Fuel Injection

A low-pressure fuel injection system is easy to recognize by looking at the injectors. If the electrical connector is on the top and the fuel appears to enter through the side or at the bottom of the

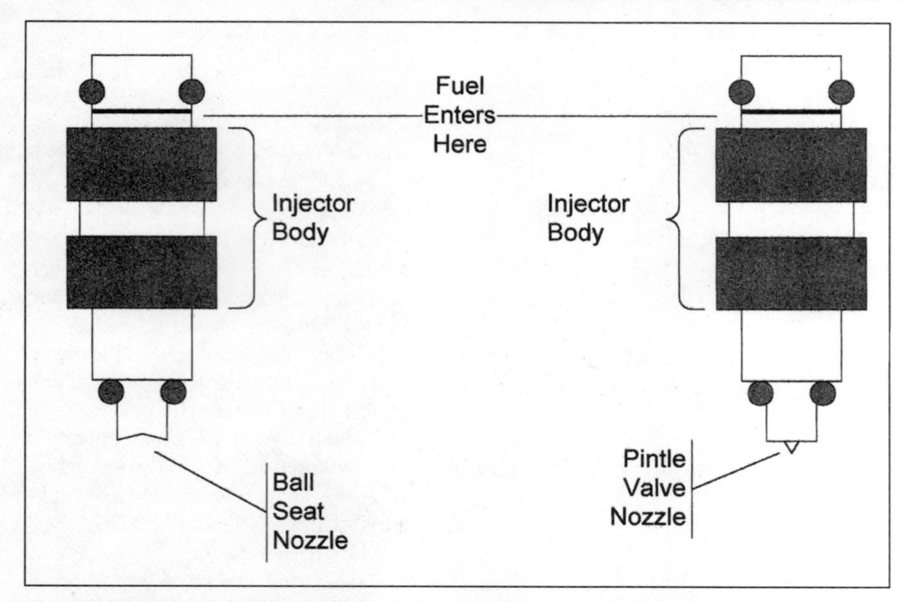

Top-feed injectors are used in the 1984 and 1985 throttle body injection systems and in all the multipoint fuel injection applications. The fuel enters these injectors through the top, passes through the injector, and is sprayed out the tip. There are two types of valves that are used on these injectors. Until recently, the most common was the pintle valve injector shown on the right. Later applications use the ball seat injector.

A high-pressure injection system is easy to recognize. The fuel enters through the top of the injector and the electrical connector is on the side.

injector, then the system is low-pressure. Low-pressure injectors are used only on throttle body applications after 1985.

High-Pressure Fuel Injection

A high-pressure fuel injection system is also easy to recognize by looking at the injectors. If the electrical connector is on the side and the fuel appears to enter the injector from the top, you are dealing with a high-pressure fuel injection system. Multipoint injection systems are all high-pressure. The Chrysler throttle body fuel injection systems on 1984 and 1985 models are high-pressure.

Turbocharged Engines

Chrysler's turbocharged engines were all four-cylinder multipoint-injected engines. Fuel system pressure is based on a specific flow rate through the injector at a given pressure differential between the fuel rail and the intake manifold. On most engines, the fuel pressure differential is kept at about 50 psi. This pressure differential is accomplished through about 40 psi gauge pressure on the fuel and about 10 psi vacuum in the intake manifold. When the throttle is opened, the pressure in the intake manifold rises about 5 psi. The fuel pressure regulator will then increase the fuel pressure by about 5 psi. Turbocharged engines will sometimes have pressure in the intake manifold increase above atmospheric. In order to maintain the proper pressure differential across the tip of the injector, the fuel pressure must be over 60 psi when the engine is at maximum boost.

Multipoint Fuel Injection

With the introduction of the 3.0-liter Mitsubishi engine in the 1987 minivans, Chrysler began using multipoint fuel injection on non-turbo applications. In these systems, the fuel is introduced into the intake manifold just ahead of each intake valve. This method of fuel distribution virtually eliminates the imbalances in air/fuel ratio that are inherent with centralized fuel delivery systems.

These systems have a typical fuel pressure of 35 to 40 psi when the engine is idling and 40 to 45 psi when the engine is not running or when the engine is accelerated. This configuration is now found on the V-6 and V-8 and V-10 applications found on the light-duty trucks.

Sequential Multipoint Fuel Injection

It might be easy to criticize Chrysler for being slow to introduce sequential fuel injection, however, there are few inherent advantages to a sequential fuel injection system over a non-sequential system. It is true that

This is what the tip of the ball seat injector looks like. Notice that the fuel will be sprayed into the engine through four small ports. The pintle valve injector has a tendency to become restricted by deposits that build up on the tip of the injector. The ball seat injector has nothing protruding though the tip into the air induction system. Deposits do not tend to build up on this injector.

The low-pressure injector sprays fuel on top of the throttle plate through multiple orifices at the lower end of the injector.

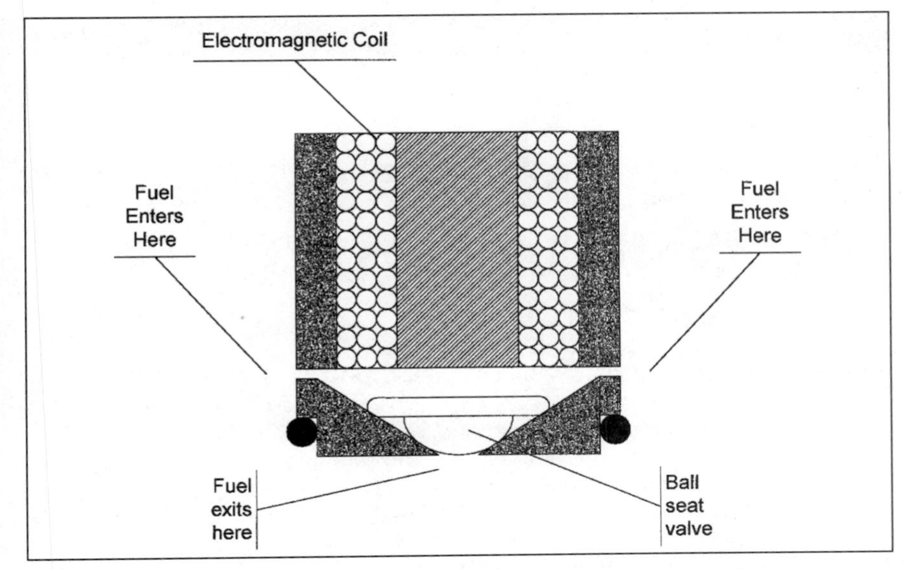

Low-pressure injection, which accounts for throttle body injection after 1985, also uses a ball seat injector. These injectors had an extremely high failure rate during 1986 but have been extremely dependable since. In these low-pressure injectors, the fuel enters through the side of the injector and is sprayed out through the bottom.

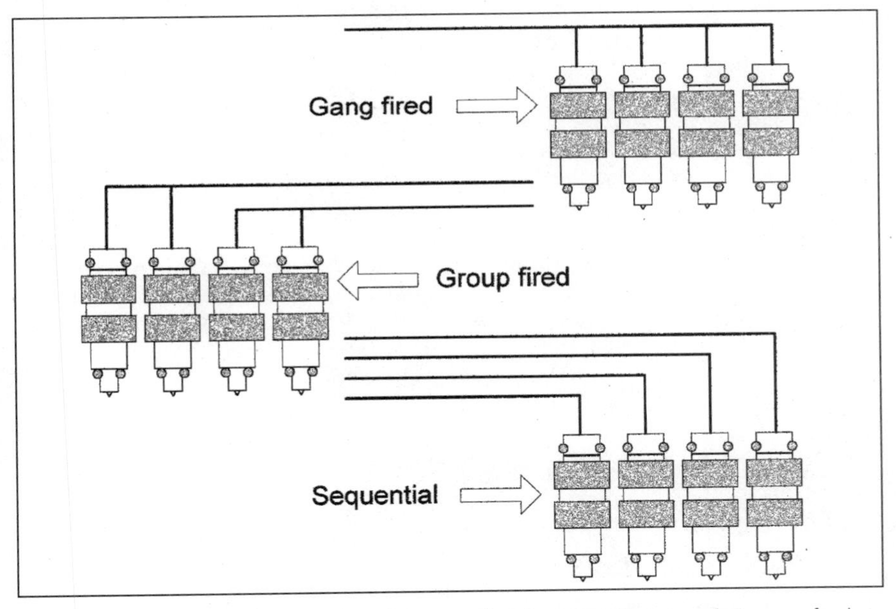

On multipoint injection systems, there are three ways the injectors can be opened. On gang-fired applications, all of the injectors are opened simultaneously. A more common method for Chrysler is to open the injectors in groups. Group-fired applications open the injector in two groups of two or in three groups of two. Sequential injection opens the injectors one at a time. The opening of the injectors is synchronized to the rotation of the camshaft.

sequential fuel injection systems tend to idle smoother and tend to have a smoother transition from idle to cruise. Non-sequential systems tend to perform better at high speeds.

An extra sensor is required for sequential fuel injection systems. The non-sequential systems need only know when one of the pistons is at a consistent though arbitrary point in the stroke. Since a sequential system strives to inject the fuel into the intake manifold at the same point on the same stroke on every cylinder, the computer must know when each of the pistons is at that same point. This sensor is a Hall Effects sensor situated to measure and monitor the rotation of the camshaft.

Injector Flow Testing and Cleaning

During the 1980s, restricted fuel injectors were a chronic problem on fuel-injected engines. Some applications had worse problems than others. About 1989, I attended an SAE workshop on the topic. The car manufacturers were blaming the fuel companies, who in turn were blaming the car manufacturers. There were several causes to the problem that have since been corrected to a large extent. The fuel companies changed their blends of additives. One additive, diolifin, a waxy substance, was attributed with sticking to the pintle of the injector and providing a medium for dirt and soot to stick to the tip of the injector. The amount of this additive was decreased and another was added to help clean the injector as the fuel flows through it. Injector and intake manifolds were modified. By the early 1990s, the problem was greatly reduced, but unfortunately not eliminated.

Performing the injector flow test requires a special tool known as an "electronic fuel injector tester" or "injector pulse tester." This device costs between $100 and $300 and can be obtained through any of the "tool truck" dealers and from many auto supply stores. The tester pulses the injector between 1 an 500 times when activated.

Connect a fuel pressure gauge to the fuel rail. Connect the battery leads of the injector tester to the battery and the number 1 injector. Cycle the key once or twice in order to pressurize the fuel system to between 16 and 19 psi. If the pressure is too high, bleed it down to the proper range before testing the injector. Set the tester for 100 pulses of 5 milliseconds and press the button. The pressure indicated on the fuel pressure gauge will drop; record the amount of pressure loss. Repeat this procedure with each injector, recording the pressure drops. Retest any injector that has a large variation from the others. Any injector which shows more than 1.5 psi difference in pressure drop, either higher or lower than the rest, is either restricted, leaking, or not closing properly. Clean the injectors with one of the injector cleaning systems that are available. These systems cost about $100 and up (one is nearly $5,000), or simply replace the offending injector(s).

CHAPTER SIX

BASIC DRIVEABILITY TROUBLESHOOTING

Driveability troubleshooting is not so much a skill as it is an art. There is a real "gut level" element to discovering the cause of these problems. I have taught, or at least tried to teach, the techniques of troubleshooting these problems to hundreds, in fact thousands, of journey level technicians over the past 15 years. Let us begin with a few truths about troubleshooting:

1. There is only one rule to which there is no exception: there is an exception to every rule.
2. The cause of a driveability problem will always be the last thing tested.
3. The fuel pressure will always be correct...unless you do not test it.
4. The problem will never be electronic...until all the mechanical is proven good.
5. The computer will never be bad... until all other possibilities have been ruled out.

Consider the Symptoms

There are symptoms that are associated with specific components and with specific defects in those components. There is a temptation to use trouble codes early in the diagnostic procedure. This temptation should be fought for as long as possible. Real world experience with these systems indicates that the system defects that cause a driveability problem are seldom system defects that also set trouble codes.

Low Power

There is a common misconception that low power is the result of a lack of fuel. There is a great deal of truth to this when talking about diesels. There is little truth to the statement when talking about spark-ignition engines. Low power in a spark-ignition engine is usually caused by a lack of air volume entering the

Canons of Electronic Fuel Injection Troubleshooting

There is only one rule to which there is no exception: there is an exception to every rule.

The cause of a driveability problem will always be the last thing tested.

The fuel pressure will always be correct...unless you do not test it.

The computer will never be bad... until all other possibilities have been ruled out.

The problem will never be electronic...until all the mechanical is proven good.

In spite of what is common belief among most people, power does not come from getting more fuel into the engine. Power in a gasoline engine comes from the air inhaled by the cylinder while the piston is on the intake stroke. When the complaint is a lack of power, check the air induction system thoroughly. This inspection should, of course, start with the air filter.

Inspection of the air induction system should include the throttle bore. Multipoint fuel injection systems can suffer from a buildup of soot behind the throttle plate. The throttle bore should be shiny and free of buildup.

engine. Sometimes the low power is caused by inadequate heat.

Lack of Air

Air, or lack of it, is the most common cause of poor power. Restrictions to airflow can occur on the intake side or on the exhaust side of the engine. The intake side is obvious: dirty air filters are often responsible for poor power. Less obvious are things that restrict the movement of the throttle pedal or the throttle plate.

Many years ago, I had a regular customer who owned a Volvo. Not too long after completing a tune-up, he returned to complain that I had done a poor job. I checked all the work I had done. Everything seemed fine. During the test drive I did notice that the car seemed to be low on power. The test drive also told me exactly the location of the problem. This was in the 1970s and heavy-grained floor mats known as "cocoa mats" were popular. Apparently, the customer had celebrated the tune-up by installing cocoa mats. They were so thick that they inhibited the movement of the throttle. The throttle never went beyond its half-open position. Excessive throttle cable slack can also prevent the throttle from going to its full open position.

To carry the air requirement to a negative extreme, I was once working as a technician in an import car repair shop near Seattle, Washington. We had a customer come in with the complaint that their Subaru had a top speed of less than 30 mph. The car had already been into several dealerships and independent shops in the area. An amazing list of what had been done was quoted to me. These things included rebuilding the carburetor, replacing the spark plugs, replacing the fuel filter, and even servicing the transmission. The problem was Mount St. Helen's volcanic ash clogging the air filter. Without the air flow, there was no power; without the power, the top speed was extremely low. Air flow is power.

Spark

Poor spark delivery can also reduce power. As stated above, power comes from air; but to be more specific, power comes from expanding the air. If there is an adequate amount of air being drawn into each cylinder, and if there is an approximately appropriate amount of fuel being carried into the cylinders by

Most driveability problems that are thought to be fuel injection related actually end up being ignition related. Most of the time, the best place to start a driveability troubleshooting is by doing a thorough tune-up.

Although a rather unsophisticated tool, there are spark testers that can be purchased rather inexpensively at most auto parts stores. This tool has a spark tester, a file, and a plug gapper all built into one tool. When the broad end of this tool is laid against a firing plug wire with the engine running, a neon light located in the center of the tool will flash. This is not a test of spark quality but is a useful "quick and dirty" tool to determine if a plug is firing at all.

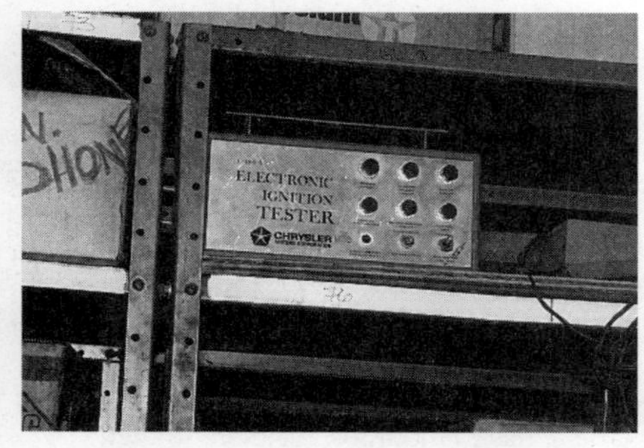

Engine analyzers are useful for testing both the primary and secondary sides of an electronic- or computer-controlled ignition system. For us real people who cannot afford a $40,000 engine analyzer, there is other equipment available. All of the automotive tool distributors offer ignition system testing equipment. The problem with these special testers is that they spend much more time in the tool box or on the shelf than actually helping their owner. The manufacturers also make testers to perform specific tasks available to their dealers and often to the public. I found this unit gathering dust on the parts room shelf of a shop in Hawaii.

the air, then the power will be created by igniting the air/fuel charge with a spark. The absence of this spark, or poor-quality spark, can prevent the proper ignition of the air/fuel charge and therefore reduce power.

Testing the spark or ignition can be a very expensive undertaking and an undertaking that can yield results that are only as good as the person running the tests. A good ignition diagnostic test will usually cost between $50 and $100. This is justified from the perspective of the repair shop owner by the fact that the test equipment can cost up to $50,000. In the mid 1990s, there appeared on the market ignition/engine diagnostic equipment of professional quality that cost less than $5,000. This did not lower the consumer cost of the tests significantly. Let's think about this realistically. A compression gauge can be purchased for about $20. A set of spark plugs for a V-8 engine is about $25. The distributor cap and rotor will cost about $25. A typical set of plug wires is about $35. Add these ignition components together and the price of the compression gauge and you have the price of the diagnostic test. Now on the surface it may seem that there may be components replaced that do not need to be replaced if the diagnostic test is not done. This is true, however, replacing the components almost eliminates the possibility that these components are defective. There is a good chance that the $50 test would have told you to replace $50 worth of parts. Either way you are out $100.

The secondary ignition system is responsible for sending spark to the cylinders. The spark plug wires have a tendency to deteriorate even under the most favorable driving conditions. Remove the plug wires one at a time and measure their resistance. A reading of 5,000 to 10,000 ohms per foot is acceptable.

With my suggestion, the entire $100 was devoted to providing new components. With the diagnostic test, half of the money was spent on questionable information.

Compression

One of the primary jobs of compression is to raise the temperature of the air/fuel charge in the combustion chamber. Low compression reduces the temperature of the charge at the point in the compression stroke where the spark plug fires. If a cylinder has low compression, the quality of the combustion process will be reduced. Low combustion quality will result in high emissions, of course, but also will cause a lowering of power from that cylinder. Low compression can cause low power, the most commonly described symptom resulting from low compression is rough engine operation, especially at idle.

When you remove the spark plug wires they should come off the distributor cap fairly easily. If they do not, remove the distributor cap and inspect the electrodes inside of the cap. On many of the four-cylinder engines, the plug wires are held in the distributor cap by clips that also serve as the electrodes. These plug wires must be removed by unlatching the clip from the inside of the distributor cap.

Ignition Timing

Ignition timing is critical for proper production of power from the engine. Many of today's engines feature distributorless ignition systems that do not make allowances for changing or adjusting the ignition timing. Firing the spark plug too soon will ignite the air/fuel charge before the compression has increased the temperature to the point where the spark can properly ignite the charge. Additionally, the charge, once ignited, will attempt to drive the piston back down the cylinder before it passes over top dead center. In other words, the expanding gases resulting from the ignition of the air/fuel charge tries to make the engine run backwards. Power output of the engine is reduced and the emissions go through the roof.

After checking the resistance of the plug wire, make sure the clip holds the plug wire firmly into the cap.

Carefully inspect the distributor cap for evidence of cracking and carbon tracking.

Actually, the proper adjustment of ignition timing is even more critical than that. The typical backyard technician and even most professional technicians think of the adjustment of ignition timing as being the point at which the air/fuel charge is ignited. This misunderstanding is understandable since that is the aspect of the proper control of ignition timing that we are able to adjust. In reality, when we adjust timing, we are adjusting the point at which the

On most applications, the plug wires are not held in place with clips that have to released from the inside of the cap. This is the cap of a 3.0-liter V-6 engine. To remove the plug wires from this cap, simply pop them out.

The 3.0-liter V-6 distributor cap is a little unusual. None of the spark plug wire towers is located near its required position for the firing order. The electrodes are therefore connected to the proper position by conductive strips. The channels seen running across the underside of the cap contain these strips.

Inspect the rotor carefully for signs of damage or burning.

burn ends, not the point at which it begins. During the combustion process, the fuel and air are drawn into the combustion chamber during the intake stroke. At the end of the intake stroke, the intake valve closes. The piston then begins its journey up the cylinder to compress the air/fuel charge. Just before the piston reaches top dead center, the ignition system bridges the gap across the spark plug with electrical energy. This ignites the air/fuel charge. As soon as the air/fuel charge is ignited, the burning mixture begins to expand. Soon the piston crests over top dead center and the expanding air/fuel mixture begins to drive the piston down the cylinder. When the piston reaches a position corresponding to crankshaft rotation position, about 10 degrees after top dead center, something unfavorable happens. The expanding gases begin to be spread apart to the point where the temperature of the gases drops to the point where combustion can no longer be supported. At this point, the fire goes out. Any fuel that is not burned by this point will be wasted.

The complete burn of the air/fuel charge takes about three milliseconds. For practical purposes, this burn time is a constant regardless of engine speed or engine load. In order to take full advantage of the power in the air/fuel charge, the burn must begin three milliseconds prior to the point after top dead center where the fire goes out. We set the ignition system to deliver a spark at that point which is three milliseconds prior to the point where the fire in the ignited air/fuel charge temperature drops below the temperature where the burn can be sustained.

To explain this concept more thoroughly, let us imagine a fictional engine where the flame front will extinguish at about 10 degrees after top dead center. Let us also say that at 800 rpm the crankshaft rotates 10 degrees in three milliseconds. The timing at 800 rpm must be adjusted at top dead center. When the speed of the engine increases to 1,600 rpm, the crankshaft will rotate 20 degrees in three milliseconds. Now the timing must be set at 10 degrees before top

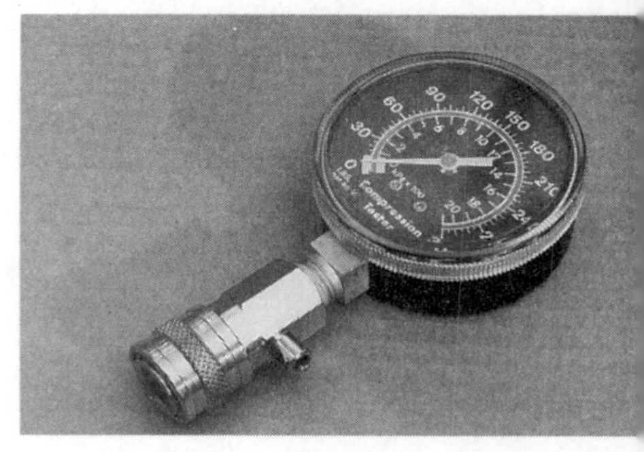

In order for the fuel injection system to do its job, the engine must be in good condition. A compression test should be performed. There are many variables that will determine exactly what the compression should be. An engine that turns over faster during the test will yield a higher compression reading than one that turns over slowly. If you hold the throttle open while performing the test, you will get a higher reading than if you let the throttle remain closed. Basically, you should find the compression to be even on all cylinders. Typical readings are between 150 and 180 psi.

dead center. When the speed of the engine is 2,400 rpm, the crankshaft will swing 30 degrees in three milliseconds. Now timing must be set at 20 degrees before top dead center.

Adjusting the ignition timing makes sure that the burn of the air/fuel charge is complete before the dropping piston expands the heated gases to the point where they are too cold to continue to support combustion. The end result is low power.

Summary

In my experience as an automotive technician, and as a fuel injection specialist in particular, one of the most common complaints I have heard is that of low power. Power is a relative thing. It seems that another car always has more power, that your car *had* more power, or that your car should have more power.

The fuel injection computer and the system associated with it is primarily designed to deliver fuel and to control the ignition timing. Power and loss of power is most closely associated with the ability of the engine to breath. Although fuel- and timing-related problems can cause a loss of power, the place to start a troubleshooting routine is with looking for restrictions in the air induction system and with the exhaust system.

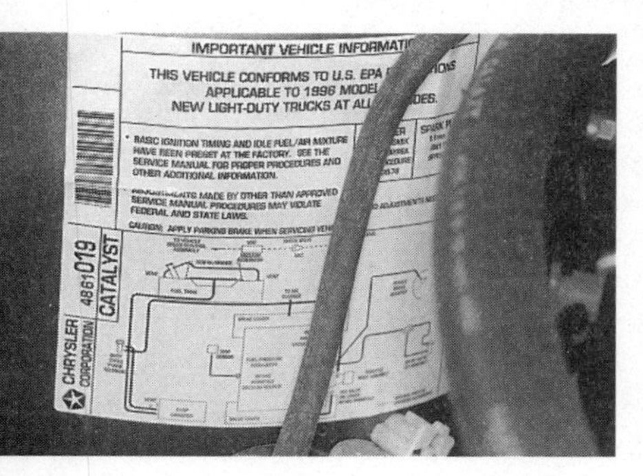

For the proper instructions on checking the ignition timing, be sure to refer to the emission control label located under the hood. This decal was found on the McPherson strut tower.

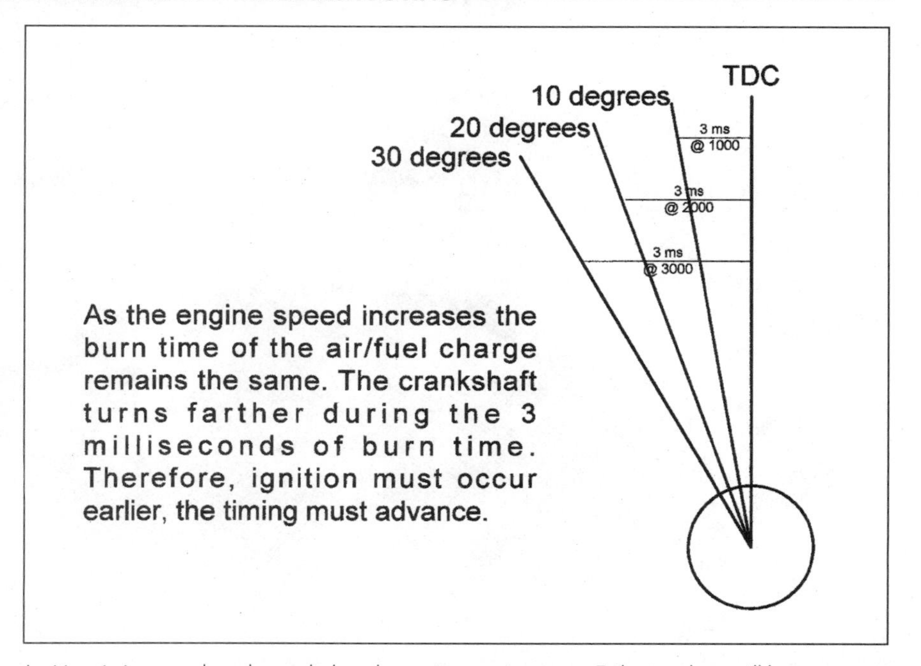

Ignition timing must be advanced when the engine rpm increases. Failure to do so will have a severe effect on exhaust emissions and power. Even distributorless ignition systems require timing advance. Although you cannot adjust the ignition timing, you can still confirm that the timing advances. Connect a timing light to a spark plug wire. Make a reference mark in three places around the harmonic balancer and start the engine. Shine the timing light on the balancer. One of the three marks you created should be visible. Rev the engine. The timing mark should appear to move.

Engine Runs Rich

I cannot remember a customer ever coming to me and saying, "My car is running rich." Instead, they speak of the problems symptomatically. Typical symptoms when the engine is running rich are: brown or black smoke, a fuel smell, a rotten egg smell, or poor fuel economy.

Although the most common user complaint is poor fuel economy, there are much more serious consequences if the problems are left unrepaired. Two of these potential problems rank pretty close to one another in terms of seriousness. The first is overheating the catalytic converter.

An engine that is running rich produces high levels of carbon monoxide (CO) and hydrocarbons (HC). When these enter the catalytic converter, it tries very hard to convert these gases to carbon dioxide (CO2) and water. This process generates heat. The more CO, the more heat; the more HC, the more heat. Soon the temperature of the catalytic converter will be high enough to begin melting the substrate. Melting of the substrate reduces the cross-sectional area of the converter and this causes temperatures to increase even more. Soon the catalytic converter is ruined, the exhaust system becomes restricted, and there is a noticeable loss in power.

The second problem with an engine that runs rich is engine damage. When the fuel flow is excessive, the gasoline, an excellent solvent, tends to wash the oil off of the cylinder wall. This wash-down reduces the lubrication of the cylinder walls and

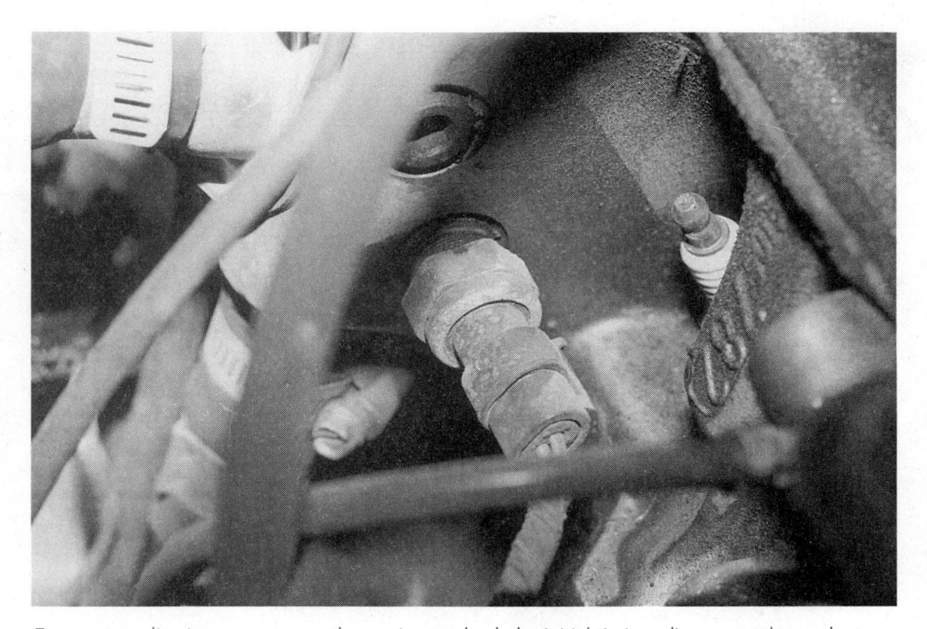

For most applications, to prepare the engine to check the initial timing, disconnect the coolant sensor and start the engine. The absence of the coolant sensor signal will cause the computer to go into a limp-home mode. In this mode, the timing becomes fixed at initial start-up. When the timing test is completed, reconnect the coolant sensor. On pre-SMEC applications, it will be necessary to turn the ignition switch off and restart the engine to get the computer to exit the limp-home mode. On SMEC and later applications, the computer will return to normal operation as soon as the sensor is reconnected.

When an engine is running rich, it produces high levels of carbon monoxide and hydrocarbons. In addition to harming the atmosphere, there is the potential of damaging the catalytic converter.

that can result in engine deterioration. Engine overhauls that are instigated by wear of the cylinder walls are always expensive.

Now a rich running condition is a problem that truly is likely to be caused by the fuel injection system. The term "rich" means that there is an imbalance in the air/fuel ratio. Either the engine is getting too much fuel or not enough air. Check the air filter for restrictions caused by dirt and oil. Check the exhaust system for restrictions as well. If the engine is unable to remove the spent exhaust gases from the cylinders, then it will be unable to draw in fresh air for combustion.

If there is no evidence of restriction to air flow, then it is time to get into the system that controls fuel flow. The first thing that should always be checked when the fuel injection system is suspected of being the cause of a driveability problem is the fuel pressure.

High fuel pressure can cause the engine to run rich. The fuel injection computer has been programmed to assume that the fuel pressure is a specific amount. Flow rate is directly linked to the pressure of the fuel. Low fuel pressure

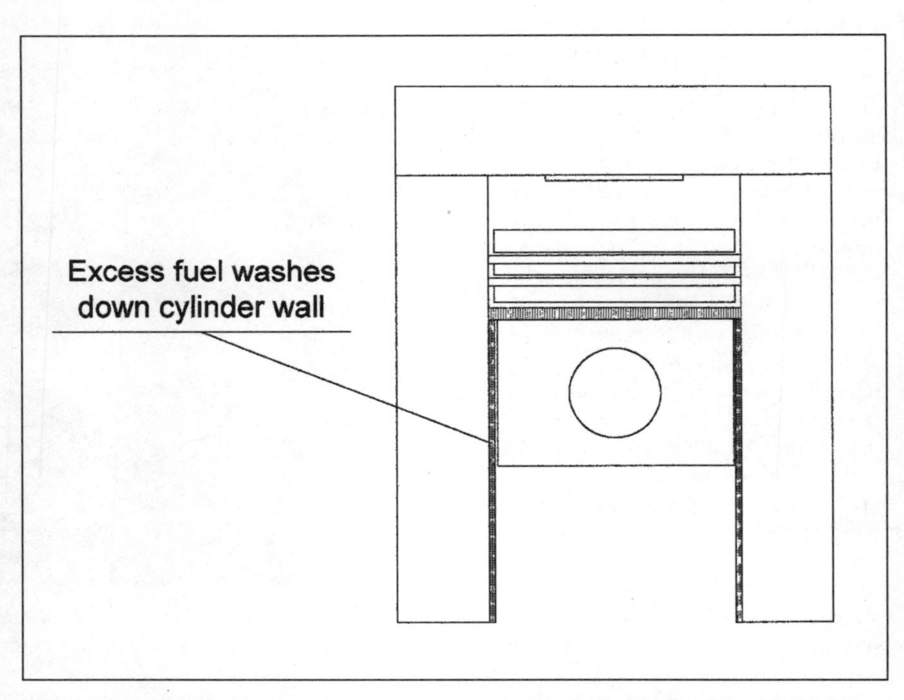

When the engine runs rich, excess fuel may cling to the cylinder walls. This fuel dilutes the oil that should lubricate the cylinder wall. The long term result will be excessive wear to the engine.

When the engine runs too lean, it becomes a polluter, just like when it is running too rich. The primary emission when the engine is running too lean is unburned fuel.

can cause an insufficient supply of fuel while high fuel pressure will cause an excessive flow of fuel. Causes of high fuel pressure are restrictions in the fuel system return line to the tank or a defective fuel pressure regulator.

Engine Runs Lean

Lean-running engines are usually observed by the operator or owner as a stumble, a hesitation, or a bog. While fuel is not the actual power of the engine (remember, it is the expanding gases), an insufficient amount of fuel can cause an insufficient amount of heat. When there is not enough heat to properly expand the air, there is a loss of power. With many ignition systems, especially the older ignition systems, the coil may not have enough energy to fire the spark plugs in lean-running cylinders. In fact, this can cause a misfire that can cause HC (hydrocarbon) emissions to elevate.

Check the fuel pressure. Low fuel pressure can cause the engine to run lean. Check the injectors. During the mid-1980s, restricted injectors were very common. These restrictions were caused primarily by sediments and carbon building up on the external tip of the injector. Reformulation of gasoline and changes in injector and intake manifold design have greatly reduced this problem.

When looking for the causes of low fuel pressure, remember that the fuel pump is a device designed to deliver a volume of fuel to the injection system. Pumps do not create pressure; they only deliver a volume. Therefore, low fuel pressure is almost always a low-volume problem. The most notorious volume restrictor is a dirty fuel filter. Low pressure can also be caused by kinks in the fuel supply (inbound to the fuel injection system) lines, a bad fuel pressure regulator, and a worn fuel pump. Of these three, the least likely

The TPS is a potentiometer. A potentiometer is a piece of metal (the wiper) that moves back and forth across a carbon-based paint. Eventually the wiper will wear a hole in the paint. When the wiper moves across this hole, the voltage will drop to zero, telling the computer that additional fuel for acceleration will no longer be needed. The result is a severe hesitation or stumble.

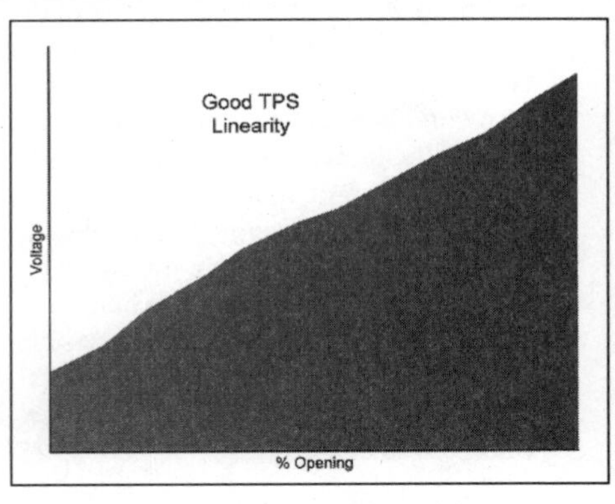

Problems with the TPS can cause stumbles and hesitation. Connect an analog voltmeter to the signal wire of the TPS. Slowly move the throttle to the wide open position while watching the voltmeter. The voltage should change smoothly and proportionally to the amount that the throttle is opened.

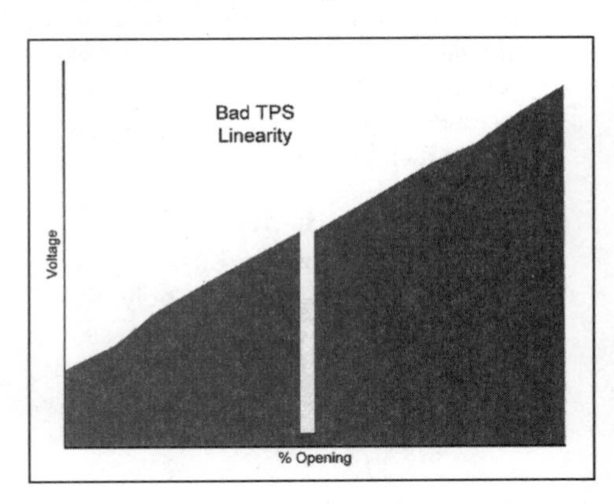

to result in low pressure is a defective fuel pump. Typically, when a fuel pump goes bad, it fails completely.

Hesitation

Next to fuel economy problems, the most common driveability complaint probably has to do with hesitations. The term hesitation is used to describe the sensation that occurs when you move your foot toward the floor and the engine says, "You want me to do what?" before the car accelerates.

Hesitation is caused by an insufficient supply of one of the three basic ingredients needed to get power from an engine: air, fuel, and spark. I have had customers complain of a hesitation when the real problem was a bind in the movement of the throttle linkage. Technically, they are right in their description because this bind interferes with air flow. A weak spark can affect combustion quality, especially when the air/fuel ratio is incorrect. Upon initial acceleration, the throttle has just been opened and there is a large increase in the mass of air entering the combustion chambers. This causes at least a momentary lean condition. A lean cylinder is much more difficult to ignite than a normal or rich cylinder. The spark plug sparks, but not with enough energy to completely burn the air/fuel mixture in the combustion chamber. The result is a momentary loss of power that we call a hesitation. Even if the air/fuel ratio is correct, low power from the ignition system can cause a hesitation.

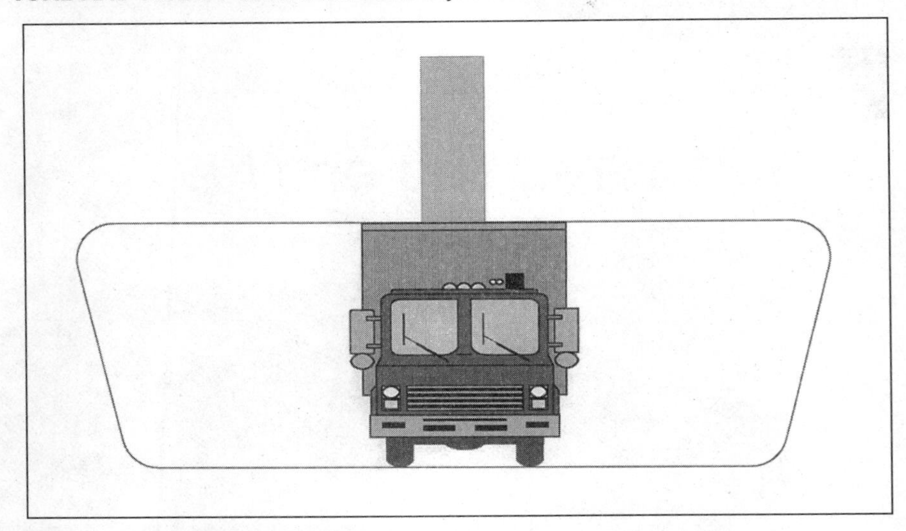

A stalling problem can be very embarrassing when there is a Kenworth in your rearview mirror.

Stumble

The terms hesitation and stumble are often used interchangeably. That really is okay since the causes of each of these problems are pretty much the same. Technically, a stumble occurs when the engine is unable to deliver enough power to provide for the change in engine demand requested by the driver. Causes of stumbles, like hesitation, include: an interruption in air flow, inadequate fuel delivery, or a weak spark.

Car Fails Emission Test

At this writing, most jurisdictions in the United States test only three gases. These are carbon monoxide (CO), hydrocarbons (HC), and carbon dioxide (CO_2). Carbon monoxide is tested to ensure that the air/fuel ratio is correct. Hydrocarbons are tested to verify the quality of combustion, and carbon dioxide is tested to make sure the exhaust system is in good condition. Although problems with the fuel delivery system can cause any of the three to be incorrect, it is far more likely that the problem comes from malfunctions in the air induction, exhaust system, or ignition system.

Stalling

This may just be the most annoying of all problems to most drivers, and certainly one of the most dangerous.

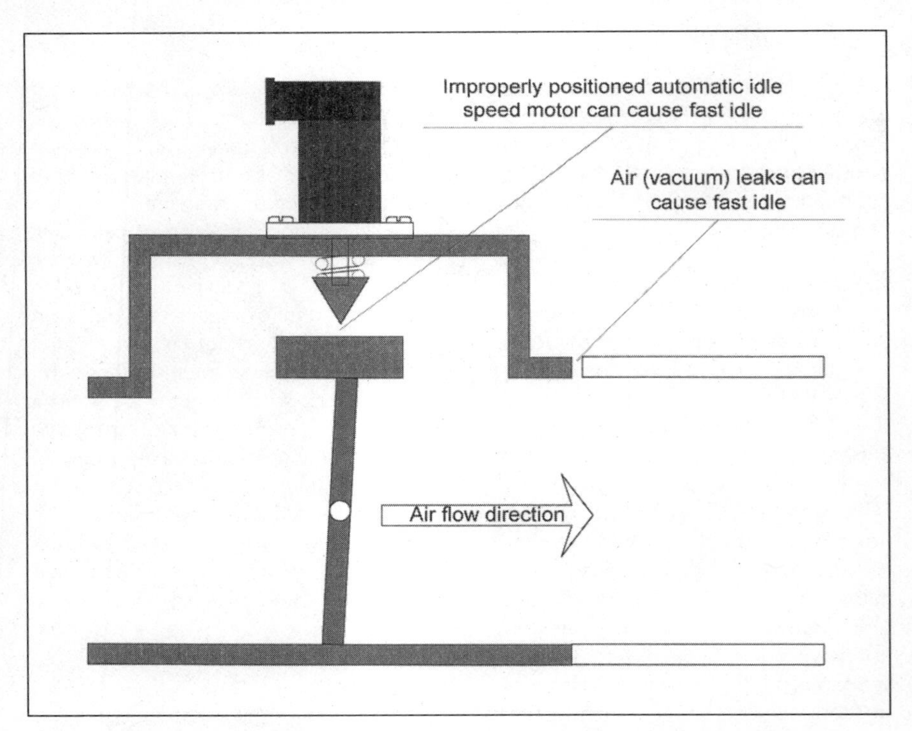

Assuming that the throttle plates are closing properly, there are two other likely places where extra air could be leaking into the air induction system, causing a high idle. The first of these is the idle air stepper motor. This device controls idle speed based on computer demand. Located in a bypass channel to the main stream of air flow, it can be in one of 256 different positions. If the computer is issuing incorrect commands, or if the stepper motor is failing to obey the commands, too much air can flow into the engine. Any air (vacuum) leak in the induction system after the throttle plates can also cause a high idle.

There are few things more embarrassing than pulling onto a busy freeway in your LeBaron and having the engine die just as you are looking in your rear view mirror to see a nice big Kenworth with a rebel flag on the grille barreling down on you at 85 miles per hour. Stalling is caused by the loss or sudden degrading of ignition spark, an interruption of fuel flow, sudden excessive fuel flow, or a sudden blockage to the air flow. Although there are a couple of simple things that can cause stalling, the most common things are the high-tech items.

The first low-tech thing to check would be the fuel level. When the vehicle accelerates, the fuel will often slosh away from the fuel pickup. The engine may be starved for fuel long enough to stall. On older, carbureted applications, this was not a problem because the carburetor stored fuel in the fuel bowl. On acceleration, the fuel might slosh away from the fuel pickup but will slosh back again before the fuel bowl goes dry. You do not have the luxury of a fuel bowl in a fuel-injected engine; any loss of fuel volume through the pump will cause an instant loss in fuel pressure and therefore an immediate decrease in the amount of fuel being delivered to the engine.

Another low-tech cause of stalling is shorts in the wiring harness of the ignition system. This includes both the primary and secondary sides. Look for breaks in the wires. Look in particular for damage to the insulation of either the primary or secondary wires to the ignition coil. These inspections are tedious but essential because even the smallest crack in the insulation can allow the ignition system to ground and cause the engine to stall.

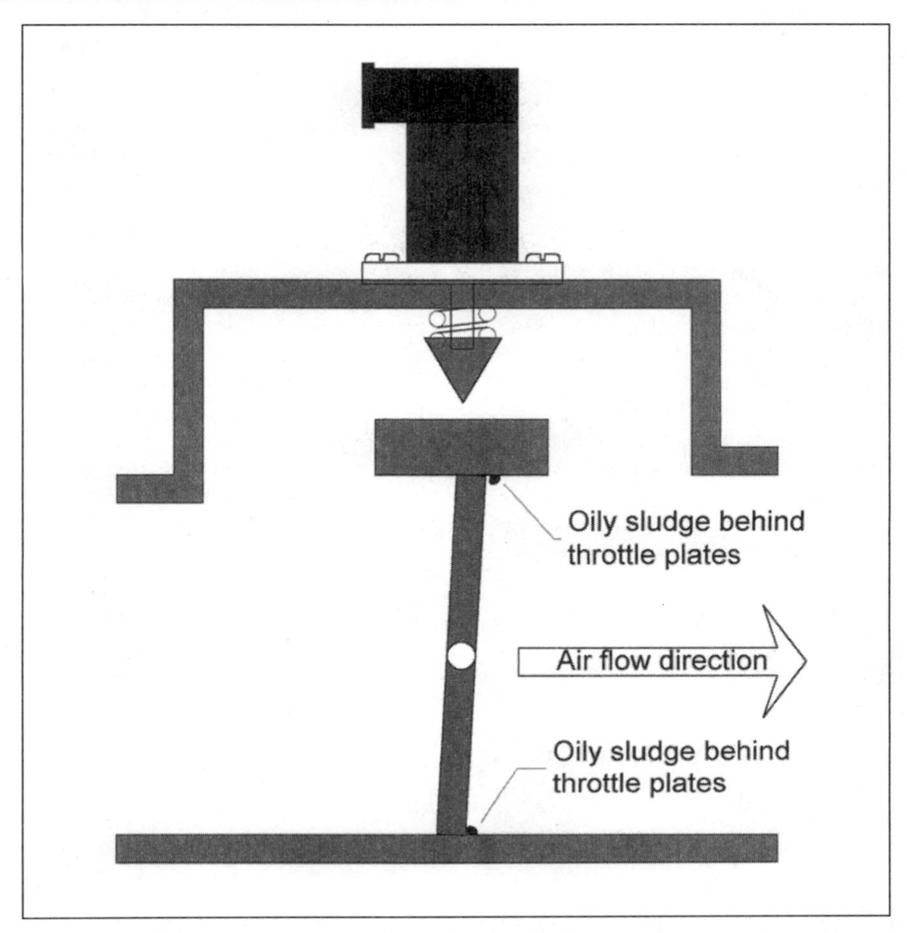

One of the leading causes of idle problems is the build-up of sludge behind the throttle plates. This can cause several problems. First, the sludge impedes air flow past the throttle plates when they are closed and therefore the automatic idle speed motor must be solely responsible for controlling idle. Second, the throttle plate may fail to close properly, which can cause incorrect closed throttle voltages from the TPS, which means the computer may fail to recognize when the throttle is closed and therefore fail to properly control idle speed.

Engine Idles Fast

High idle speed problems can never be caused by a problem with fuel delivery. The speed of a gasoline spark-ignition engine is determined almost entirely by the mass of air entering the engine. If you open the throttle, you increase the air mass; close the throttle, and you decrease the air mass. Begin by checking the throttle linkage adjustment. Make sure that the linkage is not adjusted so tightly that it partially holds the throttle open. Make sure that the linkage, or cable, does not bind. Also look for sources of extra air entering the engine. A common place for this extra air to leak into the engine is where the throttle assembly is mounted on the intake manifold.

Like everything else talked about so far, the idle speed of the engine is controlled by the computer. After the basic checks referred to above have been completed, there are some computer-related items to test. Those procedures will be described later.

Engine Will Not Idle

Again, engine speed is determined by the air mass entering the engine. If the engine does not idle but runs at all other speeds, it is usually an indication that there is an insufficient mass of air entering the engine when the throttle is closed. There are three major possibilities here. First, the throttle stop may not be properly adjusted; second, the throttle bore may be coked; and third, the idle control device may be inoperative.

There is an adjustment, called the minimum air adjustment, that is essential for the proper operation and proper idling of a fuel-injected engine. The minimum air adjustment determines the amount of air allowed to pass across the throttle plates when they are closed. Incorrectly adjusted minimum air can result in tip-in hesitation, stumbling, and stalling on deceleration.

To begin the minimum air flow adjustment, it is necessary to check for throttle body coking. When a port fuel-injected engine (an application with one injector per cylinder) is shut off, hot crankcase vapors rise through the positive crankcase ventilation (PCV) system into the air intake. As they rise, they carry oil and soot which rests in the intake system, coating the connector tube between the power module, SMEC, or SBEC, and the throttle plates. When the engine is started, incoming air picks up this oil and soot, depositing it at the first low-pressure area it comes to just behind the throttle plates. Evidence of coking is a ridge of soot built-up behind the throttle plates which can be felt with the tip of the index finger.

Coking behind the throttle plates forms a seal which reduces the amount of air passing across them. Minimum air flow is affected, causing hesitation and stalling. The Idle Air Control passage can also become coked, causing erratic idle speed control problems.

The best way to clean throttle bore coking is to remove the throttle assembly from the intake manifold. Using solvent and an old toothbrush, scrub the area in front of and behind the throttle plates. Remove the Automatic Idle Speed valve from the throttle assembly, dip a rag in solvent, and clean the pintle and spring of the AIS. With a small toothbrush, or several pipe cleaners twisted together, scrub out the AIS passage. Reassemble the throttle assembly and install back on the intake manifold. In order to extend the interval for this cleaning procedure, also scrub the corrugated rubber hose that connects the computer to the throttle assembly.

It should be noted that throttle body coking is almost never a problem on TBI cars. On TBI, there is a constant flow of fuel through the throttle bore which keeps any build up from occurring behind the throttle plates.

After all that, you can finally adjust minimum air. Minimum air is the amount of air passing across the throttle plates when the throttle is closed. This specification is given in rpm. Connect a tachometer to the negative side of the coil. If your engine is distributorless, then you may have to use an inductive tach or the tachometer on the instrument panel if the car is equipped with one. A scanner can also be used. Force the AIS motor to close off the air passage as much as possible. This can be done by just barely opening the throttle and waiting for the computer to bring the idle speed down. Then repeat the procedure until

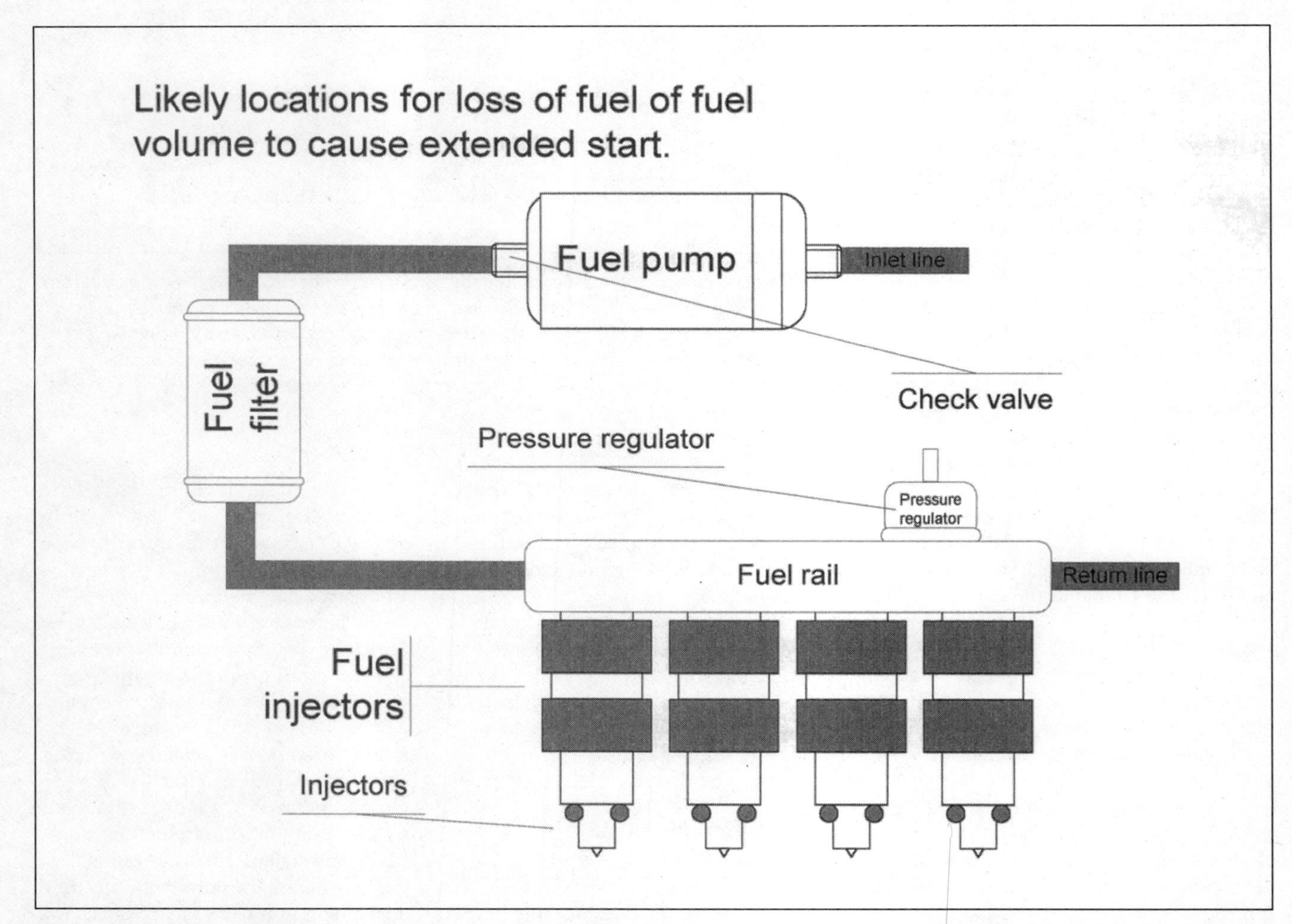

Another common starting complaint is the extended start. Fuel-injected engines that are in good condition and tuned properly start almost the instant the key is rotated to the start position. When they fail to do so, it is often a fuel supply-related problem. The fuel pump check valve, the fuel pressure regulator, and the injectors should hold a volume of fuel in the system when the engine is not running. If any of these items leak, there will not be a supply of fuel ready when the engine is cranked. Worst of the potential leak points is the injectors, as they will also flood the engine when they leak.

the computer can no longer compensate for your movement of the throttle. This is at best a flaky procedure and definitely not the most precise method. A far better way is to use a scan tool. There is a test mode in the Chrysler program of most scanners that will force the AIS to close off the idle air bypass. This minimum air idle speed should be considerably less than the curb idle speed. Typical minimum idle speeds range from 500 to 700 rpm. For most applications, this rather large spec range is adequate.

Engine is Difficult to Start

For most people, one of the most noticeable advantages of electronically fuel-injected engines is the ease with which they start. If the engine runs well once started, it is unlikely that the problem is fuel delivery oriented. The same is true of the ignition system. The most likely cause of difficulty in starting is a problem in the starter motor system. The starter motor requires 100-300 amps, depending on engine size, to crank the engine at sufficient speed for it to start. This equates to about 1,000 to 3,000 watts of power at minimum cranking voltage. That in turn translates to about 1.34 to 4.02 horsepower. The bigger the engine, the more horsepower it takes to crank the engine over at proper speed. Resistance in an electrical circuit reduces current flow and therefore reduces the amount of power available to crank the engine over. The result will be low cranking speed. Low cranking speed will mean that the engine will be creating less compression in the cylinders. Less compression means less heat from compression. The bottom line is that the ignition system has to work extra hard to ignite the air/fuel charge. A tune-up will provide new ignition components, and therefore might make the engine start easier, but will not cure the problem.

Beginning with the battery, test all the components of the starting system. To test the battery, disconnect the primary side of the ignition coil. Connect a voltmeter across the battery terminals. Observe the open circuit voltage. Crank the engine for 15 seconds. Observe the battery terminal voltage while the engine is being cranked. The voltage should remain above 10 the whole time the starter is engaged. After the key is released, the voltage should rise within about five seconds to about 95 percent of the original open circuit voltage. If the voltage fails to rise rapidly, the battery is defective and must be replaced. If the voltage drops below 10 volts at any time during the 15 seconds, recharge the battery and repeat the test. During the second test, if the voltage drops below 10 volts or if the voltage fails to recover rapidly, replace the battery.

To eliminate resistance in the starter circuit as being the cause of the starting problem, or to isolate resistance in the starting circuit that is the cause of the starting problem, perform the following test. Connect the red lead of a voltmeter to the positive battery terminal. Connect the black lead of that voltmeter to the biggest post of the starter solenoid. Crank the engine while observing the voltage. If the voltage displayed on the voltmeter is greater than 0.2 volts, then the cable has high resistance and could cause the current flow to be reduced enough to cause difficult starting. Repeat this procedure for each of the cables in the starter system. Replace any cable that shows a voltage drop greater than 0.2 volts.

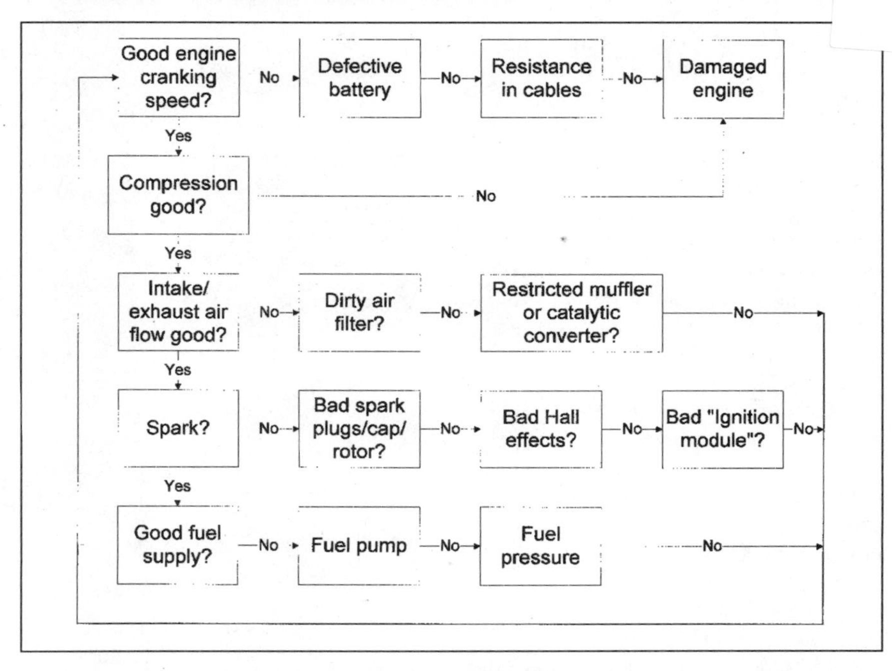

A quick and dirty chart for troubleshooting no-start problems. Note that emphasis is on the mechanical and electrical. Most no-start problems remain related to these basic areas.

Extended Start Problem

An extended start problem is where the engine cranks at the proper speed and starts, but it must crank for several seconds first. The most common cause of this symptom is a loss of fuel volume in the system while the engine is shut off. The fuel pump is equipped with a check valve on the outlet side. This valve prevents fuel in the system from draining back into the tank. If the check valve should become defective, it will cause the extended start symptom. After the car sits for a while, the engine will have to be cranked for several seconds before it will start. The cure for this problem is replacing the pump. Keep in mind that a loss of residual fuel volume is not the only possible cause of an extended start symptom.

No-Start Problem

The following is a logical procedure in testing for a no-start condition.

1. Check Engine Cranking Speed

Probably one of the quickest and easiest (although not very scientific) ways of checking engine cranking speed is by sound. This is especially effective if you are working on your own car or a model with which you are very familiar. The use of a tachometer for this test is good since it will not only tell you what the cranking speed of the engine is, but if you do not get a tach reading, that will tell you the primary ignition is not working. Traditionally, the tachometer is connected to the negative side of the coil. If you cannot find the negative side of the coil, or if you are working on a distributorless engine, connect your digital tach to the purple/white wire that runs from the ignition module to the ECM.

2. Check Compression

Compression is a function of engine cranking speed and proper cylinder sealing. Like checking engine cranking speed, the skilled technician can often tell by the sound of the engine whether or not there is sufficient compression to start. If there is any doubt as to the quality of compression, a proper compression test should be done. Low compression on one or two cylinders usually is not enough to keep the engine from starting. A no-start condition would require that several of the cylinders have low compression. Low compression in all cylinders might be caused by a jumped timing chain or stripped timing gears. If the compression is good in some cylinders but not in others, the most likely cause is a blown head gasket, worn rings, or bad valves.

3. Check Air Flow

To a large extent, the ability of the engine to bring air into the combustion chamber is dependent on the same things that give the engine compression, as well as intake system design, the condition of the exhaust system, the condition of the air filter, the idle air control valve, and the throttle plates.

Testing this system involves a visual inspection. Check the condition of the air filter; if it appears dirty or restricted, replace it. Ensure the idle air control valve and throttle plates are not coked. Also inspect the throttle for free movement.

On the opposite side of the combustion chamber from the intake system is the exhaust. In order for the intake to be able to pull air in, the exhaust must be capable to let air out. Have someone crank the engine while you hold your hand over the tail pipe. There should be noticeable pressure against your hand. If you are unsure whether or not there is enough, compare it to the amount being pushed out the exhaust while cranking a car that will start.

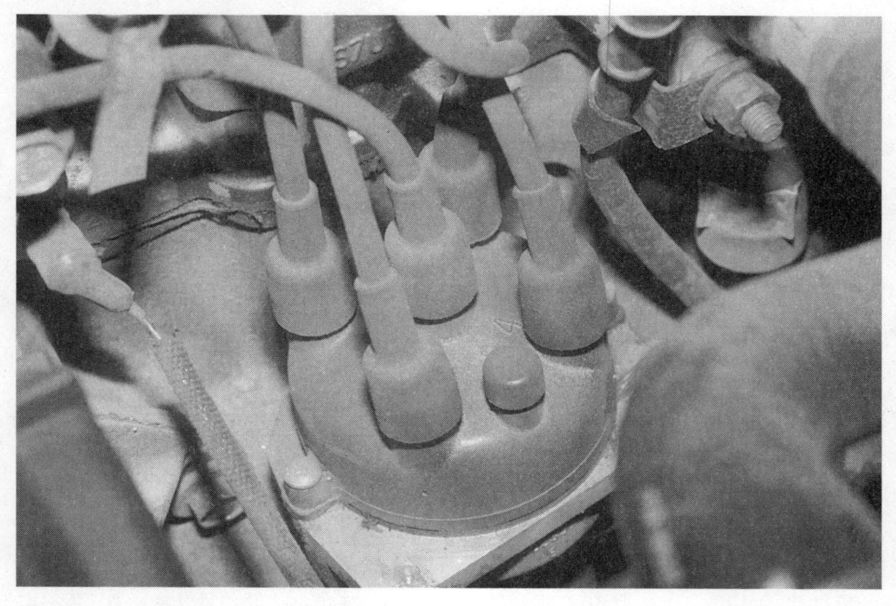

Distributor-type ignition systems have dominated the industry since the 1920s. To test the output of the distributor, remove one of the spark plug wires at the distributor cap and install a known good ignition wire. Insert an insulated screwdriver into the wire and hold the screwdriver 1/4-inch from ground. Crank the engine. If there is no spark, check the cap, rotor, coil wire, coil, and wires connecting the computer and the coil before condemning the computer. Remember that on the distributor cap style shown here, the cap must be removed and the wire unlatched from the inside.

Although seldom will the air induction system be so dirty as to cause a no-start condition, the filter should be checked as a part of any diagnosis. There are parts of the country where dust, ash, and agricultural by-products can cause severe restrictions and air induction problems.

4. Check the Spark (Distributor Type Ignition)

A. In order to test for spark, insert a screw driver into the spark plug end of a plug wire and hold the screwdriver about 1/4 inch from a good ground. Crank the engine and check for spark.

If there is a spark, remove one of the spark plugs, insert it into the plug wire, place the plug on a good ground, such as the intake manifold, and crank the engine. If there is no spark, proceed to step B.

If the spark plug sparks and does not appear worn or old, proceed to the No Start, Fuel Diagnosis section. If there is no spark at the spark plug, replace the spark plugs.

B. If there was no spark from the spark plug wire probe, check the negative terminal of the ignition coil with a test light. If you are not sure which side of the coil is negative, then insert it first on one side, crank the engine, then on the other side. One side of the coil should have a steady supply of switched ignition voltage while the other side should flash on and off as the engine is cranked. If neither has power, then you have an open wire in the voltage supply to the coil. If both sides have steady power, then the problem is the primary ignition's control of the coil. In the old days, the first thing to look at would have been the points. Today we have to look at that which replaces the points: the ignition module, pickup coil, and the interconnecting wiring.

On distributor applications, the Hall Effects is located inside the distributor. The shiny, somewhat triangular object is the permanent magnet of the sensor. The actual Hall device is located opposite the magnet.

The Hall Effects has three wires going to it. One wire is the power supply wire (orange). On a Chrysler, this is usually 8-9 volts. Another wire provides a ground (black and light blue), and the third wire (gray and black) carries the signal to the computer. Coming off the Hall Effects unit itself, these wires are in a triple "zip-cord" format. The wire colors indicated are usually the colors of the wires meeting the zip-cord.

C. Test the Hall Effects sensor. The Hall Effects sensor is the standard type of engine speed sensor that Chrysler has used since the debut of electronic fuel injection in 1984. This pickup produces an AC signal. Disconnect the connector at the pickup and connect an AC voltmeter. Be sure that nothing will get tangled or damaged when the engine is cranked. Crank the engine and watch the voltmeter. If it measures a voltage of at least .5 volts AC, then the pickup is good. Check the wiring and test the ignition module.

A Hall Effects pickup is a semiconductor carrying a current flow. When a magnetic field falls perpendicular to the direction of that current flow, part of that current is redirected perpendicular to the main current path. The semiconductor is placed near a permanent magnet. A set of metal blades, or armature, attached to a rotating shaft or other device passes between the Hall Effects semiconductor and the permanent magnet. As the armature rotates, the magnet field is alternately applied to the Hall Effects and interrupted. The result is a pulsing current perpendicular to the main current path. This frequency is directly proportional to the speed of armature rotation. Since the output is only dependent on the presence of the magnetic field, the Hall Effects unit is capable of detecting armature position even when there is no rotational speed.

Testing:

With an Ohmmeter

There is no valid test procedure on the Hall Effects using an ohmmeter.

With an Oscilloscope

Connect the oscilloscope to the Hall Effects signal lead. Rotate the armature. Depending on the number of blades and the rotational speed of the armature, the scope pattern could appear either as a square wave or a flat line that rises and falls with rotation.

With a Voltmeter

Connect a voltmeter to the Hall Effects output lead. The voltmeter should display either a digital high (4 volts or more) or a digital low (around 0 volts). Slowly rotate the armature while observing the voltmeter. If the voltmeter had read low it should now read high; if the voltmeter had read high it should now read low. If the voltage fluctuates in this manner as the armature is rotated, then the Hall Effects is good.

With a Dwell Meter

Since the signal generated by the Hall Effects is a square wave, the dwell meter becomes a natural for testing. Connect the dwell meter between the Hall Effects output and ground. Rotate the armature as fast as possible (for instance, crank the engine). The dwell meter should read something besides zero and full scale. If it does, the Hall Effects is good.

With a Tachometer

As with the dwell meter, the tachometer is also a good tool for detecting a square wave. Connect the tachometer between the Hall Effects output and ground. With the armature rotating as described in the paragraph on the dwell meter, the tachometer should read something other than zero if the Hall Effects is good.

D. Test the Ignition Module (located in the Power Module, SMEC, or SBEC). Other than replacing with a

In the newer engine designs, the Hall Effects device is located on the end of the camshaft or crankshaft.

This is the classic ignition coil. Rarely used on modern applications, this oil-filled marvel was the standard of the industry from World War I to the 1980s. Like all ignition coils, when the engine is being cranked, a test light connected to one side of the coil primary (small) wires should glow steady while a test light connected to the other side should blink.

Distributorless ignition systems are found on both V-6 and four-cylinder applications. In the coil pack shown here, there are two coils. The rear pair of secondary output leads sends spark to cylinders number one and four. The closer pair sends spark to cylinders two and three. Each time one of the coils fires, it delivers spark to two cylinders. One of the cylinders will be on the compression stroke, getting ready to fire; the other is on the exhaust stroke. The cylinder on the exhaust stroke has no ignitable gases. Therefore, the spark across that plug merely completes the circuit. This is called the waste spark.

Modern ignition coils are "fat free." The cooling oil of earlier coils is replaced with a heat-dissipating plastic. The compact design makes these new coils much more flexible from a design/location perspective. That, of course, is a polite way to say that sometimes they are very difficult to test while mounted.

known good unit, there is no accurate way to test an ignition module without special equipment.

5. Check the Spark (Distributorless Type Ignition)

Distributorless ignition systems may be found on several Chrysler V-6 applications. Like the distributor applications, if there is no high-voltage spark, the fault could lie in the coil, ignition module, or engine position sensors.

The Chrysler distributorless ignition system uses Hall Effects sensors to monitor the positions of the camshaft and crankshaft.

A. Checking the Crankshaft Position Sensor

Probe the orange wire at the crankshaft position sensor connector. This wire runs from terminal 7 of the computer to the crankshaft position sensor and provides power for the sensor. The reading should be 8 volts or more with the ignition on.

Probe the black and light blue wire on the crankshaft position sensor connector with an ohmmeter. Measure the resistance in this wire, which runs from the crank sensor to terminal engine block ground. The resistance should be very close to zero. If the resistance is close to infinity, then repair the wire.

Check the voltage with the key on at terminal 24 of the ignition module, a gray/black wire. The voltmeter should read either close to 0 volts or over 5 volts. Whichever voltage it reads when you rotate the crankshaft, the voltage reading should change. If it was five, it should drop to zero; if it was zero, it should rise to five.

B. Checking the Camshaft Position Sensor

With the key on there should be 8+ volts in the orange wire running from the cam sensor to terminal 7 of the computer. Check this at the cam sensor.

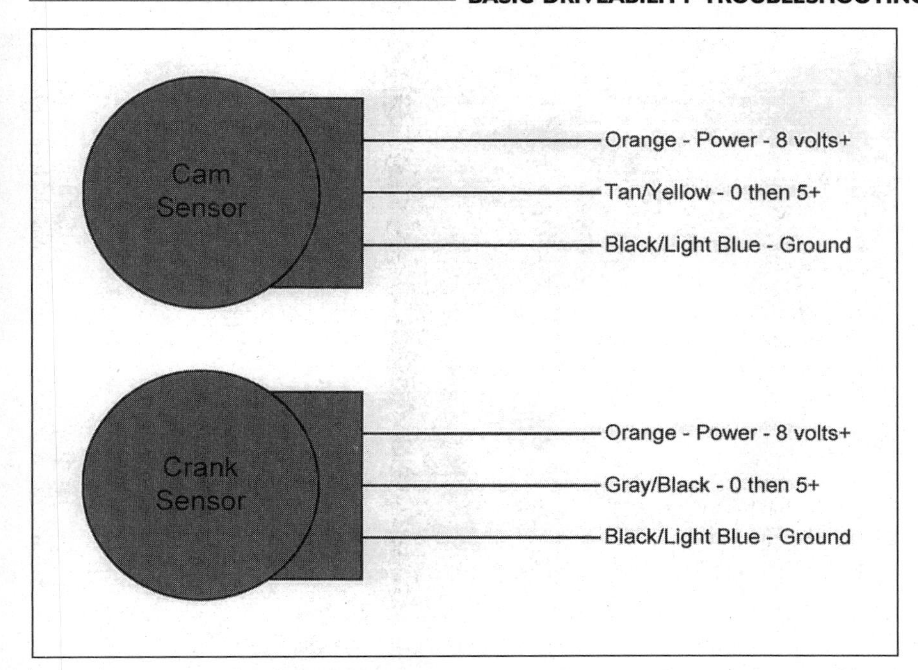

All Hall Effects sensors are tested in basically the same manner. One wire will have a supply voltage ranging from 5-12 volts. A second wire will be connected to ground through the computer, and the third wire will carry the signal.

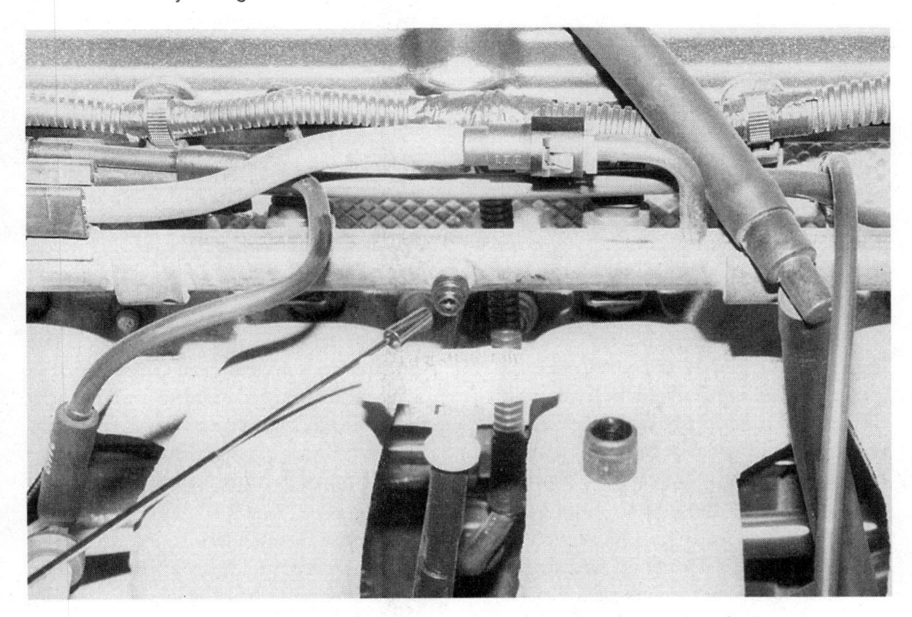

The pointer is highlighting the schraeder valve provided by Chrysler for checking fuel pressure. Although hundreds of dollars can be spent on fuel pressure gauges, not three hours before writing this caption I saw a gauge for $39.95 at my hometown parts store.

The ground for the sensor is a black/light blue wire running from the cam sensor to terminal engine block ground. Measuring with an ohmmeter from the cam sensor connection to ground should read resistance but not infinity.

The tan/yellow wire runs from the cam sensor to the 44 terminal of the computer. This wire carries camshaft position information to the module in order to sequence the injectors. The voltage should read either 5+ volts or zero. Rotate the crankshaft. If the voltage was low, it should go high; if the voltage was high, it should go low. If it does not, replace the crank sensor.

Quick Check Tips

If there is no spark, but the injector pulses, then the lack of spark is a result of secondary ignition problem such as a bad coil, distributor cap, or rotor. If there is no spark and the injector does not pulse, then the problem is likely a primary ignition problem.

A quick and easy test for pulses from cam and crank sensors is to connect a dwell meter to terminals 24 and 44 of the computer. If the sensors are good, the dwell meter will read somewhere between 10 and 80 degrees on the four-cylinder scale. If not, it will read 90 or zero.

No Start, Fuel Diagnosis (Testing for Fuel Pump Operation)

When the key is turned to the run position, but the engine is not cranked, the fuel pump will run for about two seconds, then shut off. If you can hear the fuel pump run for this few seconds, then you know that the fuel pump relay and the computer's control of the relay is operative.

Connect a fuel pressure gauge to the fuel rail on the multipoint injection systems; tee into the inbound line on the throttle body applications. Crank the engine for several seconds. The fuel pressure gauge should indicate more than 30 but less than 45 psi for multipoint and between 15 and 22 psi for throttle body systems. This pressure should hold for several minutes to hours slowly bleeding off. If this pressure drops off quickly, then the fuel pump check valve, fuel pressure regulator, or an injector is leaking.

If the pressure is low, run a volume test before and after the fuel filter. Disconnect the fuel filter on the outbound side. Using an approved fuel container, install a hose on the outbound end of the fuel filter and crank the engine for about 15 seconds. The pump should flow a minimum of a pint. If the flow is less, remove the filter and repeat the test. Install a new filter if the flow is good before the filter but reduced after the filter.

If the pressure is low, but the flow is adequate, replace the fuel pressure regulator.

If the fuel pressure is correct, check to see if the injectors are opening. There are several good ways to do this, but the best way is to use a stethoscope to listen for them to open and close while the engine is being cranked. If a mechanic's-

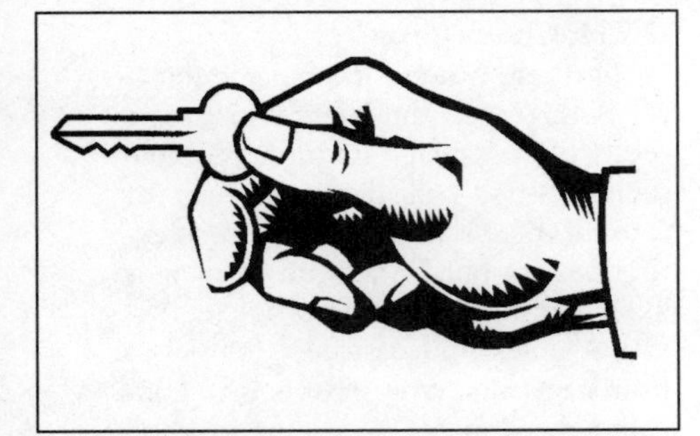

These are the only special tools required to extract the trouble codes from a fuel-injected Chrysler. Insert the key into the ignition switch. Cycle the key on and off three times, stopping in the on position. Observe the flashes of the light. The codes will be flashed once each in a two-digit format.

All of the specialized machinery and equipment in the world is not a substitute for a discerning eyeball or two.

style stethoscope is not available, a piece of 1/2-inch heater hose held near the ear and next to the injector will do almost as well. If the injectors are not clicking, and there is no spark, repair the ignition system first. If there is spark but the injectors do not click, check for voltage, while cranking, at the green/black wire which supplies voltage for the injectors. If voltage is present, connect a tachometer to the negative side of the injectors. Cranking the engine should show some sort of a reading (we do not care what the reading is as long as it is not zero). If there is no reading, check the Hall Effects sensors.

Pulling Trouble Codes

Most people, even professional technicians, think that the use of trouble codes is the most effective way to troubleshoot electronic fuel injection systems. I find this a bit perplexing. As I mentioned earlier, I have been working with electronically fuel-injected engines since the early 1970s. In those early days there were no on-board diagnostics, only logic and reasoning.

Let us talk a little about what trouble codes will and will not do for you. At the most, they will point you in the direction of the cause of the problem. At the worst, they will point you in the direction of the symptom. Keep in mind that the symptom and the problem are two different things.

To extract the trouble codes on a Chrysler, insert the key in the ignition switch and cycle the key on and off three times leaving the key in the on position. At this point, the Power Loss or Service Engine Soon light will be on. It will quickly go out and then begin to flash the codes. The codes will be flashed out in a two-digit format always beginning with the lowest code numerically and proceeding to the highest. You will know that the computer has given you all the codes when the light flashes out a code 55. Code 55 means end of message. Each code will be flashed only once. If you are unsure of your reading of the codes, simply turn the ignition switch off and repeat the three key cycles. Later we will discuss the exact meaning of each of the codes and how to troubleshoot them.

Visual Inspection

A thorough visual inspection is probably the single most important diagnostic procedure that a technician can perform in finding a driveability problem. As a former trainer with a major scope manufacturer, I have seen the human eyeball outperform $25,000 engine analyzers time after time.

During a visual inspection, the following items should be carefully examined:

Checking the Coolant Level and Proper Mix

This stuff used to be called antifreeze. In Fairbanks, Alaska, where I used to live, it still is. In the old days, one could get by without it in climates where oranges grow. Today, the antifreeze is an assumed component of the computerized engine control system. Today's engines run at temperatures that can exceed 230 degrees F. At these temperatures, water, even in a pressurized system, will boil. The antifreeze is necessary to prevent boilover at these temperatures. The antifreeze also has the ability to help spread the engine's retained heat more evenly throughout the block and head. Pure water will allow localized hot spots to build up, which can contribute to detonation.

As bad as pure water is for the operation of the engine, pure antifreeze can be worse. The principal element of antifreeze is ethylene glycol. The freezing point of water is 32 degrees F; the freezing point of ethylene glycol is only slightly below that. It is only when the water and ethylene glycol are mixed that the freezing

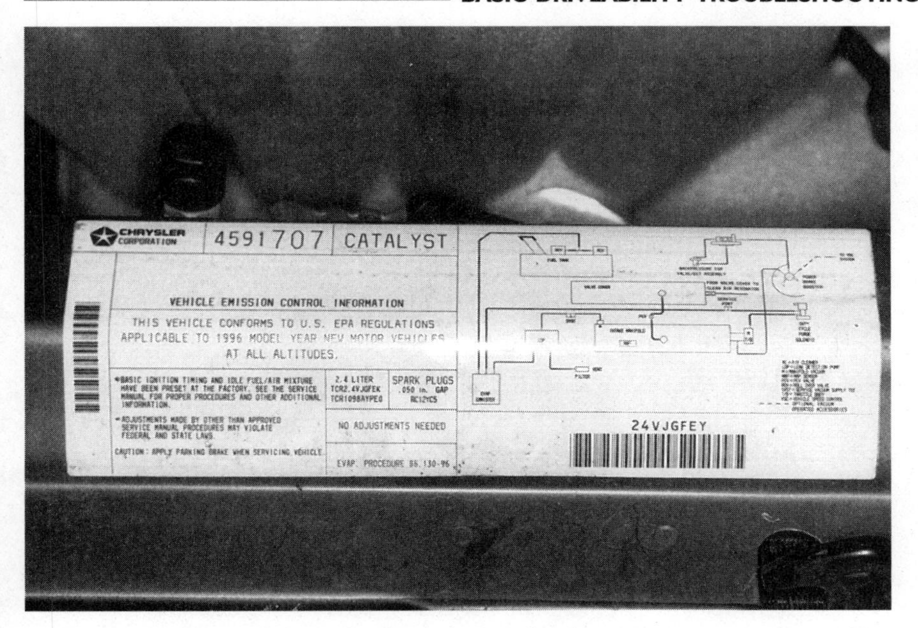

Under the hood of every vehicle produced in the last two decades is a vacuum routing diagram. This diagram can be used during the visual inspection to be sure that the vacuum hoses are properly routed. Improperly routed hoses can result in poor driveability and high emission levels.

temperature is significantly lowered. A mix of about 40 percent water and 60 percent ethylene glycol will lower the freezing temperature of the mixture to -65 degrees F.

Besides the obvious potential of overheating the engine, a car that is low on coolant may run rich, get poor fuel economy, and have detonation problems.

Checking the Radiator Hoses, Connections, and Clamps

Although defective radiator hoses, clamps, or connections will not cause a driveability problem, they can cause a loss of coolant, which in turn can cause any of the problems described above.

Checking the Battery Cables

Incorrect voltage to the ECM or the fuel pump can cause improper operation of those components. Good connections are essential for delivering the proper amount of voltage to all of the vehicle components. This is especially true as the load on the electrical system increases when lights, wipers, and rear window defoggers are turned on.

Inspect battery cables for signs of bubbling or swelling under the insulation near the cable end. This is a telltale sign of hidden corrosion.

Note: Do not remove the battery cables at this point! Doing so will erase trouble codes that may be important to the proper diagnosis of the driveability problem.

Check the cables for evidence of corrosion in the form of a white or green powder around the battery terminal end. If there is any doubt about the cleanliness of the battery terminals, pull the trouble codes. Pulling the codes at this point will ensure that the information stored in the computer's diagnostic memory will be retained for use later in the troubleshooting procedure. These codes will become the only way of determining any intermittent defects that may require attention after the "hard faults" have been diagnosed.

After writing down all of the fault codes, turn off all electrical loads on the car that can be turned off. Disconnecting the battery terminal with a load on the battery increases the possibility of damaging the computer due to an induced voltage spike. Disconnect the terminals and clean or replace them as necessary.

Checking the Vacuum Hoses

Several of the sensors and actuators require either manifold or ported vacuum in order to function properly. Inspect all the vacuum hoses for evidence of swelling, cracks, or leaking. Check for a snug fit at all connections and proper routing. Under the hood of North American-delivered cars there is a vacuum hose routing diagram. This diagram is often part of the EPA sticker. Compare the routing of the vacuum diagrams to this chart.

Checking the Wiring Harness and Routing

Thoroughly look at the condition of the main engine wiring harness as well as the computer engine control harness. Look for broken or frayed wires, loose connections, damaged insulation, and wires pinched under brackets or other engine parts.

Unfortunately, there is not a routing diagram for the wiring harness. Ensure that all of the computer or sensor wiring harnesses are at least 2 inches from secondary ignition wiring and that the wiring harness is not stretched.

Checking for Fuel Leaks

The threat of fire from a fuel leak is obvious; however, there are other considerations. A fuel leak will cause a drastic reduction in fuel economy and may lower the fuel pressure enough to cause stumbling, hesitation, or limited power at the top end.

Inspect around the injectors where they attach to the fuel rail, all of the fuel line connections at the filter, and from the filter to the fuel rail. A place that is often overlooked is the fuel pressure regulator. Remove the vacuum supply hose (some TBI engines do not have one) from the fuel pressure regulator, if there is fuel in the hose, replace the fuel pressure regulator.

Checking for Exhaust Leaks

An exhaust system that is leaking out exhaust gases is also leaking in oxygen. Oxygen leaking into the exhaust system can cause the oxygen sensor to believe that the engine is running lean. When it reports this condition to the computer, it will respond by enriching the mixture. The end result will be a very rich-running engine, poor fuel economy, fouled plugs, and all of the symptoms that go with these conditions.

Test Driving

The test drive should be more than an opportunity to get away

Symptom	Old-Tech	New-Tech	Other
Engine does not crank	Battery, starter, battery cables		Engine seized
Cranks but won't start	Distributor cap, rotor, spark plugs, coil wire, dirty fuel, no fuel, fuel leaks, restricted air filter, restricted fuel filter	Hall Effects stuck or defective injector, bad coolant temperature sensing circuit, defective computer	
Hard cold start	Bad spark plugs, vacuum leak, weak battery, wrong viscosity engine oil,	Defective TPS circuit, coolant temperature sensing circuit, defective Automatic Idle Speed control motor circuit	
Rough idle cold	Vacuum leaks, worn spark plugs	Defective Automatic Idle Speed control motor circuit, coolant temperature sensing circuit	Low compression
Stall, stumble, hesitation	Defective or malfunctioning EGR system	Fuel filter restricted, fuel pump output low, restricted fuel filter, defective fuel pump, defective TPS circuit, defective coolant temperature sensing circuit	
Hard start hot	Internal leak in carburetor, vacuum leak	Leaking injector, loss of fuel pressure	
Rough idle hot	Vacuum leak	High fuel pressure	
Stalls on deceleration	Incorrect idle speed adjustment, vacuum leak, EGR sticking open	Leaking injector, vehicle speed sensor problem, high fuel pressure	
Low top speed	Restricted intake, incorrect ignition timing, restricted exhaust		
Lack of power	Restricted intake, incorrect ignition timing, restricted exhaust		
Surge at cruise		Leaking injector, high fuel pressure	
Poor mileage	Incorrect starting procedure	Incorrect fuel pressure	
Flooding	Incorrect starting procedure		
Engine diesels	Engine idles too fast, internal leak in carburetor		
Idles too fast	Vacuum leak	Malfunction of AIS motor	

The following chart compares old technology components to new technology components. It attempts to match the functions item by item.

Old-Tech	New-Tech
Accelerator pump	Throttle position sensor
	Rise in fuel pressure on acceleration
	Manifold absolute pressure sensor
Choke	Coolant temperature sensor
	Air charge temperature sensor
Power valve	Manifold absolute pressure sensor
Vacuum advance	Manifold absolute pressure sensor
Thermal vacuum switches	Coolant temperature sensor
Mechanical advance	Hall Effects signal to the computer
Float level	Fuel pressure
Idle mixture screws	Oxygen sensor
Carburetor jet size	Oxygen sensor

Accelerator pump	**Throttle position sensor** **Rise in fuel pressure on acceleration** **Manifold absolute pressure sensor**
Choke	**Coolant temperature sensor** **Air charge temperature sensor**
Power valve	**Manifold absolute pressure sensor**
Vacuum advance	**Manifold absolute pressure sensor**
Thermal vacuum switches	**Coolant temperature sensor**
Mechanical advance	**Hall effects signal to the computer**
Float level	**Fuel pressure**
Idle mixture screws	**Oxygen sensor**
Carburetor jet size	**Oxygen sensor**

On the left side of this chart are old-technology components; on the right side are the new-technology components that roughly equate to the function of the old tech.

from the shop, garage, or shade tree. Many times I have diagnosed a driveability problem as I drove the car with no test instruments. Each component in the engine, ignition, and fuel injection system has a specific job to do. Since the job of an auto manufacturer is to provide a dividend to their stockholders by building and selling cars, they are not inclined to use two parts to do a job that could be done by one. Each symptom coincides with a limited group of tasks that are to be performed by the engine control system. Each task coincides with a limited group of components. Therefore, when a component fails to operate properly, there are a limited number of symptoms that can occur. The nearby chart is a general guide to things that should be checked for a specific symptom.

Matching Conditions

For many readers, this paragraph might seem like a "well, duh" paragraph. However, there are many professional technicians that do not follow the simple concept that I am about to mention.

There is an old joke that describes the mistake:

A man walking down the street observes a fellow that is obviously inebriated pacing back and forth under a street light with his head down. The man walks up and, looking down, asks, "What are you looking for?"

"I dropped a dollar bill," said the drunk.

After a few minutes of looking with the drunk, the man asks, "Where did you lose it?"

The drunk points to an area on the ground about 30 feet away. "Over there."

"Then why are we looking over here?" asks the man, obviously exasperated.

"Because the light is better over here."

This is what many technicians do. They hook up bulky and complicated test equipment and perform a series of tests with the vehicle sitting still. In many cases, these are adequate test conditions to find a problem. But in many cases they are not.

As I am writing this book, I have an Isuzu Trooper belonging to my son sitting in the driveway. The Check Engine light comes on only after the engine is warm and has been driven at freeway speeds for five minutes. If I were to take it to the dealership and was not specific about when the Check Engine light comes on, it is likely that the technician would do an inadequate test drive.

Gather as much information about when the symptom occurs as possible. Does it do it going uphill or downhill? When the ambient temperature is cold, or when the ambient temperature is hot? When the engine is cold or when the engine is warmed up? When it is raining? When it has been dry for days? On rough roads? On freeways? Just after refueling, etc? Now drive the vehicle under these conditions with your portable test equipment hooked up.

What To Watch for

Basically what to watch for is obvious. Watch for the complaint for which you are seeking the cause. But more than that, there are corollary problems that may not be the primary complaint but do offer clues about the probable causes of the primary complaint. Some of the other things to look for are:

1. Does the Check Engine light come on?
2. Is there a misfire?
3. How is the overall power of the engine?
4. Is there a rotten egg smell when the symptom occurs?
5. Are there any strange sounds?
6. Does the engine over heat?
7. Are there any unusual vibrations?

Scope Analysis

Let me state up front that I have worked for the analyzer manufacturers and that I have taught the use of engine analyzers for several years. I have used engine analyzers since I first got into the auto repair business over 25 years ago. All of this experience

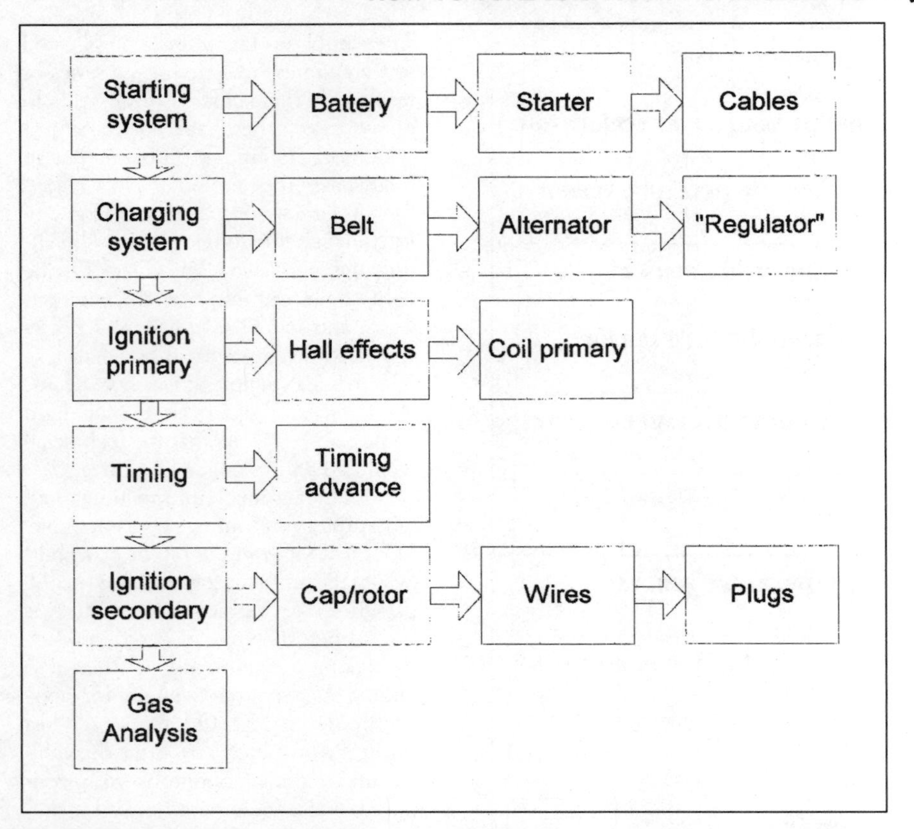

This chart shows the things that should be checked before focusing your attention on the fuel injection side of the driveability problem.

with engine analyzers has taught one thing: engine analyzers do not diagnose problems, they are only as good as the technician using them. That aside, an engine analyzer in the hands of a proficient operator can work diagnostic miracles.

The job the engine analyzer does best is to ensure that the fuel injection and computer system has a solid platform to operate on. They are very good at detecting problems with the mechanical operation of the engine. They are good at detecting problems with the ignition system, the air induction system, and the exhaust system. What they do not usually do well is analyze the fuel injection system. What follows is a brief overview of the systems that should be tested thoroughly during the diagnostic process. Absolutely essential for both the technician and the enthusiast to remember is that there is no need for a $35,000 engine analyzer to make these tests. Every test that most engine analyzers can do can be done with hand-held equipment.

Starting System

The importance of the starting system is probably obvious. If you cannot get the engine started, everything that controls the operation of the engine is irrelevant. Most modern engine analyzers take a single sample of the battery condition and display that for the technician to read. There are three battery parameters that should be checked.

The first is the state of charge. The best way to check the state of charge in a battery is with a hydrometer. This does, of course, assume that the battery cell caps are removable. Insert the hydrometer into each cell and draw in the electrolyte. Read the point where the float stabilizes. Be careful that the float does not stick to the side of the hydrometer barrel or rise to the bulb area. This can distort the readings. If the reading on any cell is below 1.200, that cell needs to be recharged. If there is more than 0.050 difference between any two cells, the battery needs to be replaced.

Many batteries do not have removable cell caps. Even those that do have removable cell caps need additional testing. The second parameter to check is battery capacity. The battery capacity can be tested using the 15-second starter load test mentioned earlier. Many engine analyzers base their diagnosis of the battery purely on this parameter.

The third test is a test of the internal resistance of the battery. Some engine analyzers are designed to measure the internal resistance of the battery to determine the capacity and power potential of the battery. The best way to do this without an engine analyzer is with the three-minute high amp charge test. First you must obtain a battery charger capable of charging at a rate of at least 40 amps. Disconnect the battery from the vehicle. Connect a voltmeter across the terminals of the battery. This voltmeter may be either digital or analog. Connect the battery charger and set it to charge at a rate of 40-60 amps. Observe the voltmeter. If the voltmeter reading exceeds 15 volts at any time during or at the end of the test, the battery is defective and should be replaced.

Note: Charging at high rates as described above can accelerate the electrolysis of water into oxygen and hydrogen. This is a very explosive combination of gases. Protective clothing and eye wear should be worn when performing this test.

Charging System

The charging system is the primary source of electrical power when the engine is running. Every component of the fuel injection system and other electronic systems assumes that its supply voltage is between 12.0 and 14.9 volts. When the voltage is outside of this range, the system can begin to malfunction and can even be damaged.

As important as the voltage is the quality of that voltage. The fuel injection computer and other electronic control modules are designed to be powered by and operate on direct current voltage. Alternating current can affect the proper operation of these modules. The alternator is actually an alternating current generator. Before the AC current leaves the alternator, the current passes through a network of rectifiers commonly referred to as diodes. The diodes convert the AC current into something at least resembling

DC. One of the more common defects that can occur in an alternator is problems with these diodes. When they are defective, they allow AC to pass through to the battery and the electronic modules. At the very least this can cause the modules to make bad decisions and could even damage them.

Testing the output voltage is simply a matter of connecting a voltmeter to the output terminal of the alternator and reading it. Checking the diodes is a bit trickier. The best way is with an oscilloscope. Connected to the output terminal of the alternator, you should see a relatively smooth pattern on the screen. If the pattern has deep valleys with an amplitude of several volts, then the alternator should be repaired or replaced.

If you do not have an oscilloscope handy, there are a couple of other effective ways to check the DC quality of the alternator output. Second best for the non-professional technician is to take a trip to the local full-line auto parts store. In their tool section or tool catalog there is a tool which consists of a small box with four light-emitting diodes and two wires. This charging system tester is supposed to be able to test the state of charge of the battery and all other components in the system. Most of its functions I consider to be marginally accurate at best. However, its ability to test alternator diodes is very impressive.

The third way to test the diodes is with an AC voltmeter. Since the alternator is supposed to put out DC, an AC voltmeter should not be able to detect anything on the output terminal of the alternator. Begin the test by connecting the voltmeter set on the AC scale across the terminals of the battery. Do not start the engine yet. Observe the reading. It will read either almost zero or in the neighborhood of 18 volts. Some AC voltmeters react in an odd manner when connected to DC. They will display a reading approximately 50 percent higher than the DC voltage to which it is connected. Only the greater power of the universe and the race of space aliens we call electrical engineers understand why. Now that you know what to expect from your voltmeter, connect it to the output terminal of the alternator and start the engine. The AC voltmeter should read either 0.00 or about 18 volts. If AC is present, the voltmeter will read one or more volts.

Note: Be very careful when using a digital voltmeter. Many of them may display a reading of, for example, 8.55. A closer look at the display will show an "mv" usually in a not-so-obvious location on the screen. This reading therefore is not 8.55 volts but rather 8.55 millivolts or 0.00855 volts.

Primary Ignition System

The primary ignition system on modern engines is very difficult for even a professional technician to test thoroughly. Our principle concern should be the condition of the Hall Effects sensors. These are used to inform the computer about the position of the crankshaft, and in the case of the distributorless engines, the position of the camshaft. The crankshaft position sensor is located in the distributor on the engines that have one and on the crankshaft on the engines that do not. This sensor is used by the computer to initiate the injectors, fire the ignition coil, and control the ignition timing.

The best way to test a Hall Effects is to use an oscilloscope. Connect the oscilloscope to the output lead of the sensor. Start or crank the engine. The Hall Effects signal should consist of a series of square wave forms with crisp edges. If there is no wave form, follow the test procedures described in the code 11 section later in this book. If the vehicle has a history of stalling problems, look carefully at the corners of the wave forms. If the corners are not crisp 90-degree turns, then the sensor may be defective and should be replaced.

Ignition Timing

Ignition timing was covered thoroughly in Chapter 2. However, I feel that I should emphasize how ignition timing must be tested and must be correct before serious diagnosis of the fuel injection system can proceed. Also confirm that the timing advance system is operating properly.

The only part of the secondary ignition system that seems to get routine maintenance is the spark plugs. When I first got into this business, it was customary for most of my customers to change their spark plugs every 15,000 miles. There have been many improvements since ZZ Top played in the basement of a parking garage in Fort Worth, Texas. Today's plugs can be expected to easily last 50,000 miles, and we are on the verge of an era where 100,000 will be considered the norm. However, just because they usually last that long does not mean they will always last that long. I have made it a habit over the years to begin most driveability diagnoses with a fresh set of spark plugs.

Secondary Ignition System

The secondary ignition system should be tested to ensure the condition and quality of the spark plugs, plug wires, distributor cap, and rotor. Realistically for the non-professional, and in some cases even for the professional technician, the best way to ensure that these components are not the cause of the driveability problem you are looking for is to replace them.

Gas Analysis

Exhaust emission gases were covered earlier in this book. At this point we are going to briefly define readings and what an incorrect reading might mean.

Gas	Typical Good Reading	What To Look For If Incorrect
Carbon monoxide (CO)	Less than 1%	Engine is running rich
Hydrocarbon (HC)	Less than 200 ppm	Combustion quality is reduced
Carbon dioxide (CO_2)	Greater than 10%	Combustion quality is reduced
Oxygen (O_2)	Less than 5%	Combustion quality is reduced or engine is running

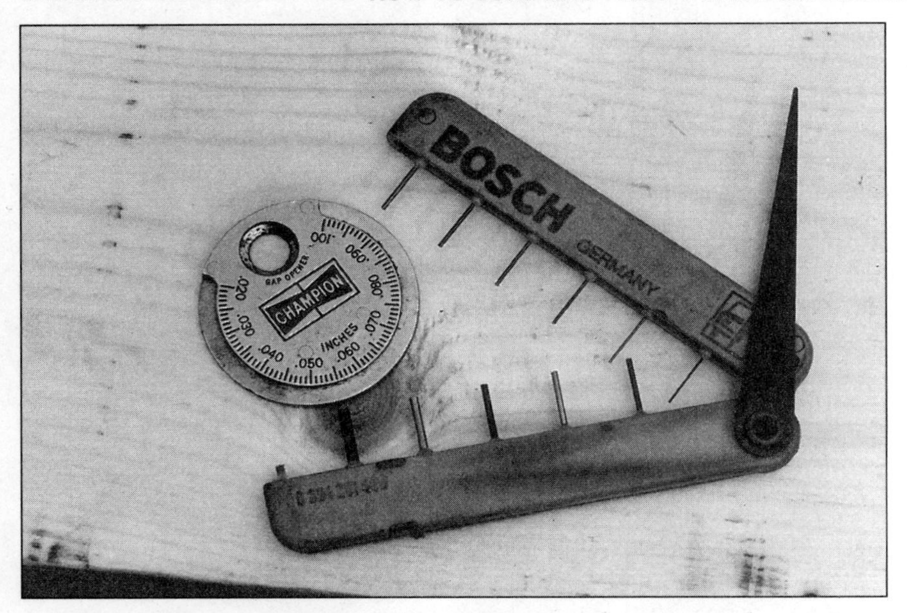

Several years ago, I got into a bad habit of assuming that the spark plugs always come out of the box properly gapped. Then I changed brands. The brand that I started putting in customers' engines were never gapped properly. After several comebacks, I decided that I had better find my plug-gapping gauges.

Oxygen Sensor Gas Analysis

It is highly likely that most readers of this book do not have a four-gas analyzer in their garage. In fact, most repair shops only have one gas analyzer and that machine is often dedicated to state emissions testing. If there is no four-gas analyzer, there is still a valid way to get some information about what is going on in the combustion chamber.

With the engine at operating temperature and the throttle open about 5-10 degrees of rotation, observe the oxygen sensor voltage. If the voltage remains constantly above 0.45 volts, it indicates that the engine is running rich. If the voltage remains below 0.45 volts, it means that the engine is running lean or that there is errant oxygen getting into the exhaust system ahead of the oxygen sensor. Keep in mind that these readings are an "instant" reading and are not a valid indication of what is going on with air/fuel ratio over the long term. A little later in this book we will be addressing the interpretation of the serial data information and in that information we can get information about the long-term air/fuel ratio conditions.

Fuel Pressure Testing

Connecting the Gauge

To check fuel pressure, you will need a gauge that can read up to 75 psi. Connect the gauge by teeing the gauge in between the fuel filter and the fuel pressure regulator. Remember, wherever you connect the gauge, the fuel must be able to flow through the system as the readings are being taken. On most Chrysler applications, the inbound hose to the throttle body or fuel rail slides onto a fitting and is held in place with a clamp. This is an ideal place to tee in the gauge. In general, the fuel pressure for multipoint-injected engines should be 35-45 psi at an idle. Turbocharged engines have a typical fuel pressure of about 45-60 psi. Throttle body-injected engines come in two categories. Cars produced during the 1984 and 1985 model years with throttle body injection will have a fuel pressure of about 35 to 45 psi. These are referred to as high-pressure throttle body systems. Cars produced from 1986 to the present that are equipped with throttle body injection have a fuel pressure between 15 and 20 psi.

From personal experience I should warn you that although the textbook fuel pressure on the low-pressure throttle body systems is less than 20 psi, I have observed that the actual fuel pressure on these applications is often between 20 and 24 psi. My first experience with this was in Vancouver, British Columbia. An acquaintance of mine had asked me to look at a vehicle that was putting out black smoke and had poor fuel economy. We checked the fuel pressure first and discovered that it was 23 psi. This was the first 1986 model and the first low-pressure Chrysler with which I had ever worked. I came rapidly to the appropriate but incorrect conclusion that the fuel pressure was the cause of the problem. We removed the

These two plugs have different heat ranges. Generally speaking, you should always use the heat range that is recommended by the manufacturer for that engine. Be cautious when changing brands of spark plugs. Heat ranges that are supposed to match often do not. I always recommend replacing spark plugs with the original brand.

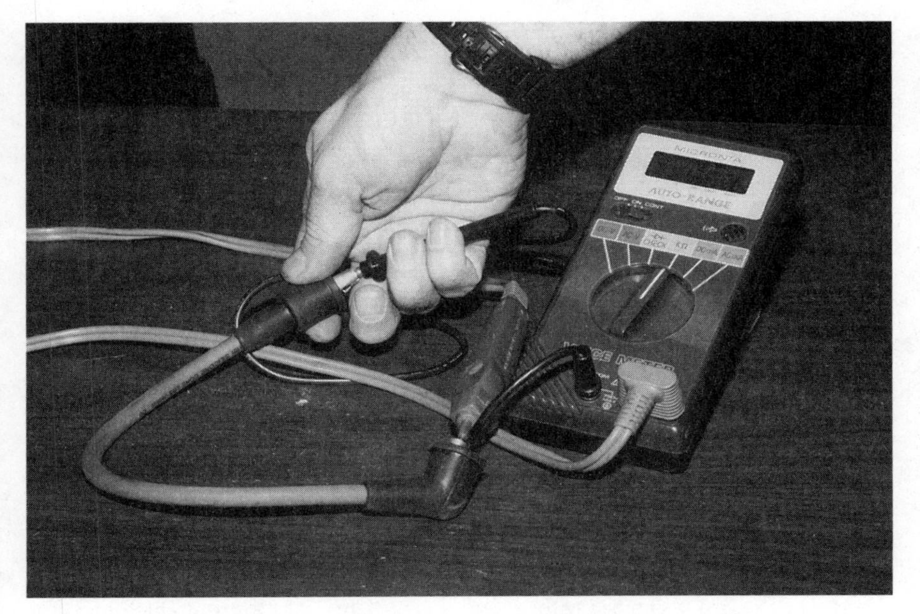

Here a spark plug wire is being tested for proper resistance. Wires that have a resistance greater then 10,000 ohms per foot should be replaced.

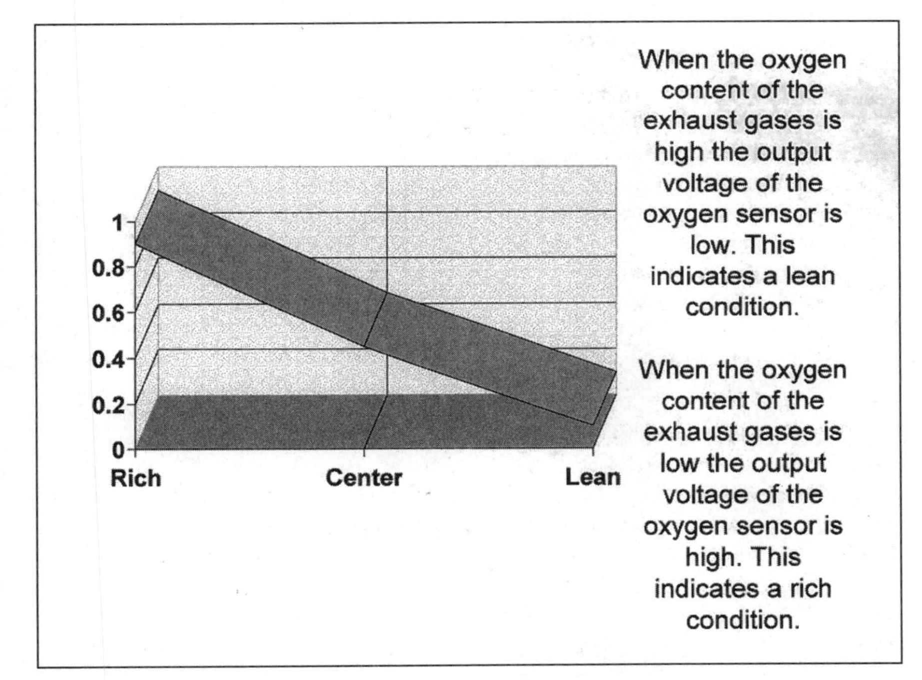

Oxygen sensor voltages greater than 0.450 volts indicate that there is a low oxygen content in the exhaust. The computer will perceive this as a rich-running condition and will respond by decreasing the delivery of fuel to the engine. A low voltage indicates that the engine is running lean. An oxygen sensor voltage that is consistently low can indicate a vacuum or exhaust leak.

return line, allowing the fuel to flow through an auxiliary hose into an approved fuel receptacle. When we checked the fuel pressure again, we found that it had not changed. The conclusion was that the fuel pressure regulator was defective. However, when the fuel pressure regulator was replaced, the pressure went from 23 to 24 psi and the symptom was still there. The actual cause of the problem ended up being the injector.

Since this event, I have revised the point at which I will condemn fuel pressure for being too high. If the fuel pressure is not greater that 50 percent higher than the correct fuel pressure, it is probably not the cause of the problem you are trying to find. This is not to say that if the fuel pressure is supposed to be 35-40 psi and you find that it is 48 psi it should not be corrected. It only means that you have probably not found the actual source of the problem and you need to keep looking. You still need to eventually figure out why the fuel pressure is not right.

Test Conditions

Whenever possible, check the fuel pressure under the conditions where the problem occurs. Do not forget about looking for the dollar bill under the street light referred to earlier in this chapter. If the fuel pressure drops at the time the symptom occurs, there is a restriction to fuel flow in the inbound side of the fuel supply system.

CHAPTER SEVEN

USING TROUBLE CODES

C O N T E N T S

Electrostatic Discharge

Before talking about the aggressive attempts to analyze and repair trouble codes, a caution about static electricity. This is something that people in some locations and climates around the world hardly ever think about. In other climates, it is very noticeable. In 1959, my father was transferred to Amarillo, Texas. For those not familiar with the location of this town, it is in the center of the Texas panhandle. Although I was only in the lower grades of school, several things about the environment of the city stand out in my mind. First there was the smell of the refineries, then there was the smell of the feed lots, and then the sound of sonic booms on almost a daily basis from the nearby Air Force base. Finally there was the fact that every time you touched something metal that provided a path to earth, you would get zapped by a spark of static electricity. I remember being fascinated by the fact that when I lifted the blanket from my bed, the room would light up from the sparks.

As I said, Amarillo sits in the center of the Texas panhandle, a place where straight-line winds are so fierce that light aircraft have been known to land with almost a zero ground speed. During the winter, these winds are very dry and accompany temperatures that are lower than 20 degrees F. This cold, dry, windy climate is ideal for static build-up in your body. Discharging this build-up into a piece of electronic componentry can damage delicate electronic devices.

On the other side of the planet, and definitely the other side of the static problem, is the U.S. territory of Guam. Guam during the winter has temperatures that average over 90 degrees F with humidity factors between 50 percent and 90 percent. Basically, the weather never changes. Here the concept of getting bit by door knob static is virtually unheard of.

Since the early 1980s, much has been said about the effects of static electricity on electronic components. Some components are more sensitive than others. Semiconductor devices are what all the furor has been about. There are several different types of semiconductors used in automotive electronics, each with a different degree of sensitivity.

Chrysler electronic fuel injection systems can generally withstand 5,000 to 15,000 volts of electrostatic discharge. At first this looks like a very large amount of voltage to be sent into an electronic device accidentally, but it is not. Typical everyday activities can generate voltages between -2,500 volts and +500 volts.

Protection from Electrostatic Discharge (ESD)

1. Store components in their original packages.
2. Do not touch component leads or pins.
3. Before handling the computer, discharge the static from your body by touching a grounded metal surface.
4. Do not slide the computer or device across any surface.
5. Keep static-generating materials away from the work area. Things such as plastic, cellophane, paper, candy wrappers, and cardboard are common static generators.
6. Keep clothing away from static-sensitive devices.
7. Never connect static-sensitive devices with power applied.

How much precaution should you take? Well, I prefer to operate an electronic device as though I am in

Amarillo even when I am in Guam. However, if your climate is closer to that of Guam, then you can be a little less anal-retentive about it.

Using a Digital Voltmeter

The most important tool in working with and troubleshooting modern Chrysler fuel injection systems is one of the cheapest and easiest to acquire. Digital voltmeters that are adequate to do the job range in price from $25 to $500. There is an advantage to some of the more expensive meters. They combine many of the functions that we will later describe using tachometers and dwell meters for into one package.

Code number	Turn Check Engine light on?	Definition
88	No	Start of test (usually only seen with flash test)
11	No	Engine not cranked since battery was disconnected; no distributor reference signal (Note: distributor reference signal does not apply to 2.5-liter Dakota pickup)
12	No	Memory stand-by power lost
13	Yes	Slow or no change in MAP sensor signal between KOEO (Key On-Engine Off) and idle
14	Yes	MAP sensor voltage excessively high or excessively low
15	Calif. Only	Vehicle speed; distance sensor circuit
16	Yes	Loss of battery voltage (does not apply to engines equipped with a knock sensor)
16	No	Knock sensor
17	No	Engine running too cold
21	Calif. Only	Oxygen sensor circuit
22	Yes	Coolant temperature sensor circuit voltage excessively high or excessively low
23	No	Throttle body; air temperature sensor circuit voltage excessively high or excessively low
24	Yes	Throttle position sensor circuit voltage excessively high or excessively low
25	Calif. Only	Idle speed control (ISC) motor driver circuit
25	No	Automatic idle speed motor driver circuit
26	Yes	Peak injector current has not been reached or injector circuits have resistance
27	Yes	Fuel injector control circuit or injector output circuit notresponding
31	Calif. Only	Canister purge solenoid circuit failure
32	Calif. Only	Exhaust gas recirculation (EGR) system open; short or failure in EGR control solenoid; power loss to PCM during diagnostic test
33	No	Air conditioning clutch cutout relay circuit
34	No	Speed control (cruise control vacuum or vent control solenoid circuits)
35	No	Cooling fan relay, high speed fan or low speed fan control relay(s)
35	No	Idle switch circuit; cooling fan relay circuit
36	Yes	Air switching solenoid circuit (non-turbo) or the wastegate solenoid on turbocharged models
37	No	Part throttle unlock solenoid driver circuit (lockup converter automatic transmission only) or shift indicator light circuit
41	No	Charging system field current too high or too low
42	No	Automatic shutdown relay (ASD) driver circuit
43	No	Ignition coil control circuit or spark interface circuit
44	No	Loss of FJ2 to logic board/battery temperature out (1987) or failure in the SMEC/SBEC
45	No	Overboost shut-off circuit on MAP sensor reading above overboost limit detected; overdrive solenoid (A-500 or A-518 automatic transmission)
46	Yes	Charging system voltage too high
47	No	Charging system voltage too low
51	Calif. Only	Oxygen sensor detects lean
52	Calif. Only	Oxygen sensor detects rich
53	No	Module internal problem; SMEC/SBEC failure; internal engine controller fault detected
54	No	Problem with the distributor synchronization circuit
55	No	End of code output
61	Yes	BARO solenoid failure
62	No	Emission reminder light mileage is not being updated
63	No	EEPROM write denied—controller failure
64	No	Flexible fuel (methanol) sensor indicates concentration sensor input more/less than the acceptable voltage
65	No	Manifold tune valve solenoid circuit open or shorted
66	No	No message from the transmission control module (TCM) to the powertrain control module (PCM)
66	No	No message from the body control module (BCM) to the powertrain control module (PCM)

The digital voltmeter boasts a high impedance input (10 million ohms or more). This allows the voltmeter to be connected to very small current flow circuits without affecting the voltage reading. Voltmeters with a low input impedance rob power from the circuit being tested. This causes the voltage readings to be lower than they really are. For this reason, the digital meter is what should be used any time precise voltage readings are required.

There is a disadvantage to the use of digital voltmeters. Since it is digital, it merely samples voltage and displays the sample reading. There are major gaps between these samples. Transient fluctuations are completely missed. Devices such as the throttle position sensor potentiometer create a steadily increasing voltage as the throttle is opened. As the TPS wears, there may be places where the wiper no longer contacts the carbon film strip; this would result in a sudden drop in voltage. If the digital voltmeter's sampling did not match up with the voltage fluctuation, the cause of a major driveability problem could be missed. For this reason the analog voltmeter is a better tool for measuring variations in voltage.

Using an Analog Voltmeter

Where the digital voltmeter displays its reading as digits, the analog voltmeter uses a needle moving across a scale to display its readings. The benefits of the analog voltmeter have been ignored since the introduction of the electronic engine control systems at the end of the 1970s. This is because the majority of inexpensive analog meters have a low input impedance. As mentioned previously, a low impedance meter can distort readings. Rumors about technicians ruining computers and other components by using analog meters to take measurements are largely exaggerated.

The analog meter will detect fluctuations in voltage much better than the digital voltmeter. When a transient voltage change occurs, it will show up in the analog meter as a fluctuation in the needle.

Use the analog meter when you are looking for fluctuations in voltage. Use the digital when you are looking for precise readings.

Note: Because of the extremely low current output of the oxygen sensor, most analog voltmeters will ground

There is a little inside joke in the auto repair industry. Each manufacturer must have an engineer who is in charge of hiding essential components. In the diagnostic workshops I teach, I tell my students, "The diagnostic connector is always located in the same place: between the front and rear bumpers and between the left and right door handles." Well, on Chrysler products, it was under the hood until the mid-1990s when it was moved under the dash to come in compliance with new government regulations.

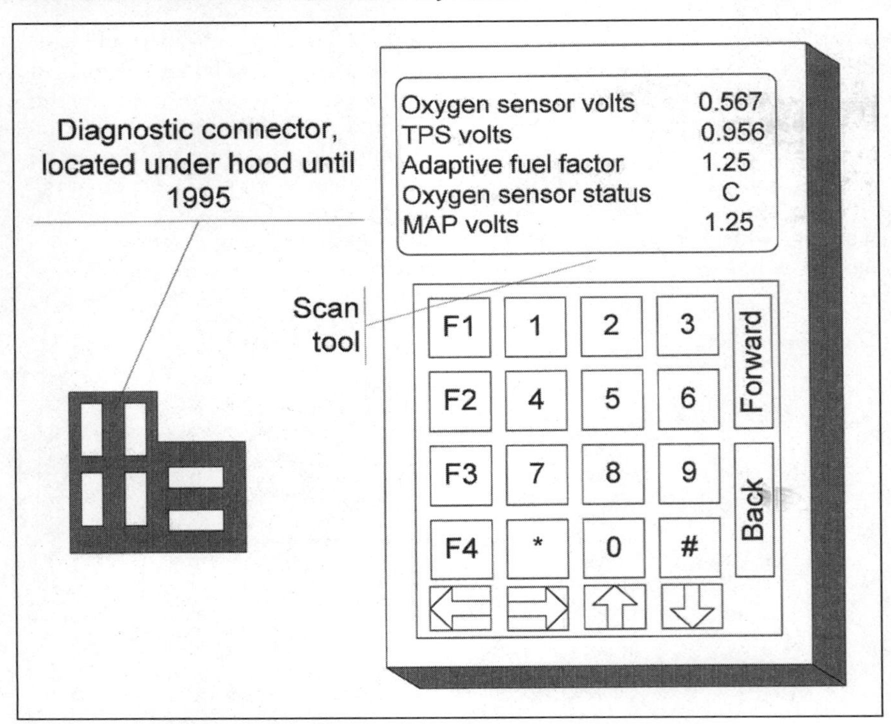

Although codes can be acquired by cycling the ignition switch on and off, using a scanner is much easier. Not only will the scanner give you the codes themselves, but it can be very useful in the troubleshooting of the codes.

Code 11 will only set if the computer has never seen a pulse from the crankshaft rotation sensor. When the crank sensor fails, the engine will die but no code is set. When you have a no-start condition, locate the connector shown here. It is near the battery, and disconnect it for about 15 seconds. Reconnect and crank the engine for 7 seconds. Now pull the codes again. If you get a code 11, it means that the crank Hall Effects sensor is defective.

out the oxygen sensor; reading the meter will display 0 volts. Always use a digital voltmeter or a 10-megohm input impedance analog meter when taking oxygen sensor readings.

Trouble Code Explanations

Code 88

Code 88 is actually a holdover from the days of the DRB (Digital Readout Box). This was the original Chrysler-dedicated scanner. Introduced in the early 1980s, the unit had only two seven-element light-emitting diode (LED) displays with which to communicate to the technician. When the DRB was plugged in and the ignition turned on, the unit powered up. As it was powered up, the DRB would light up all of the LED elements. This caused an 88 to be displayed on the screen. So as you can see, an 88 is not actually a code at all, just a test procedure to ensure the technician that all the LED elements are working.

Code 11

This code is unique in the industry. Code 11 means that the computer has never, in its entire life, ever seen a pulse from the Hall Effects sensor. Now think about it, you are driving down the road and the engine dies due to a failure of the Hall Effects circuit. Will there be a code 11? The answer is no; the computer had seen pulses from the Hall Effects before it failed. No code 11 will be stored. "So what good is it?" you ask.

When you begin to troubleshoot any no-start condition, you should begin by checking to see if there is spark at the ignition coil and if the injectors pulse. Testing to see if the injectors pulse can usually be done simply by touching them as the engine is being cranked. If you are not sure if you feel the injectors, then get a piece of heater hose. Hold one end of the heater hose to your ear and the other next to an injector. If either the spark plug sparks or the injector pulses, then the problem is not the Hall Effects. Once you have determined that neither the injector nor the spark plugs are working, proceed by pulling the trouble codes. There may be several, but there will not be a code 11. After recording the codes for future reference, locate the main power lead to the computer. On many Chrysler Corporation applications, this is a large red

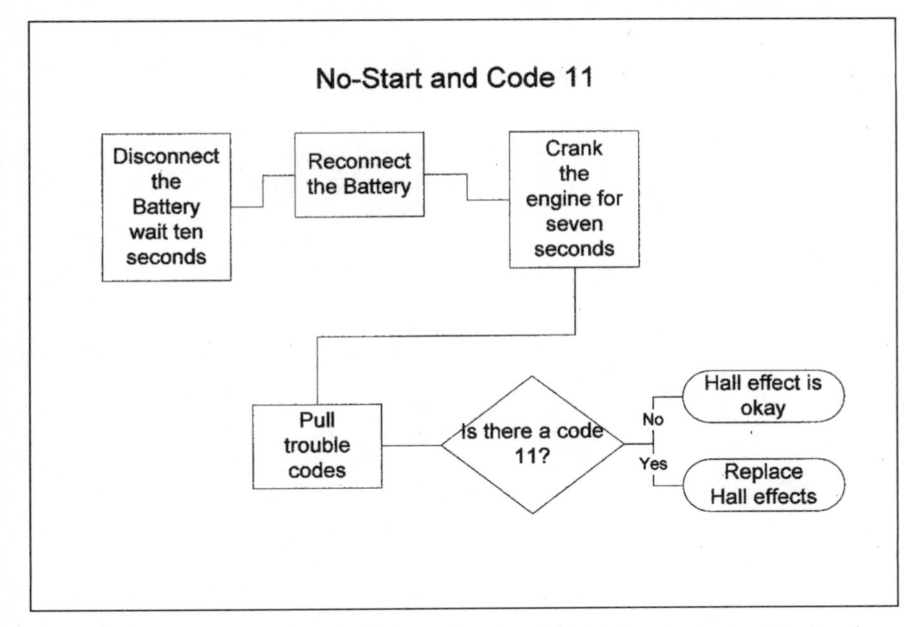

This is the diagnostic routine for code 11. Remember that when this flow chart states, "Replace the Hall Effects," consider the possibility that the real problem is in the wiring.

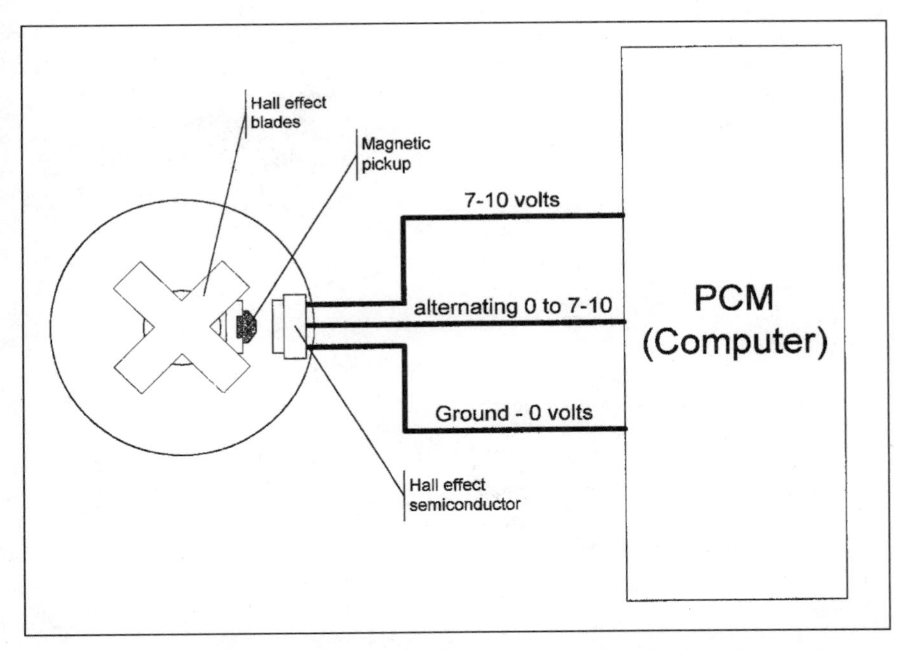

One wire on the Hall Effects should have 7-10 volts constantly. Another wire should have a voltage that alternates between 0 and 7-10 volts as the Hall blades rotate. The third wire should be ground.

cable connected to the positive battery cable. It is not the positive battery cable itself, but rather a smaller cable about half its diameter. Once you locate the cable, follow it until you come to a connector. This connector will only be about 6 inches from the battery terminal. If you cannot find this connector, then simply disconnect the battery. Leave the power disconnected from the computer for about 10 seconds. Reconnect the battery power to the computer and crank the engine for a minimum of 7 seconds. Now pull the codes again. If one of the codes is a code 11, then the Hall Effects or the wiring to the Hall Effects is defective.

To distinguish between a Hall Effects problem and a wiring problem, locate the wiring harness connector closest to the distributor or crankshaft position sensor. With the ignition switch in the off position, disconnect the three-wire connector. Now turn the ignition switch back on and measure the voltage on the three wires at the computer end of the wiring harness. At least one of the wires should have between 7 and 10 volts. This is the power supply wire for the Hall Effects. One of the other two wires should have a voltage that is a little lower than the power supply wire. The third wire, the one without the voltage, should have continuity to ground. Verify this with a voltmeter. Remember that the ground wire, although it will be continuous with ground, will have some resistance in it. Expect the resistance to be 0 to a few hundred ohms. If the resistance reads infinity, this wire is open and must be repaired. If the wiring harness checks out good, then replace the Hall Effects unit.

Code 11 (2.5-Liter Dakota)

The computer in this application is not programmed to log a code 11.

Code 12

This code means that the battery power to the computer has been lost sometime within the last 20-40 restarts of the engine. An odd thing about this code, odd but logical when you think about it, is the fact that every new car arriving at the dealership has a code 12 in its memory. This is because the engine on that car has not yet been started 20-40 times.

Codes 13 and 14

The MAP sensor is a piezo-resistance device which changes a 5-volt reference voltage in response to changes in manifold pressure. As manifold pressure increases, the voltage from the MAP sensor also increases. Before going any further with this explanation, it should be noted that most technicians and car hobbyists are used to thinking in terms of manifold *vacuum*. The scientific fact is that there has never been a vacuum in the intake manifold. What is in the manifold is a pressure that is lower than atmospheric pressure. We are also used to thinking that when the throttle is opened, the manifold vacuum decreases; the reality of this is that the manifold pressure is increasing and

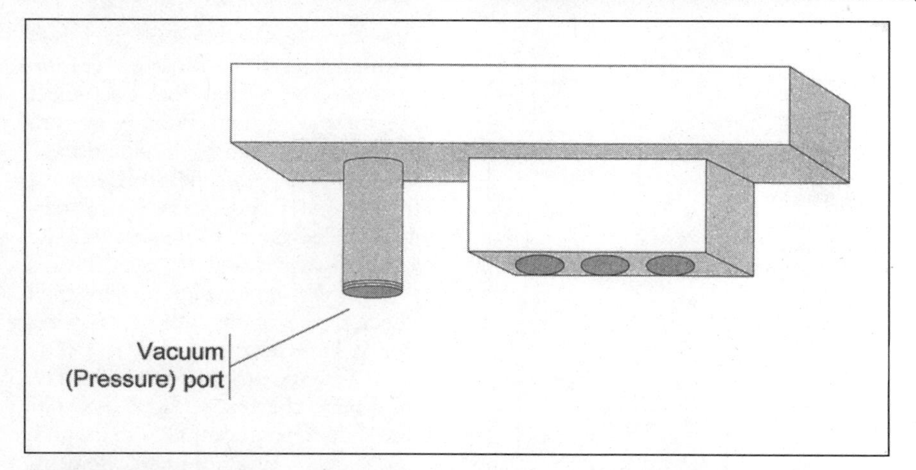

The MAP sensor has three wires and a vacuum port. The purpose of the MAP sensor is to monitor manifold pressure (vacuum) so that the computer knows the load on the engine and air flow rate.

On the 1984 models, the MAP sensor is located above the computer in the passenger's kick panel. The 1985 model has the MAP sensor located on the logic module in the passenger's kick panel. The style shown here was introduced in 1986 and is found in the engine compartment. If you determine that the MAP sensor needs to be replaced, be sure to record any numbers written on the sensor. The parts store may need these numbers.

On the 3.0-liter applications, the MAP is located near the alternator. Before condemning the sensor, check the hose that connects the MAP to the intake manifold.

because the pressure is increasing the reading on a vacuum gauge drops.

With the key on and engine off, the MAP sensor sends a voltage to the computer of about 3.5 to 4.8 volts, depending on the altitude and current atmospheric pressure. When the engine is started and allowed to idle, the voltage will drop to approximately 1/3 of the key-on, engine-off voltage. This is because the pressure in the manifold at idle is about 1/3 the pressure in the manifold when the engine is not running. When the throttle is opened, or the engine is put under a load in some other fashion, the manifold pressure will increase. As the manifold pressure increases, the voltage output from the MAP sensor also increases. At or near sea level an idling engine will cause the voltage signal from the MAP sensor to be between 1.2 and 1.9 volts, usually averaging about 1.5. As long as the idle voltage is in this neighborhood, and the voltage increases as the engine load increases (such as when you snap the throttle), and decreases as the engine load decreases, then the MAP sensor is functioning normally.

All TBI applications use a MAP sensor as the primary means of detecting both engine load and airflow into the engine. Most of the multipoint-injected engines use a MAP; the exception is the Mitsubishi Chrysler products. These applications use a Karman Vortex sensor.

On the turbo applications it should be noted that all of the output voltages for a given pressure are cut in half. This would make the idle voltage about .75 volts, and the key-on, engine-off voltage would be about 2.25 volts. This is because the MAP sensor on a turbo not only has to measure pressures below atmospheric, but also has to boost pressures above atmospheric.

The computer uses the MAP sensor information for the control of two major systems. First, the MAP sensor signal is used to measure the flow of air into the engine. Secondly, it is used to measure engine load in order to retard the ignition timing when the engine comes under a load.

The MAP sensor senses air flow by measuring the difference between barometric and manifold pressure.

Many late-model applications bolt the MAP sensor directly to the intake manifold. This eliminates the need for hoses and restriction problems associated with these hoses.

When the engine is being started, the computer senses the manifold pressure as the key is rotated through the on position heading toward start. This sample will be barometric pressure. It is stored in a memory and referred to by the computer until the engine is turned off. The pressure begins to vary as the engine runs. The closer the engine-running manifold pressure is to the barometric sample, the greater the load the engine is under.

The MAP sensor in 1984, 1985, and a few 1986 models is located on the logic module, which is located in the passenger's kick panel of the car. On all other models the MAP sensor is located under the hood. Please take extreme

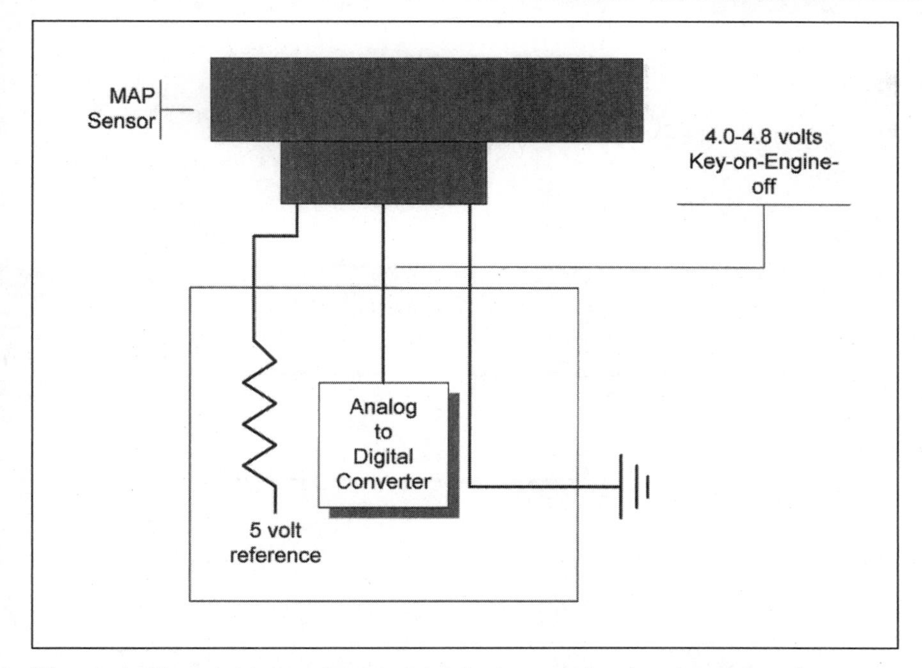

When the ignition switch is turned on, but the engine is not running, the voltage sent to the computer should be between 4.0 to 4.8 volts. If you are at an extremely high or low altitude, the voltages may be a little higher or a little lower. This is also an approximation of what the voltage will be when the engine is at wide-open throttle and full load.

care when removing the MAP sensor from the logic module to run the tests that are described below. The MAP is connected to the logic module by long and surprisingly thick wires. When the MAP is removed from its mounting, these wires are drug across the printed-circuit board. Damage may occur to the board or to the components on the board if caution is not used.

Also unique in the industry is Chrysler's ability to distinguish between vacuum problems with the MAP sensor and electrical problems with the MAP sensor. Code 13 means that the computer feels there is a vacuum problem causing incorrect readings from the MAP sensor. Code 14 means there is a perceived electrical problem with the MAP sensor. The way the computer distinguishes between them is a rather clever manner. When the key is rotated through the on position heading toward start, the computer takes a sample from the MAP sensor. Almost immediately after the engine starts the computer takes another sample. These two samples are compared and if they are not significantly different, a code 13 is logged.

Code 14 is logged when the reading from the MAP sensor is either above 4.9 volts or below 0.02 volts. These readings should never occur if the MAP sensor circuit is working properly. Therefore, the computer knows the circuit must have an open, short, or ground.

If the circuit does have an open, short, or ground, it is likely that the voltage will not change when the engine is started. Therefore, if a code 13 appears by itself, it is a safe bet that the problem is in the vacuum routing to the MAP sensor. If a code 14 appears by itself, the problem is in the wiring to the MAP sensor or in the MAP sensor itself. If code 13 and 14 appear together, the problem is also likely to be electrical. The reason for this is the open, short, or ground is present whether the engine is running or

A hand-held vacuum pump to simulate a running engine can be handy when testing a MAP sensor. Apply about 18 inches of mercury vacuum to simulate a running engine.

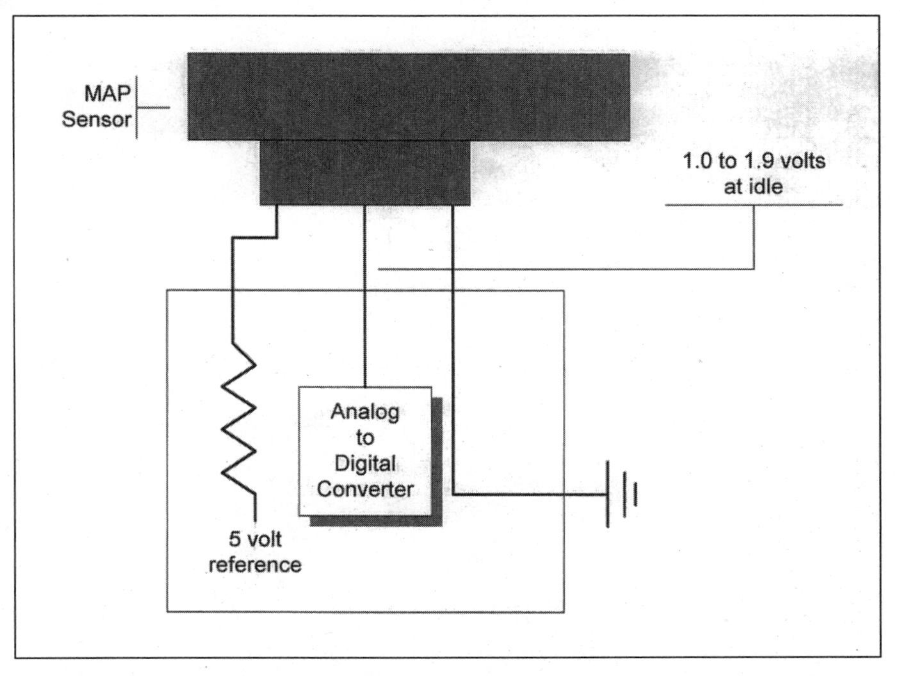

When the engine is idling, the MAP signal should be between 1.0 to 1.9 volts. If the voltage does not drop when the engine is started, it can cause a code 13 to be generated. Inspect the hose or hoses that connect the MAP to the intake manifold.

This is a top view of a late-model MAP sensor. This type mounts directly on the manifold. Note that the pins are very tiny and are easily damaged by ham-handed technicians. Be careful when disconnecting and reconnecting the harness connector.

not. Therefore, when the engine starts, there is no change in the voltage.

To troubleshoot a code 13 that is by itself, tee a vacuum gauge into the line that runs from the intake manifold to the MAP sensor. Note the reading on the vacuum gauge. It should, of course, read zero inches of mercury right now. Start the engine. If the vacuum rises significantly, replace the MAP sensor. If the vacuum does not rise, repair the vacuum hose to the MAP sensor.

To troubleshoot a code 14, or a code 13 and code 14 together, begin by using the test described in the preceding paragraph to confirm that vacuum is getting to the MAP sensor. If there is vacuum, shut the engine off and disconnect the MAP sensor electrical connector. Measure the voltages of the wires in the electrical connector leading from the MAP sensor to the computer with the voltmeter's black wire attached to a good engine, chassis, or battery ground. The voltage readings should be about 5.0, about 4.8 (just a little less than the 5.0 wire), and 0.0 volts. If the readings are good, connect the voltmeter from the 5.0-volt wire to the ground wire. You should get a reading of 5.0 volts. If you read 5.0 volts when the red lead was in the 5.0-volt wire, and the black lead was attached to ground but read considerably less or zero now, repair the open ground wire. If the ground wire is not open, then replace the computer.

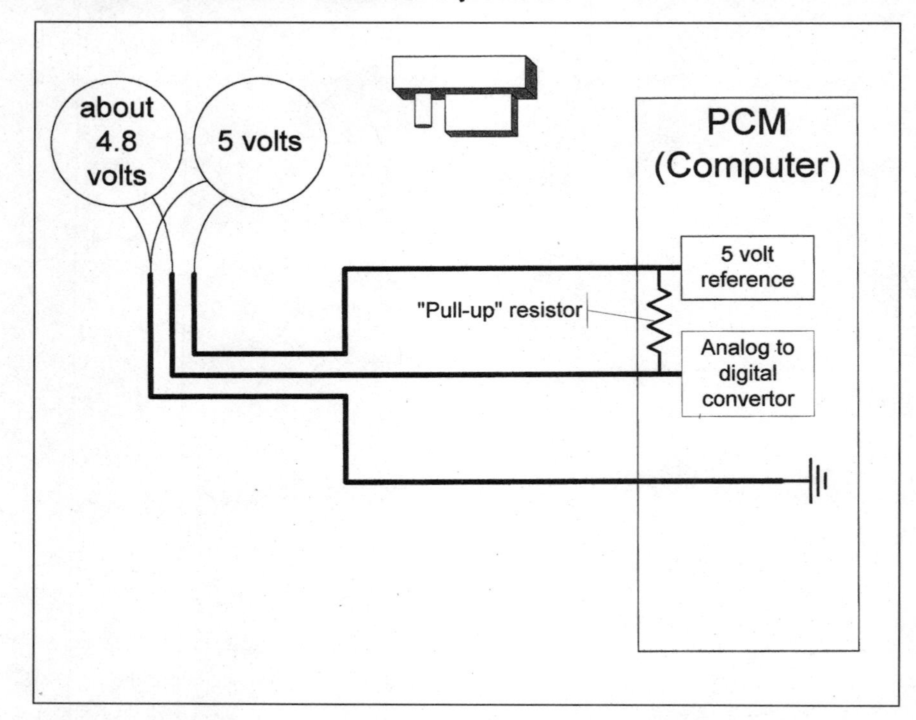

Checking the MAP sensor wiring harness for open circuits is very easy. There is a pull-up resistor between the 5-volt reference wire and the signal wire. Disconnect the MAP sensor. Turn the key on. Connect a digital voltmeter between the 5-volt reference wire (often violet and white) and the ground wire (often black and light blue). The voltage should be very close to 5 volts. Then check the voltage between the signal wire (usually dark green and red) and the aforementioned ground wire. The voltage will usually be less than the reading between 5 VREF and ground but still very close to 5 volts.

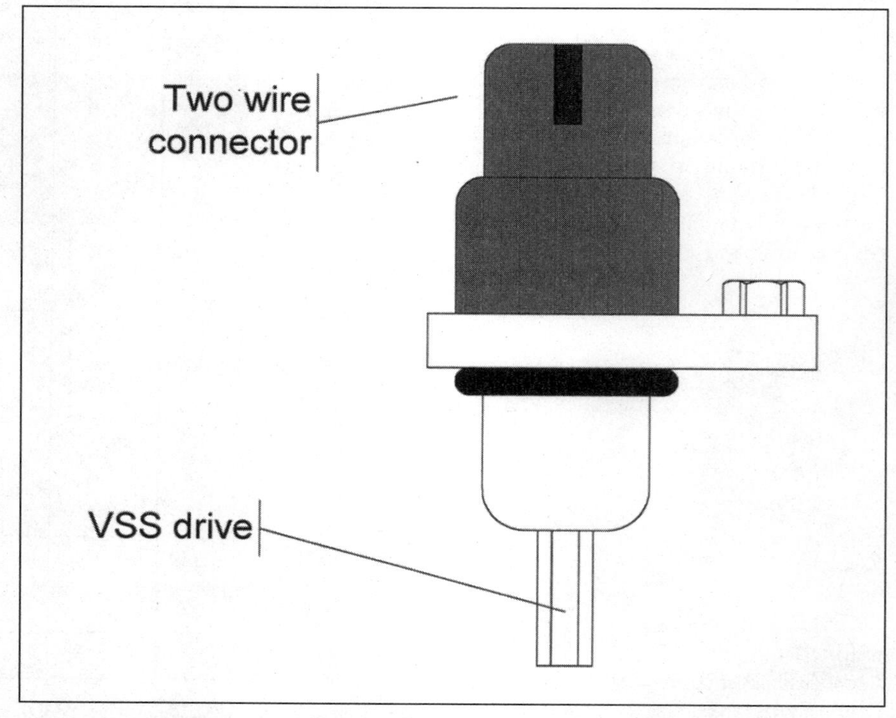

The vehicle speed sensor is located in the transmission or, on earlier applications, in the speedometer cable.

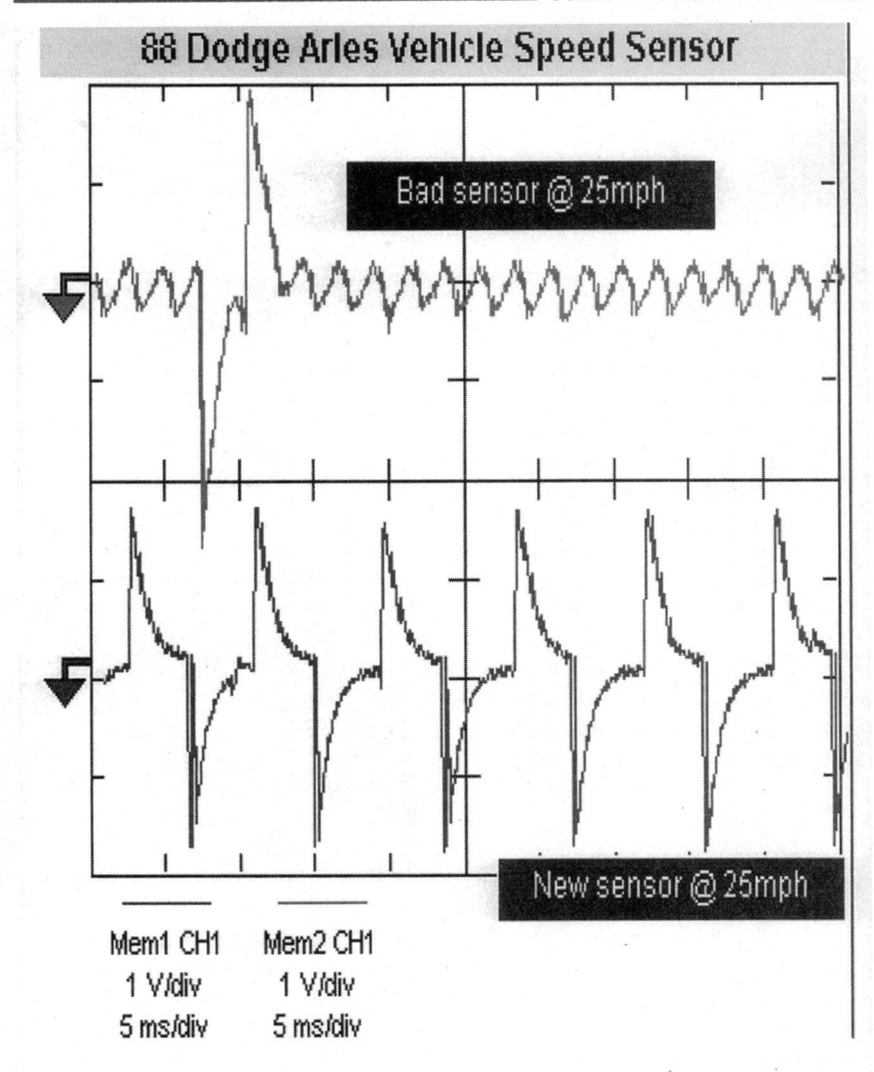

On top is the signal put out by a defective speed sensor. The lower part of the drawing shows the signal from a good speed sensor. Many Chrysler publications refer to the speed sensor as a "distance sensor."

If none of the wires has 5.0 volts when measured with the black lead of the voltmeter connected to engine, chassis, or battery ground, then this wire is defective and must be repaired. In the case of the applications where the MAP is mounted on the logic module, this means for most readers that the logic module must be replaced. If, however, your vocation or avocation involves the inner working of electronic equipment, the logic module can be disassembled and the damaged MAP wires replaced or repaired. Remember to assume that you are in Amarillo even when you are in Guam.

Once the 5.0-volt wire has been identified and/or repaired, test the ground wire by connecting the voltmeter between the 5.0-volt reference wire and the ground wire. You should read approximately 5.0 volts. If you read approximately 0.0 volts, then the ground wire or computer is defective. Check the resistance in this wire from the MAP to the computer. If the resistance is infinity, then repair the wire. If the wire is good, then the computer must be defective. The same rules and cautions apply to repair this ground wire on the 1984-1985 models as was discussed earlier.

Code 15

Code 15 relates to the vehicle speed sensor. I remember this as one of the most common codes that came up during my time in the Dodge dealership. Now in reality it may be that it was no more common than any other code. Sometimes time has a way of distorting our perceptions on such things. The job of the vehicle speed sensor is to inform the computer about the speed of the vehicle so the computer can calculate engine load and prepare the engine idle speed control system while the vehicle is decelerating. On deceleration, the idle speed control system will allow additional air flow into the engine as the vehicle rolls to a stop. The best way to test the vehicle speed sensor circuit is with a scanner. If the scanner detects a signal from the vehicle speed sensor, it will display a reading other than zero. Refer to the serial data stream section for more detail.

Code 16

Code 16 is used to inform the technician that the battery voltage as sensed by the computer has dropped below 4 volts, or has dropped between 7.5 and 8.5 for more than 20 seconds. Now in most cases, code 16 itself is rather insignificant compared to the most likely cause of the code, which is a defective battery. Defective batteries can, of course, keep the vehicle from starting. Somehow, if the engine does not start, the fact that there is a code 16 seems almost irrelevant. The computer does not begin sampling until about one minute after the engine starts. Therefore, the problem cannot be the battery voltage dropping low while the engine is being cranked.

If there is a history of starting problems, then the battery, and especially the cables leading to and from the battery, should be checked thoroughly. Repair any problems found with these cables.

The actual and precise cause of code 16 is a drop in the voltage on the battery voltage sense wire as it connects to the computer. Connect the voltmeter at the voltage sense terminal of the computer. The voltage should be within 1 volt of battery voltage. If it is, check for loose connections, loose terminals, and loose wires. If there is not, repair the wire that runs from the battery sense ter-

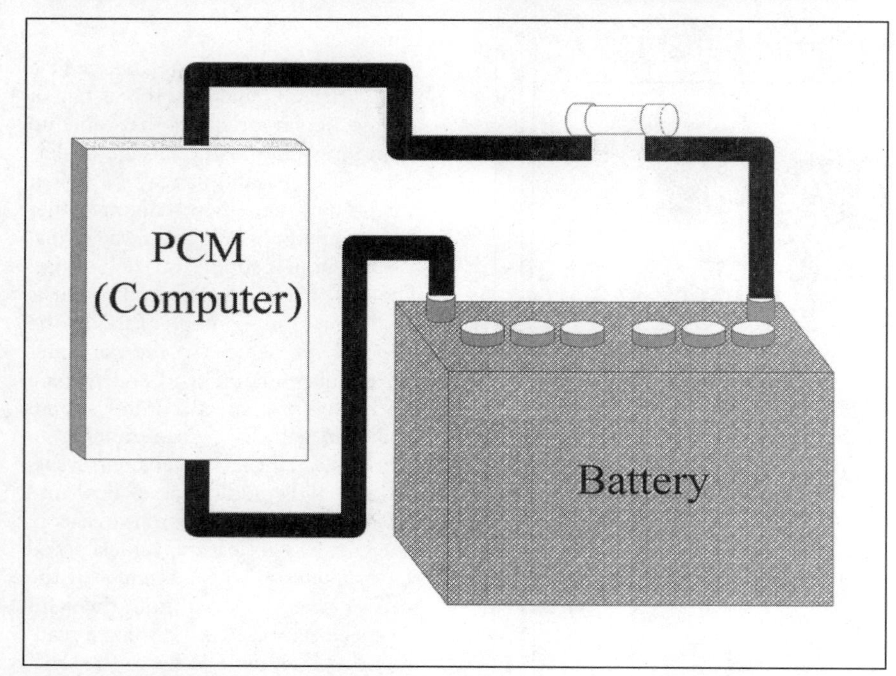

Although navigating the wiring harness of any modern automobile is as complicated as navigating Los Angeles freeways, the circuit that can cause code 16 is quite simple. The most likely location for the problem that results in this code is the connection at the PCM. Also check the cables, fuses, and the battery itself.

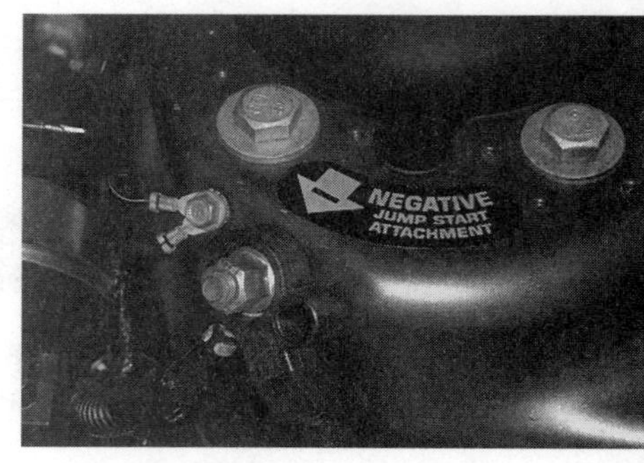

On many applications, the computer is grounded to the body. Sometimes these grounds are subjected to dirt, salts, acids, and other corrosive substances. Make sure that the attachment point for these grounds is at, or near, the original location.

On this application, a large battery cable connects voltage to a junction point. The computer's battery power source is often at that point rather than at the battery.

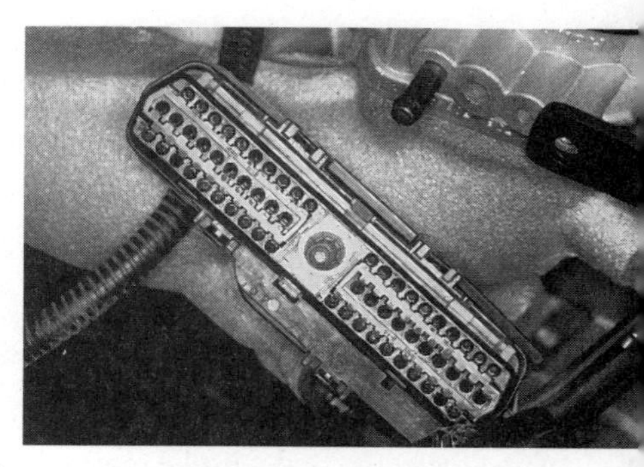

My colleagues and students have repeatedly questioned Chrysler's wisdom in placing the computer under the hood, especially those applications where the computer is near the battery. Remove the connector and inspect each pin carefully. Look for evidence of corrosion and expanded terminals. This is an important task for most codes, not just code 16.

minal to the direct battery feed terminal of the computer.

Code 17

On non-turbo applications, code 17 means that the engine is taking too long to warm up. Basically, it means that the engine temperature was still below 174 degrees after 8 minutes of operation. There are several obvious things to check first. When you spit, does the spittle hit the ground as a solid piece of ice? This might be a clue that the ambient temperature is too cold for the heat caused by combustion to overcome. I used to live in Fairbanks, Alaska. I remember several mornings leaving the warmth of the house to get into a car whose block core temperature had fallen to about 50 degrees below zero. I would start the engine, more than a minor miracle in itself, and drive about 3 miles to work. The temperature gauge would not even come off the low peg. When a scanner was attached to measure the temperature after getting to work, the temperature was not yet above freezing. Now, in that kind of climate, code 17 might be caused by the ambient temperature. In Oklahoma City, I doubt that low temperature would cause code 17.

Another possible cause is a defective or missing thermostat. The job of the cooling system thermostat is to reduce—in fact, virtually cut off—flow of coolant through the radiator until the engine almost reaches operating temperature. Thermostats come in different temperature ratings. I can remember the first time I had to make a decision about what temperature thermostat to use. I was 16 years old, and my dad was making the world safe for democracy by maintaining B-52s in Guam. I was suddenly the man of the house. I went to the local K-Mart and selected from a rack of thermostats. I chose one that had a temperature rating of 160 degrees, after all, this was Texas, and the cooler the engine would run the better. First of all, the temperature of the thermostat has very little to do with the final operating temperature of the engine, only with the haste at which that operating temperature is attained. As it turns out, my decision was a perfect example of the ignorant wisdom of youth.

Today's automobiles are a little different story. Manufacturers are now required to guarantee to the EPA and other government agencies that the engine will produce emissions that fol-

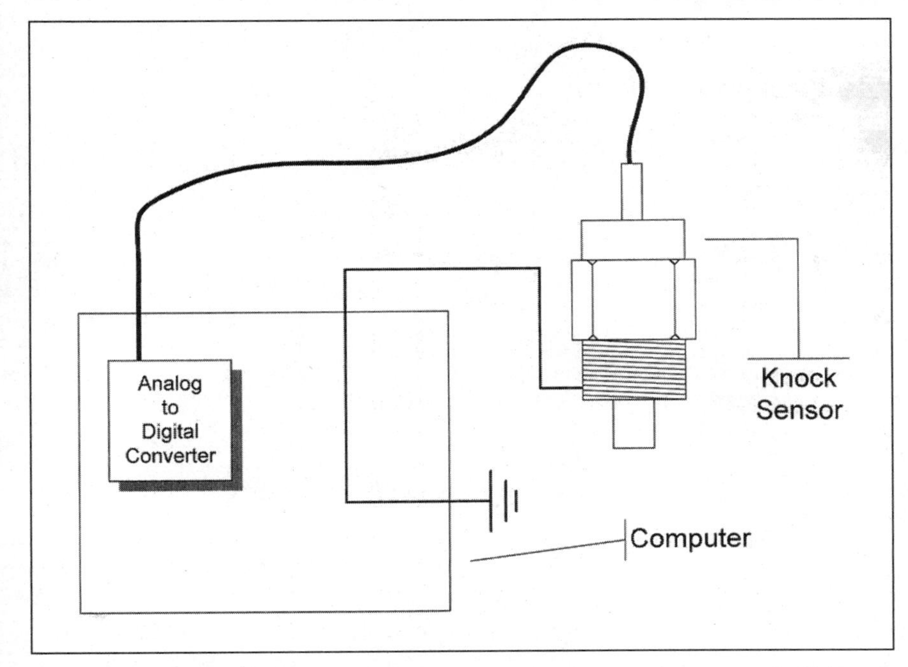

Turbocharged engines run high combustion pressures and temperatures. This increases the likelihood of potential engine-damaging detonation. The signal from the knock sensor is used by the computer to retard the timing to reduce detonation. Disconnect the harness connector at the computer and connect an AC voltmeter to the knock sensor input wire. Lightly tap on the intake manifold with a small hammer. If any voltage is registered, the sensor and the sensor wiring harness is good.

low an approved pattern and level. When the engine warms up differently than designed, this can upset emission patterns. Modern engines are controlled and monitored by computers. These computers expect to see certain things happen at certain times. They expect the engine to warm up at a certain rate. When you install a colder thermostat than is recommended for the engine, the engine will follow a different warm-up pattern. In the case of most Chrysler applications, the computer will respond by setting a code 17.

Code 17

On some applications, code 16 represents a problem with the knock sensor. Turbocharged Chryslers incorporate a microphone-like device in the intake manifold or cylinder head called a knock sensor. The knock sensor detects vibrations which result from knocking or pinging. Back in the "big fin days," knocking and pinging were controlled by rich mixtures, leaded gas, and low compression ratios. Today's lean-running engines which run on unleaded gas at gradually increasing compression ratios have to depend on electronics to guard against detonation. Since turbocharging effectively increases the compression ratio of the engine, Chrysler chose to use knock sensors on its turbo applications to reduce detonation.

When a detonation is heard by the knock sensor, the timing retards 3 or 4 degrees each second until the knock goes away. It then begins to slowly advance the timing until, assuming that engine loads remain steady, the knock sensor begins to hear detonation. It then backs the timing up slightly and will remain at that setting until the engine load or throttle position changes.

At first glance a system to retard timing might seem to work against good performance and power, however, the opposite is true. A detonating engine is an engine that is wasting power, the air/fuel charge in the combustion chamber is being ignited too soon and is therefore trying to drive the piston back down the cylinder as the crankshaft is trying to push the piston up on the compression stroke; this robs power and wastes fuel. The knock retard system controls detonation, thereby restoring the large part of this lost power and economy to the engine.

The knock sensor consists of a piezo-electric element connected to the computer by a single wire. When a detonation occurs, the piezo-electric element generates an AC voltage that is detected by the computer.

The most effective way to test this device and its circuit is with an oscilloscope. Connect the oscillo-

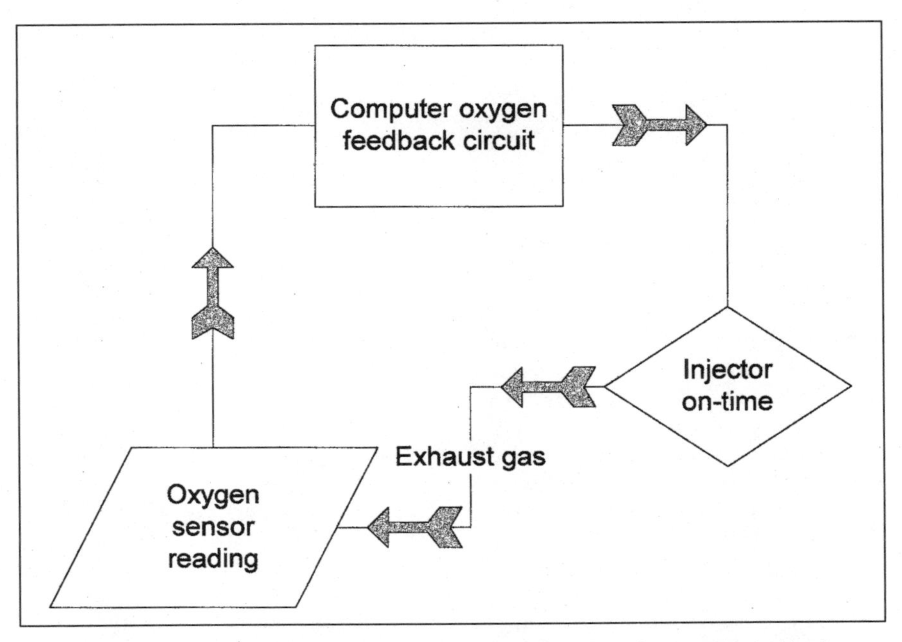

The oxygen sensor measures the amount of exhaust gas oxygen. This information is used by the computer to detect errors in air/fuel ratio. When the computer detects such an error, it assumes that the problem is caused by incorrect injector on-time. The computer then alters the injector on-time. Again, the oxygen sensor samples the gases, finds an error, and the cycle continues ad infinitum.

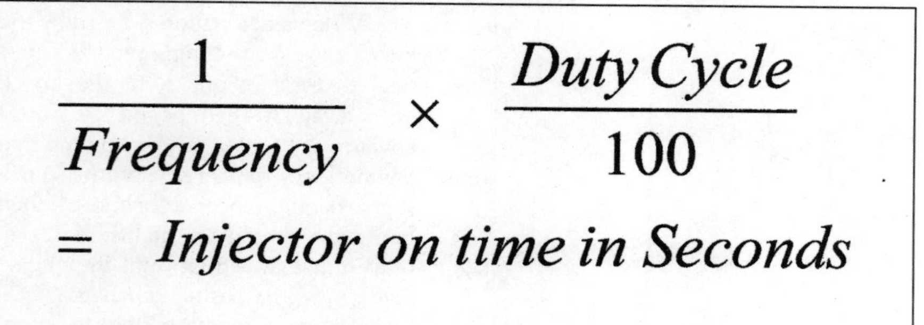

$$\frac{1}{\text{Frequency}} \times \frac{\text{Duty Cycle}}{100} = \text{Injector on time in Seconds}$$

Injector on-time can be measured with a dwell meter and tachometer. Put the tach/dwell meter on the four-cylinder scale. Do this even if the engine is a 10-cylinder. Take a tach reading on one of the injector wires. One of the wires on each injector will give a "tach" signal and the other will not. Divide the reading you get by 30; this is the frequency. Set the tach/dwell to read dwell. Multiply the reading by 1.1; this will be the duty cycle reading. Plug the frequency and duty cycle into the formula shown here; that will give you the injector on-time in seconds.

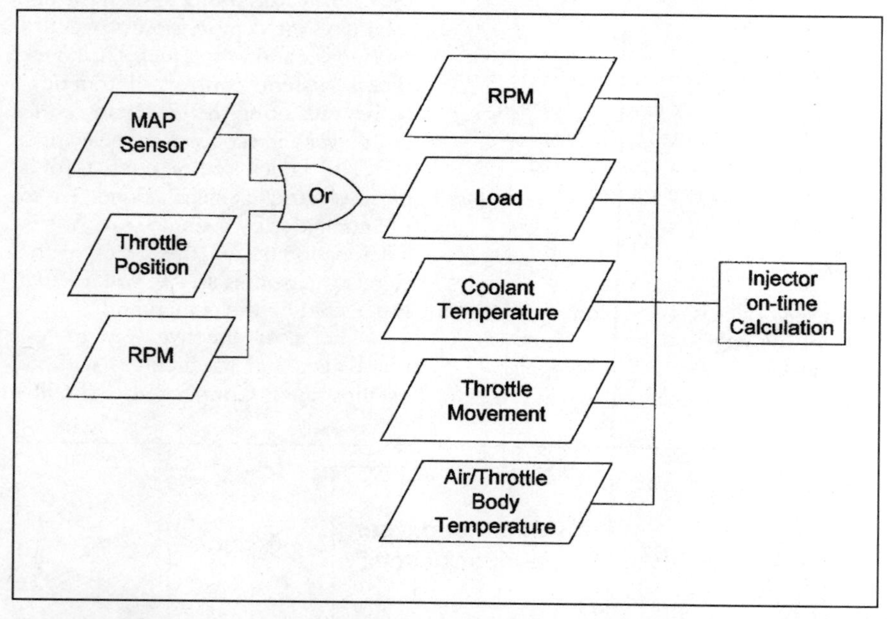

There are tools and software available from automotive tool dealers that can read and display injector on-time directly.

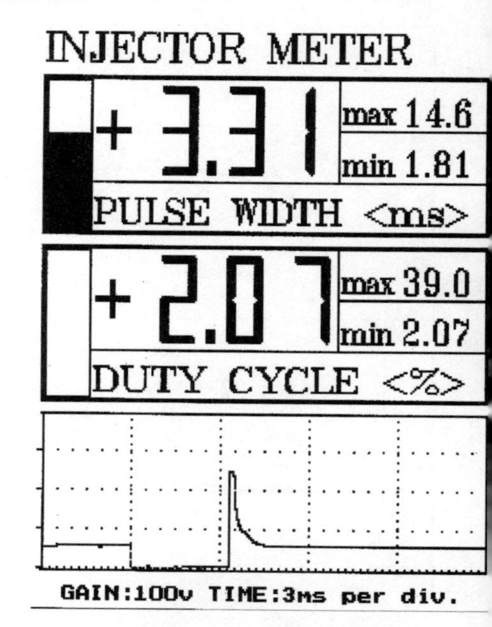

When the engine is cold, or when it is first started, the oxygen sensor is ignored. This time is called open loop.

scope to the knock sensor input wire at the computer. Tap on the engine near the knock sensor. You should see the oscilloscope pattern fluctuate. If the scope does not fluctuate, then disconnect the connector on the knock sensor and connect the oscilloscope directly to the knock sensor. Again tap on the sensor. Is there a deflection on the oscilloscope screen?

If the scope did not deflect when connected to the computer end of the wire but did deflect when attached to the sensor directly, repair the wire. If it still did not deflect when attached to the knock sensor directly, replace the knock sensor.

If you do not have a scope handy, the same tests can be made using an analog AC voltmeter that has a scale that will read in millivolts. Simply connect the AC voltmeter as was described for the oscilloscope and look for a needle deflection.

Code 21

Code 21 relates to the oxygen sensor circuit. The oxygen sensor is an electrochemical device consisting of two layers of platinum separated by a layer of zirconium oxide. One plate is exposed to ambient oxygen while the other is exposed to the oxygen content of the exhaust system. Although an electronics engineer would probably disagree, this is a lot like plates in a battery separated by an insulator. The oxygen sensor is installed in the exhaust system, usually close to the manifold. When the engine is started, the hot exhaust gases passing through the manifold begin to heat the oxygen sensor. When the temperature of the oxygen sensor reaches 600 degrees F, the oxygen sensor becomes conductive for oxygen ions. In the same way the plates of a battery begin to attract electrons from the electrolyte to create a voltage, the warm oxygen sensor attracts oxygen electrons. If the amount of oxygen electrons on the exhaust side are equal to the amount on the ambient air side, then the electrons equalize and no voltage is produced. As the oxygen content of the exhaust decreases, an imbalance occurs and the oxygen sensor begins to produce a voltage.

Normal operating voltages for the oxygen sensor range from a low of 100 millivolts to a high of 900 millivolts. The voltage produced can occasionally be higher or lower than these figures. About halfway between these voltages is 450 millivolts. This voltage is known

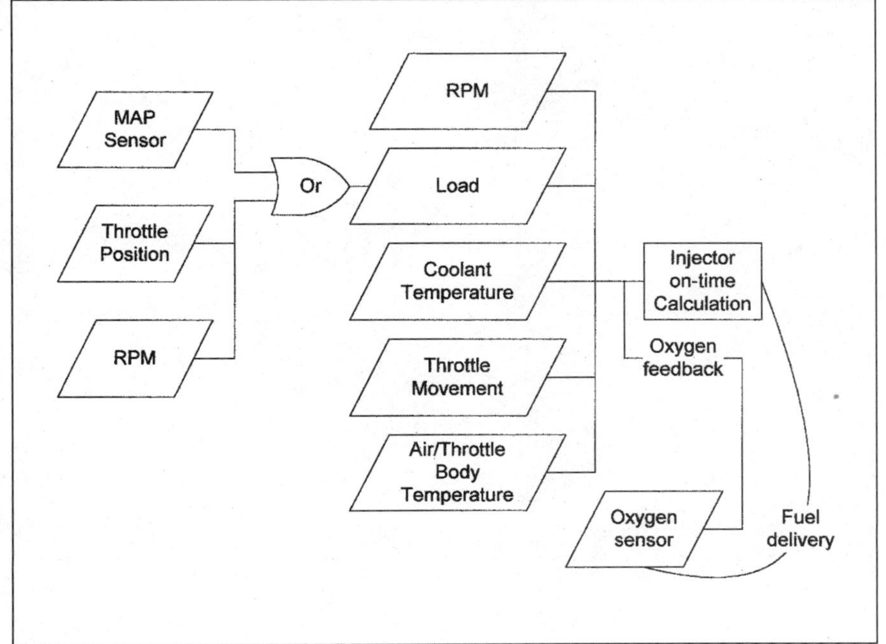

When the oxygen sensor begins to put out a signal, and the temperature of the engine has approached operating temperature, and a certain amount of time after start-up has occurred, then the computer will begin to use the information from the oxygen sensor. This mode of operation is called open loop.

as the crossover point. Any voltage less than 450 is interpreted by the computer as a lean exhaust condition; any voltage greater than 450 is interpreted by the computer as a rich exhaust condition. Anything that affects the content of oxygen in the exhaust will affect the oxygen sensor signal.

The oxygen sensor might be described as a chemical generator. When it is heated to a minimum of 600 degrees F, it will begin to produce a voltage ranging from 100 to 900 millivolts. Once operational temperature is reached, the sensor will begin to respond to changes in the content of exhaust oxygen. When the oxygen content of the exhaust is high, the computer assumes that the engine is running lean. The design of the oxygen sensor is such that it will produce a low voltage when the exhaust oxygen content is high. Oxygen content in the exhaust gases resulting from a lean combustion will result in a voltage less than 450 millivolts being delivered to the computer. When the exhaust gases result from a rich combustion, the oxygen sensor voltage to the computer will be greater than 450 millivolts. When the oxygen sensor voltage is indicating a lean condition, the computer will respond by enriching the mixture. When the oxygen sensor voltage is high, the computer will respond by leaning out the mixture. In this manner, the computer adjusts for minor errors and variations from the rest of the input sensors and controls the air fuel ratio at 14.7:1.

In spite of what you might hear from different professionals, testing the oxygen sensor is actually simple. Connect a high-impedance voltmeter to the oxygen sensor at the point where the wire from the computer connects to the oxygen sensor. Leave the oxygen sensor connected and start the engine. Allow it to run at 2,000 rpm for two minutes. With the engine idling, watch the voltage reading on the voltmeter. The voltage should be constantly changing back and forth from a voltage between 100 and 400 millivolts to a voltage of 600 to 900 millivolts.

The oxygen sensor itself is a very dependable unit. Occasionally the oxygen sensor will fail with an internal open circuit. The most common problem that

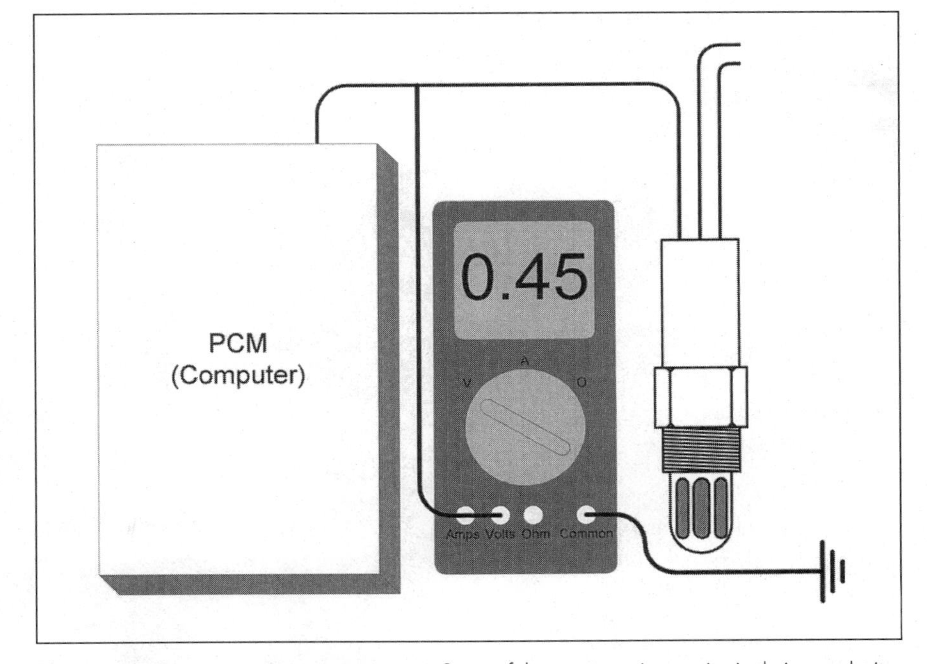

There are many ways to test an oxygen sensor. Some of them are very impressive in their complexity and showmanship. However, a simple test is the best. Connect a digital voltmeter between the signal wire and ground. Simply monitor the oxygen sensor voltage. You should observe the voltage toggling above and below 0.450 volts. If the voltage is stuck high, or in the middle, create a vacuum leak and look for a response. If the voltage is stuck low, or in the middle, goose the throttle and look for a response. If there is no response in the voltage, replace the oxygen sensor.

occurs with oxygen sensors is contamination. There are three common sources of contamination for the oxygen sensor: RTV silicone, leaded fuel, and soot.

RTV silicone is a commonly used sealer and adhesive. The fumes from this adhesive can coat the oxygen sensor, slowing its ability to respond to changes in the oxygen content of the exhaust. When the oxygen sensor becomes contaminated with silicone, it delivers a lean signal to the EEC computer.

When the oxygen sensor becomes contaminated by leaded fuel, the output voltage will be stuck above the crossover point. This is a rich exhaust indication. Although this rarely occurs by accident, be sure to check the contents of any fuel additives being used to ensure that it does not contain tetra-ethyl lead.

There are two primary sources for soot: oil and fuel. Oil soot is a result of a mechanical engine condition such as worn rings or valve guides. Fuel soot is either the result of a rich-running engine or a misfire. Either form of soot impedes the operation of the oxygen sensor. The output voltage remains constant at around 0.5 volts.

A more thorough test of the oxygen sensor circuit involves two steps.

Disconnect the connector between the oxygen sensor and the computer. Connect a high-impedance voltmeter between the oxygen sensor and ground. Start the engine and allow it to run until the upper radiator hose is hot and pressurized. Hold the oxygen lead to the computer with one hand and touch the positive battery terminal with the other hand. This delivers a small voltage to the computer, causing the computer to believe that the engine is running rich. The engine rpm and the oxygen sensor voltage should drop.

If there is no response, connect the oxygen wire that goes to the computer to ground. This makes the computer believe that the engine is running lean. The rpm and the oxygen sensor voltage should both rise.

If the system passes either test, then the system is working properly. If the rpm changes but the oxygen

On the 3.0-liter application, the coolant sensor is easy to find. On virtually all applications, the coolant sensor will be found near the thermostat housing. The lighter-colored wire is the signal wire. The darker wire is the ground.

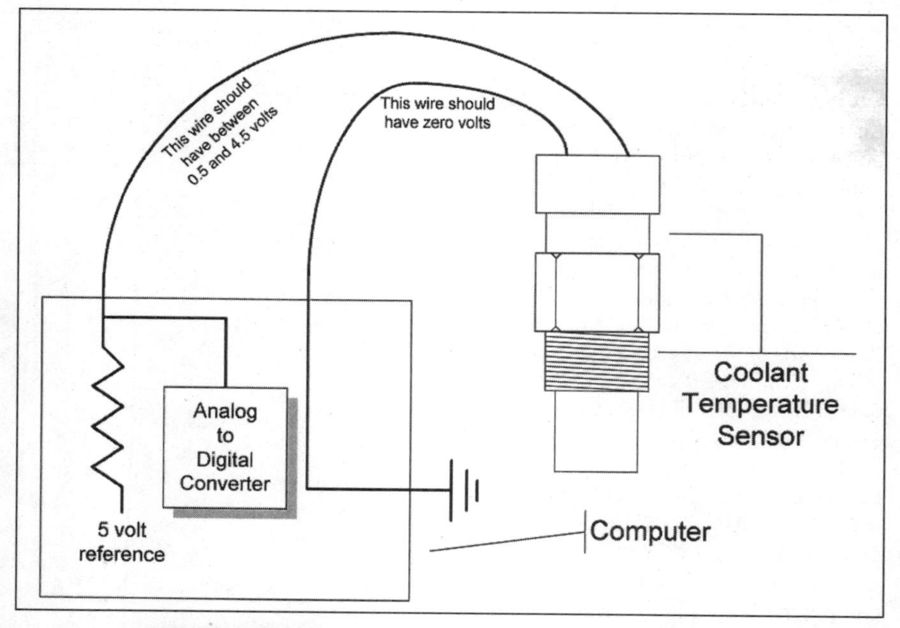

One of the wires to the coolant sensor should always have 0 volts. The other should have a voltage between 0.5 and 4.5. If the reading is in that range, the coolant temperature sensing circuit is probably working correctly. If the voltage also changes when the temperature of the coolant changes, then that proves the circuit is working.

This is looking through the intake runners on a late-model four-cylinder application. Although physically somewhat removed from the thermostat housing, a brief study of the engine design would reveal that the sensor is actually quite close in terms of coolant flow.

voltage does not, replace the oxygen sensor. If neither changes, check for continuity on the wire to the computer. If the wire is good and the connection is good, replace the computer.

Code 22

Code 22 indicates that the coolant temperature sensor circuit has sent a voltage greater than 4.96 or less than 0.51 to the computer. These voltages reflect temperatures that are impossible under normal operating conditions. The high voltage represents a temperature of less than -40 degrees F. Now I can remember the vinyl seat cracking on a cold January evening in Fairbanks. I know that it does get that cold. Under these extremely low temperatures a false code might be generated. The same goes for the low voltage of 0.51. Normal operating temperature provides a voltage to the computer of about 1 volt. The 0.51 volts represents a severely overheating engine. In fact, the temperature of the engine at this point is so high that code 22 is almost irrelevant.

Begin testing the circuit by turning the ignition switch off and disconnecting the coolant temperature sensor wiring harness. The coolant temperature sensor is located near the thermostat housing. Connect a digital voltmeter across the terminals of the harness side of the connection. Turn the ignition switch back on. The voltmeter should read 5 volts. If the voltmeter does read 5 volts, replace the coolant sensor. If it does not read 5 volts, there is a problem in the coolant sensor wiring harness.

Next test the 5-volt reference. With the ignition switch still on, connect the red lead of the voltmeter to the light-colored wire on the coolant sensor and the black lead to the battery negative terminal. The voltmeter should read 5 volts. If it does not, repair the light-colored wire between the coolant temperature sensor harness connector and the computer.

If the 5-volt reference wire tests good, turn the ignition switch off. Connect the red lead of the digital voltmeter to the positive terminal of the battery. Connect the black lead of the voltmeter to the dark-colored wire on the coolant

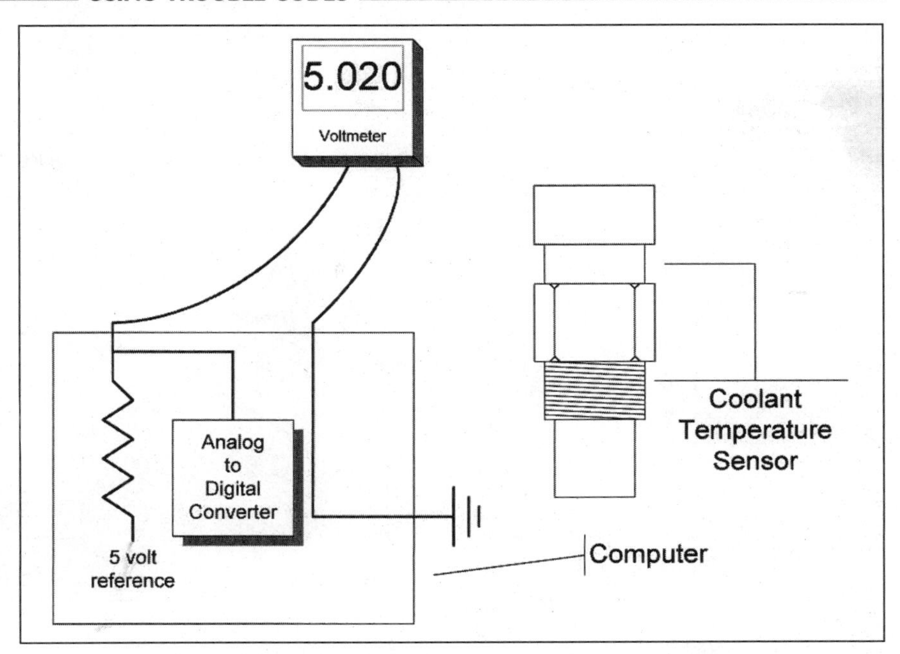

With the coolant sensor disconnected, a voltmeter connected across the two terminals of the coolant sensor harness connector should read approximately 5 volts.

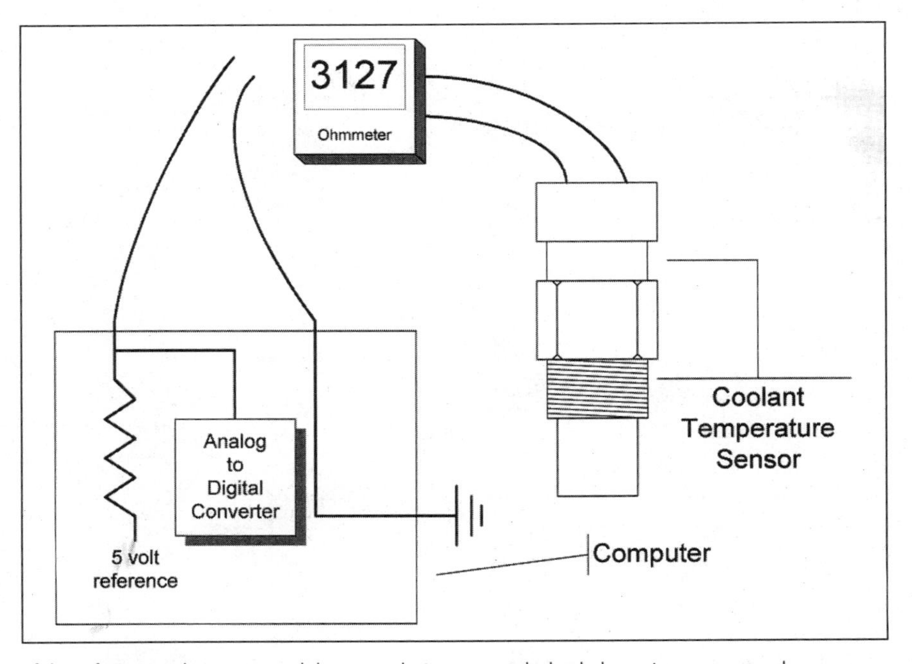

If the reference voltage wire and the ground wire are good, check the resistance across the sensor. The thermistor at the core of the sensor is one that is used widely in the industry. Cummins diesels use it. I like the specs they give you to determine if the sensor is good. "The resistance should be between 175 and 250,000 ohms." In other words, in order for a code 22 to be generated, the sensor would have to be either open or shorted.

sensor. The voltmeter should read 12 volts. If it does not, repair the dark-colored wire from the coolant sensor harness connector to the computer.

Coolant Temperature Sensor Specifications

-40 degrees F	= 100,700 ohms
0 degrees F	= 25,000 ohms
20 degrees F	= 13,500 ohms
40 degrees F	= 7,500 ohms
70 degrees F	= 3,400 ohms
100 degrees F	= 1,600 ohms
160 degrees F	= 450 ohms
212 degrees F	= 185 ohms

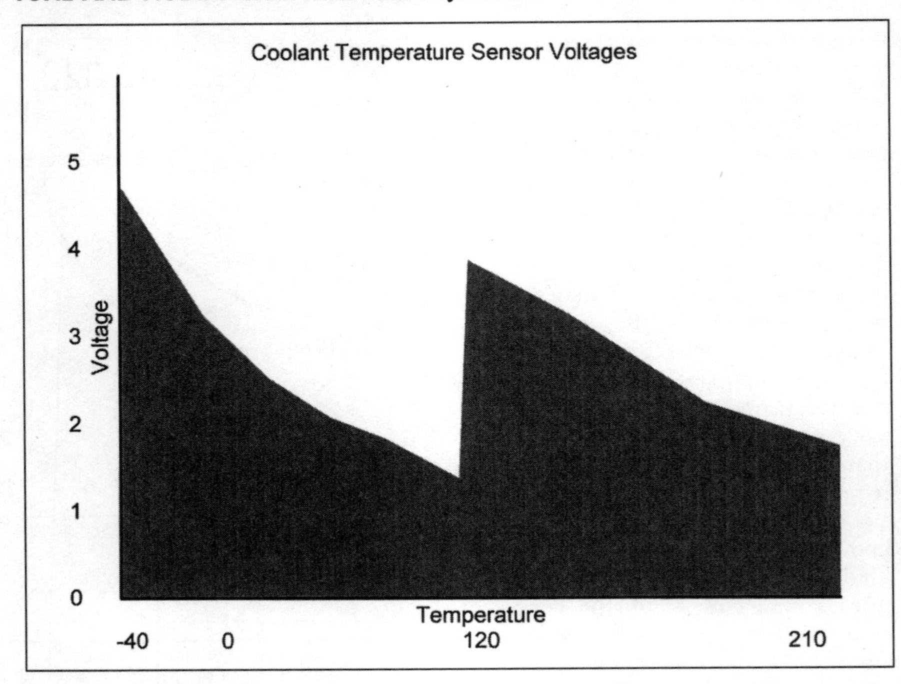

If you observe the coolant sensor voltage on late-model Chrysler applications, you will note that the voltage begins high when the engine is cold, drops rapidly, then jumps back up only to descend again. As odd as it may seem, this is normal.

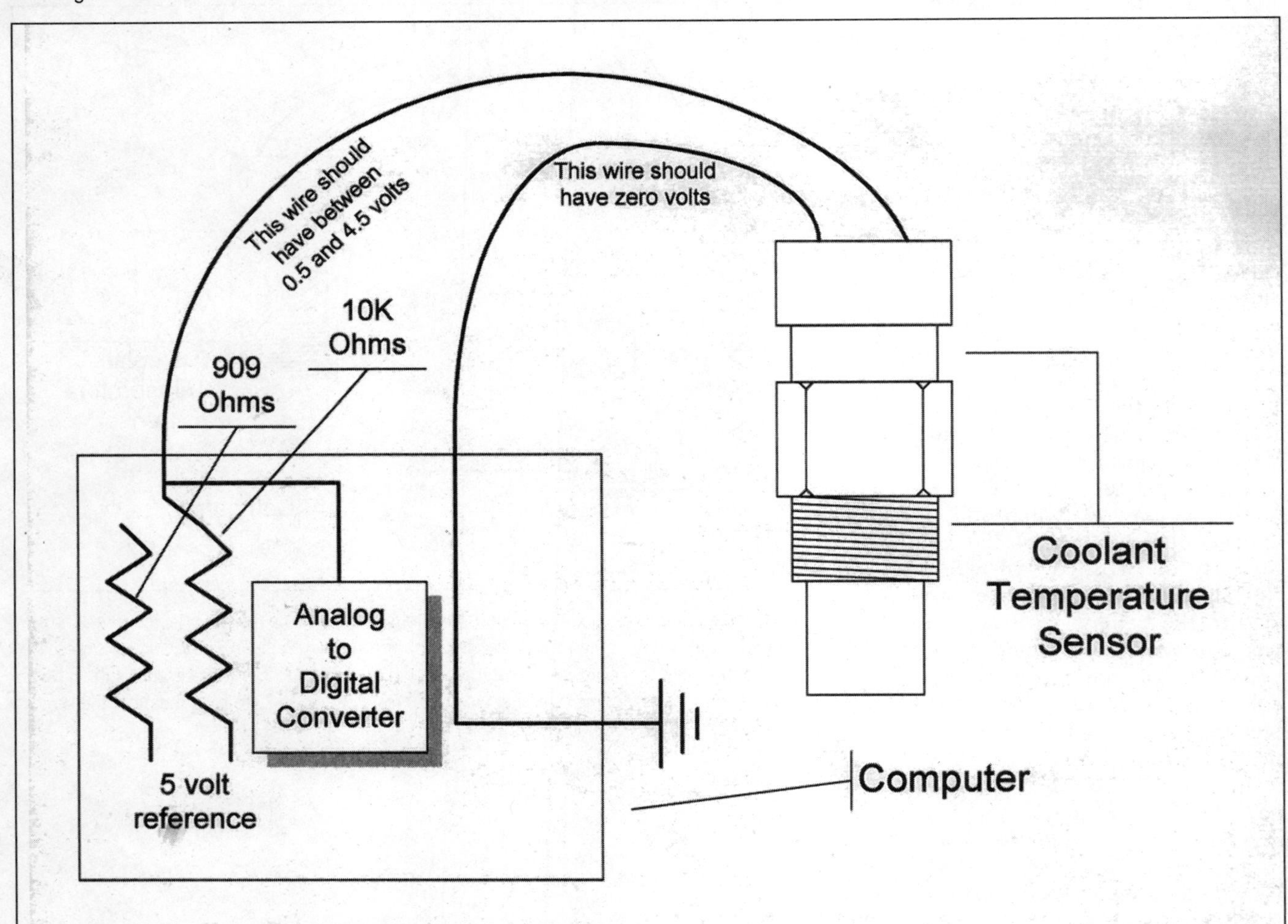

Inside the computer are two resistors used to create the voltage drop between the computer and the coolant sensor. When the engine begins its warm up, the 10,000-ohm resistor is used. When the coolant temperature exceeds about 120 degrees F, the computer switches to the 909-ohm resistor; the voltage leaps up and then begins to fall again.

Code 23

The air charge temperature sensor or throttle body temperature sensor circuit has a problem when code 23 is generated. Some multipoint Chrysler applications, especially the turbos, use an air charge to help calculate the temperature, and therefore the density, of the incoming air. The throttle body applications from 1986 on use a temperature sensor located in the throttle body to assess if the temperature of the fuel might be high enough during a hot start for the fuel to have vaporized. If the temperature of the throttle body exceeds 170 to 180 degrees F, then the computer will double the injector frequency in an effort to clear out the vaporized fuel. This code indicates that the computer has detected a signal voltage from this circuit of more than 4.98 or less than 0.06. In other words, the temperature of the charge air (air in the manifold) or throttle body is either extremely high or extremely low.

Begin testing the circuit by turning the ignition switch off and disconnecting the air temperature sensor wiring harness. Connect a digital voltmeter across the terminals of the harness side of the connection. Turn the ignition switch back on. The voltmeter should read 5 volts. If the voltmeter does read 5 volts, replace the air sensor. If it does not read 5 volts, there is a problem in the air sensor wiring harness.

Next test the 5-volt reference. With the ignition switch still on, connect the red lead of the voltmeter to the light-colored wire on the air sen-

The air temperature sensor is very similar to a coolant temperature sensor. The primary difference is that the protective cover over the thermister is open to allow the air to come into direct contact with the sensing element.

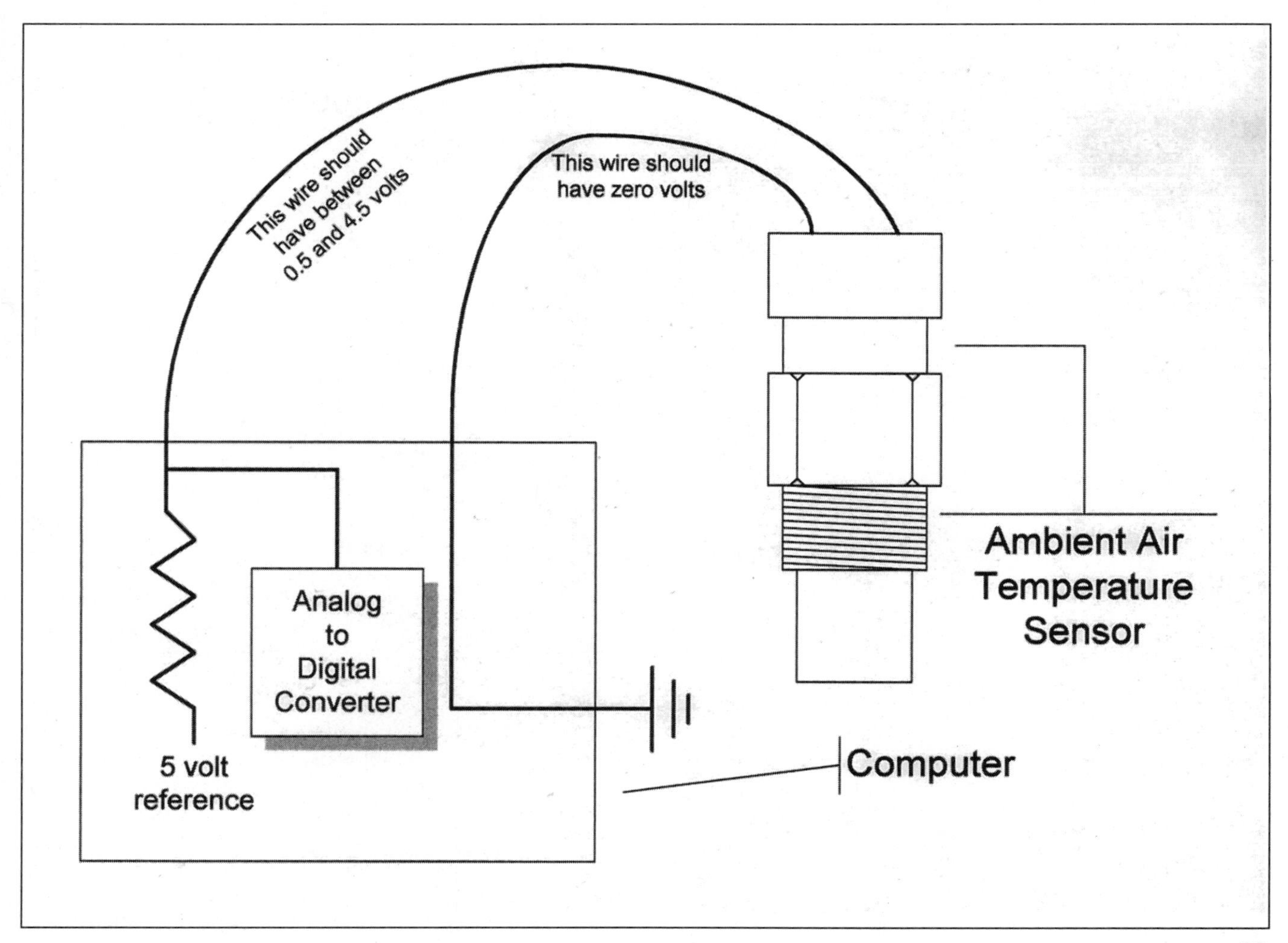

One of the wires to the air temperature sensor should always have 0 volts. The other should have a voltage between 0.5 and 4.5. If the reading is in that range, the air temperature sensing circuit is probably working correctly. If the voltage also changes when the temperature of the air changes, then that proves the circuit is working.

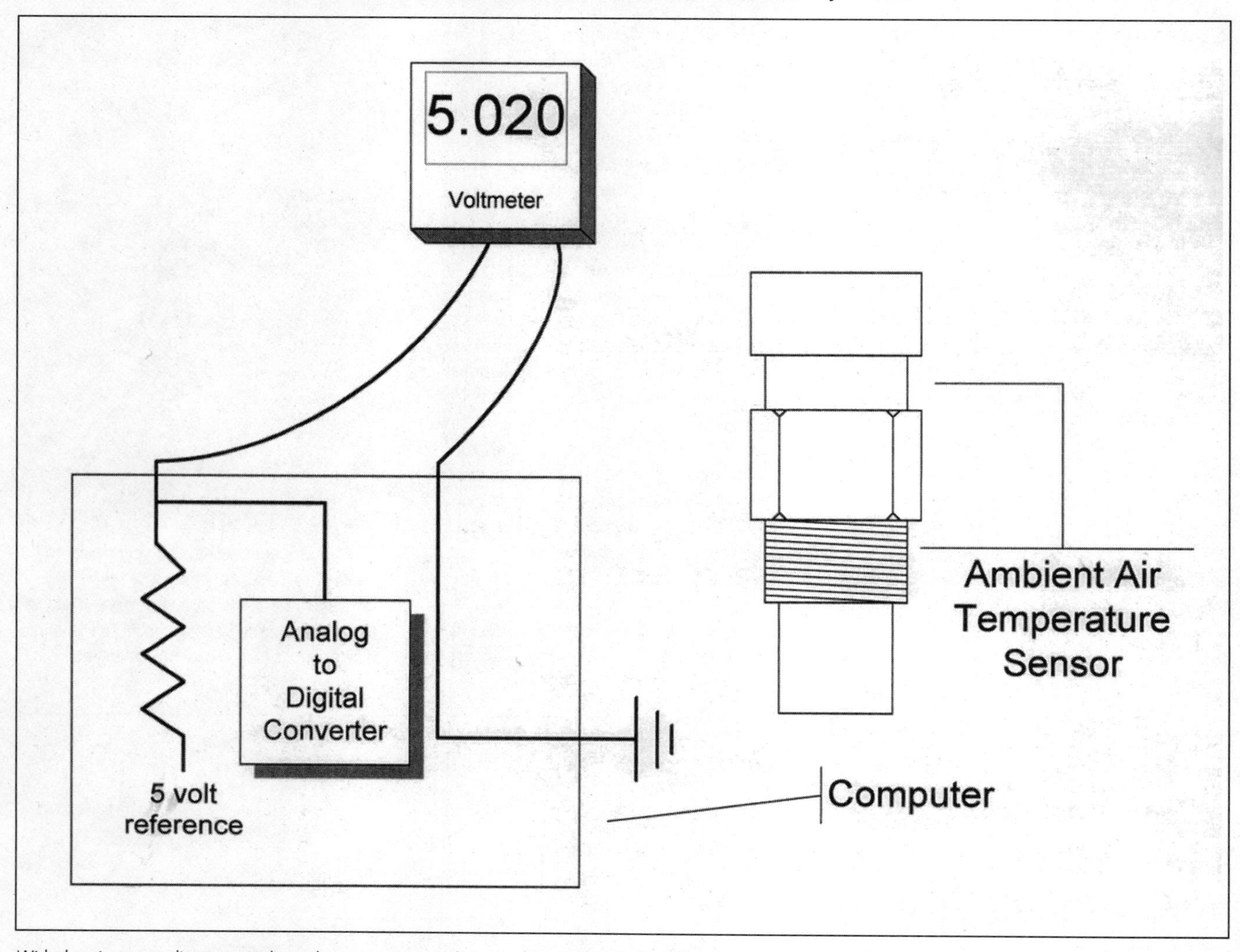

With the air sensor disconnected, a voltmeter connected across the two terminals of the air sensor harness connector should read approximately 5 volts.

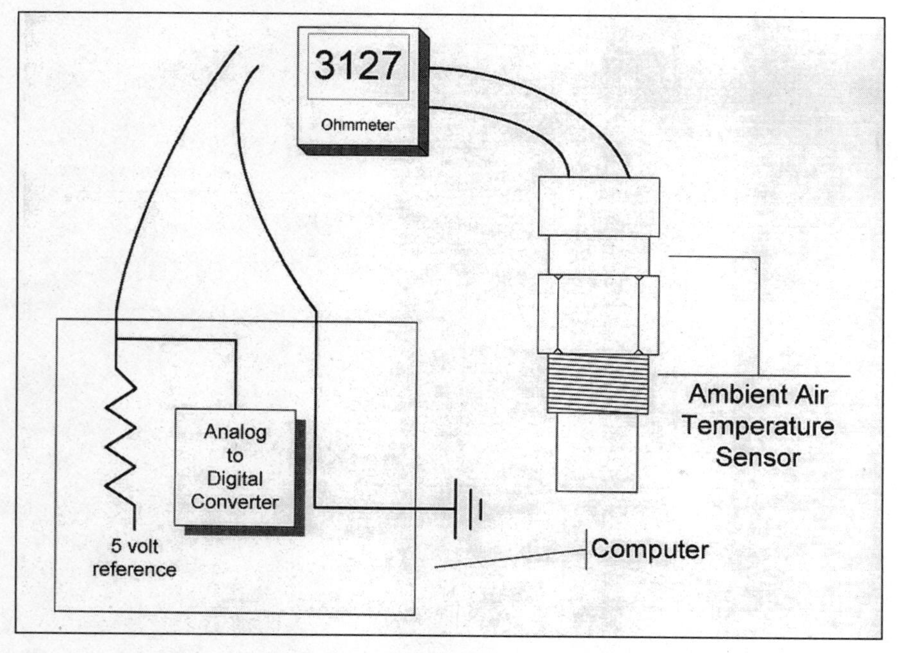

If the reference voltage wire and the ground wire are good, check the resistance across the sensor. The thermistor at the core of the sensor is one that is used widely in the industry.

sor and the black lead to the battery negative terminal. The voltmeter should read 5 volts. If it does not, repair the light-colored wire between the air temperature sensor harness connector and the computer.

If the 5-volt reference wire tests good, turn the ignition switch off. Connect the red lead of the digital voltmeter to the positive terminal of the battery. Connect the black lead of the voltmeter to the dark-colored wire on the air sensor. The voltmeter should read 12 volts. If it does not, repair the dark-colored wire from the air sensor harness connector to the computer.

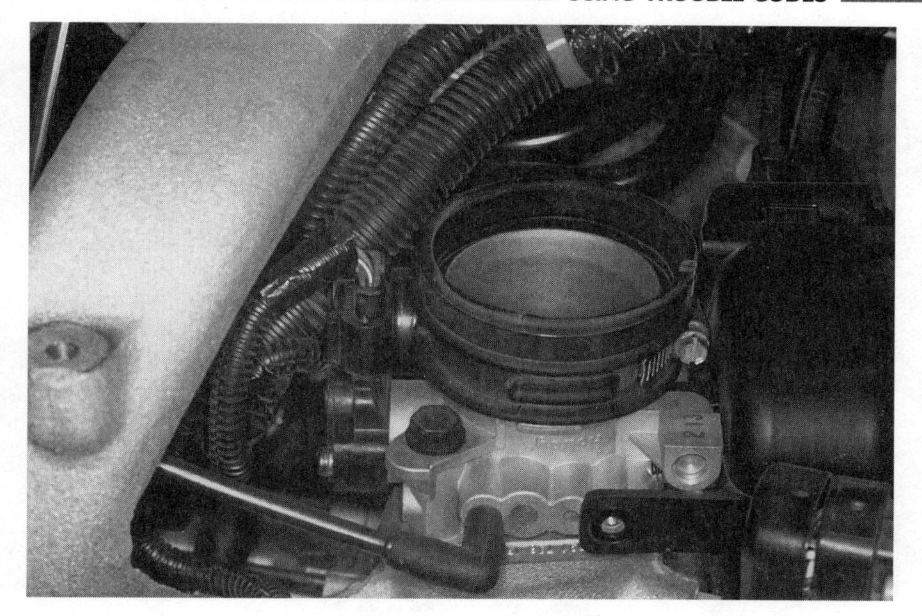

The throttle position sensor is located on the throttle body assembly. Its job is to report to the computer the position of and changes in the position of the throttle. Since this is the first electronic device to detect changes in driver demand, this is the first sensor given the opportunity to initiate acceleration enrichment; it is the accelerator pump.

Although some of the early model TPSs were adjustable, the later ones are not. The signal delivered to the computer when the throttle is closed should be about 1 volt DC. However, if the closed throttle voltage is greater than 0.5 but less than 2.0, then the code was not caused by TPS adjustment problems.

Air Charge or Throttle Body Temperature Sensor Specifications

-40 degrees F	= 100,700 ohms
0 degrees F	= 25,000 ohms
20 degrees F	= 13,500 ohms
40 degrees F	= 7,500 ohms
70 degrees F	= 3,400 ohms
100 degrees F	= 1,600 ohms
160 degrees F	= 450 ohms
212 degrees F	= 185 ohms

Code 24

The throttle position sensor essentially replaces the accelerator pump of a carburetor. Code 24 will set when the voltage returning from the TPS exceeds 4.7 volts or is less than 0.16 volts. During normal operation, the voltage range for this sensor is usually about 1 volt with the throttle closed and between 4.0 and 4.5 volts with the throttle wide open.

To find the source of the code 24 problem, begin by turning off the ignition switch and disconnecting the TPS harness from the TPS. Turn the ignition switch back on. Connect a digital voltmeter across the two outer wires of the connector. The red lead of the voltmeter should be on the wire with the white stripe. The black lead should be on the black and light blue wire. The reading should be 5 volts. If it is not, leave the red connected and connect the voltmeter black lead to battery ground. If the voltmeter reads 5 volts, repair the open in the black and light blue wire. If the voltmeter fails to read 5 volts, then repair the open or short to ground in the 5-volt reference wire.

Note: Wire colors do change from one application to another. While this text does make an effort to discuss the wires in as universal a way as possible, it may be necessary to alter the aforementioned procedure to compensate for change in wire color.

If the first test, the one with the voltmeter test leads connected across the two outside terminals, does yield 5 volts, then it will be necessary to test the center wire of the circuit. It is the center wire that delivers the signal to the computer. Connect a voltmeter with the red lead on the center wire of the TPS connector and the black lead attached to battery ground. The meter should read more than 2.5 but less than 4.9 volts. If it reads zero, repair the open or short to ground in this center wire. If the voltmeter read 5

Be sure to check the harness connector for the TPS when troubleshooting code 24. Make sure the contacts are free of corrosion, not sprung open, and not pushed back.

The throttle position sensor has three terminals. As with most of the electronic devices, these terminals are delicate. Care should be taken when connecting and disconnecting the harness.

volts of power, repair the short to voltage in the center wire.

If the first test, the one with the voltmeter test leads connected across the two outside terminals, does yield 5 volts, and the test of the center wire also yields between 2.5 and 4.9 volts, replace the TPS.

But there is another fault that can occur with the throttle position sensor that will not set a trouble code. The TPS consists of a piece of ceramic material with a carbon-based paint painted on it. As the potentiometer wiper moves back and forth across the carbon-based paint, it is not unlike taking a key and rubbing it back and forth. Eventually, the key is going to rub a hole through the paint. This very thing will happen to the paint on the ceramic material of the TPS. At that point, every time the wiper moves across the worn area, the TPS, in effect, tells the computer to decrease the power from the engine. This will result in a stumble or hesitation. Even if the vehicle does not have a trouble code 24 but does have a stumble, you should check the TPS for linearity. With the TPS connected, use an analog voltmeter to measure the voltage on the center wire. With the throttle closed, the voltage reading should be in the vicinity of 1.0 volts. Precision readings are not necessary here. Now slowly open the throttle. The voltage should slowly and smoothly rise to about 3.5+ volts. If the transition voltage does not rise smoothly, replace the TPS.

Code 25

There are three types of idle control devices that have been used by Chrysler since it began to use fuel injection in mass production. The first is a DC motor that controls a semicircular air valve. Made of plastic, this valve has a fairly high failure rate. When the computer detects that the engine is running too slow, it sends a current through the motor to rotate the valve to allow more air flow. If the computer detects that the engine is running too fast, it reverses the current through the motor and closes the valve, reducing the air flow.

The second type of idle air control valve is a stepper motor. The automatic idle speed valve (AIS) is a stepper motor controlled valve which the computer moves in order to control the speed of the engine at idle. The AIS can be moved to any one of 256 positions by the computer to ensure the correct idle speed regardless of changes in engine load due to the transmission, power steering, alternator, air conditioning compressor, or anything else. At an idle, the AIS will be at about position 20 with no loads on the engine. As engine loads increase, the rpm will tend to drop; as the rpm drops, the computer steps the AIS to a more open position (higher number). As the rpm decreases, the AIS is stepped in. AIS position is displayed through serial data and is an important piece of troubleshooting information.

Many applications use a DC motor on the throttle linkage to control the idle speed. When the throttle

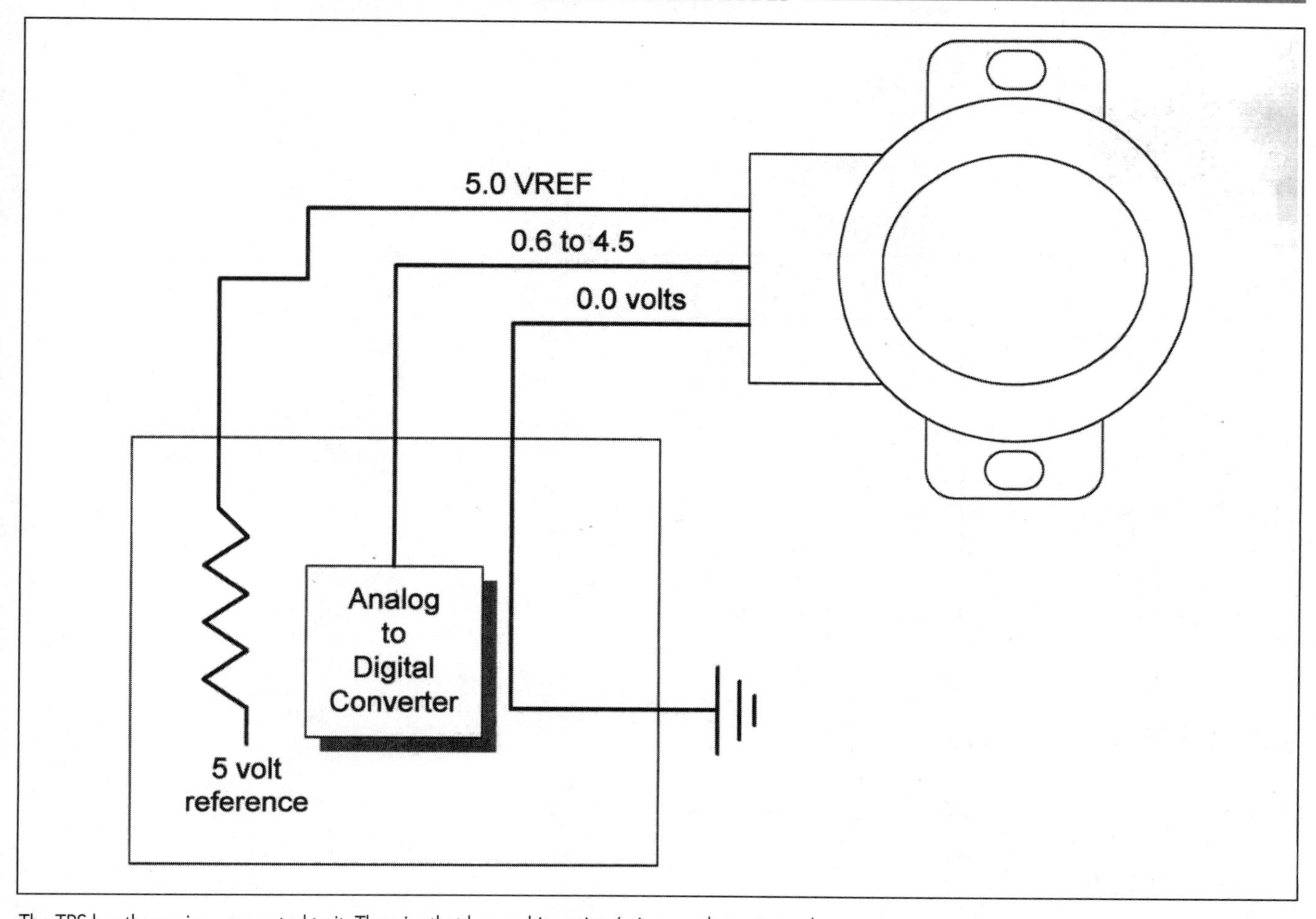

The TPS has three wires connected to it. The wire that has a white stripe (primary color may vary by application) carries the 5-volt reference. The black and light blue wire is a ground. The orange and dark blue wire carries the signal to the computer.

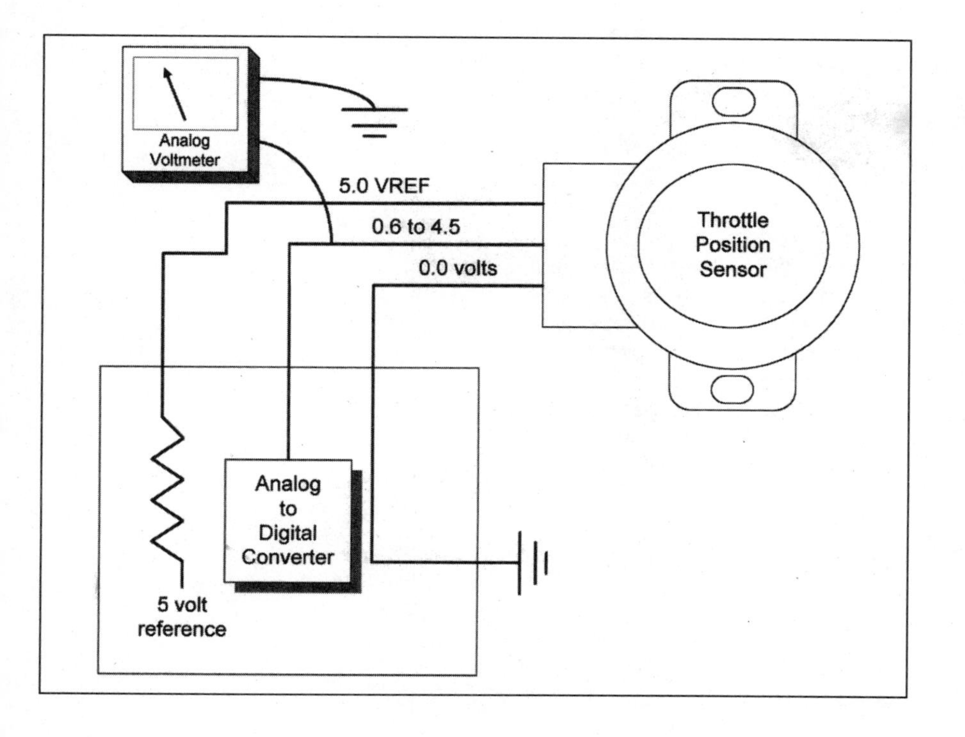

Connect an analog voltmeter to the signal wire of the TPS. This wire is usually orange and dark blue. Slowly open the throttle while observing the TPS signal voltage. The voltage should rise smoothly and evenly as the throttle is opened.

comes to rest on the end of the automatic idle speed control, a switch closes. This signals the computer to control the idle speed. By applying and reversing current through the automatic idle speed control motor, the computer can control the position of the throttle, thereby controlling the idle speed of the engine.

Code 25 is set because of an incorrect voltage in the idle speed control device circuit. Basically, code 25 responds to an extremely low resistance in the circuit. Locate the computer end of the wires that control the

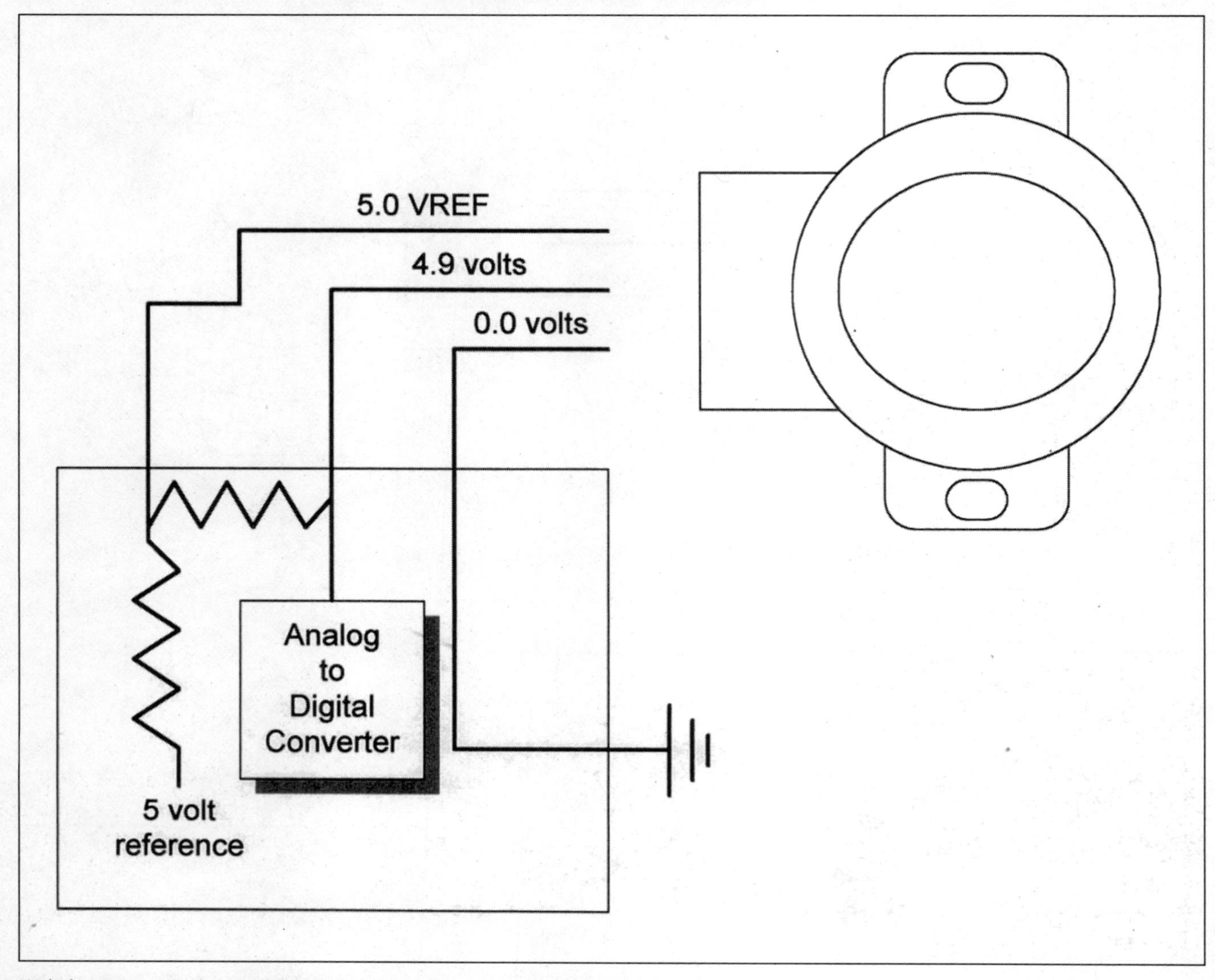

With the ignition switch on and the wiring harness disconnected from the TPS, the voltages will measure about 5.0, a little less than 5.0 volts, and 0.0 volts. For the wires that should have voltages at or near 5 volts, this test will prove that the wires are in good condition. Check the ground wire with an ohmmeter. There will be some resistance, but it should not read an open to an engine block ground.

To make sure that any closed-throttle readings you take are accurate, make sure that the throttle bore is clean. A coked throttle can keep the TPS from closing completely and deliver an incorrect signal. Clean the bore with solvent on a shop towel or use a tooth brush. A dirty throttle bore can also cause idle control problems (see code 25).

	Old Tech	New Tech
Cold engine starting	Choke	Coolant sensor
Cold engine warm-up	Choke	Coolant sensor
Off-line accelleration	Accelerator pump	TPS
Rolling accelleration	Power valve	MAP
Idle speed	Throttle stop screw	AIS
Idle load compensation	Nothing	AIS

There are no new technologies; there are only new ways of doing old things. On the cars of the 1960s and 1970s, the engine idle speed was controlled by altering the air flow past the throttle plates. Today's fuel-injected engines use a throttle bypass to control the idle speed. In the throttle bypass channel, there is a valve controlled by the computer. When the valve is opened, more air is allowed to pass through the bypass, and the idle speed is increased. When the valve is moved toward the closed position, the air flow is decreased, and the idle speed is decreased.

idle motor. With the harness disconnected from the computer but still connected to the idle motor, measure the resistance through the wiring harness and the motor. Connect an ohmmeter to the two terminals of the computer connector that are connected to the idle motor. The resistance should be greater than zero. The exact resistance is unimportant; it simply must be greater than zero. For the applications that use a stepper motor, there will be two pairs of wires that must be measured. Note that an open in the motor, connection, or wiring will not cause a code 25 even though it will cause an incorrect or erratic idle speed.

Code 26

Code 26 relates to the injector driver circuit. The exact conditions necessary to set a code 26 vary a little by application, but the bottom line is always the same. The code indicates that the resistance of the injector circuit is incorrect. The test procedure for this code should begin with identifying the injector connectors on the computer. With the ignition switch off, disconnect the injector connector on the computer. Measure the resistance through the injector. The resistance should be greater than 4 ohms but less than 20 ohms. Chances are that if the resistance is incorrect, it will be almost zero ohms or almost infinite ohms.

If the resistance through the injector circuit is incorrect, then check the resistance through the injector by itself. Like the resistance check through the entire circuit, the resistance through the injector should be between 4 and 20 ohms. If it is not, replace the injector. If it is, check the resistance through each of the wires connected between

The idle control device on late-model Chrysler products is a stepper motor. The stepped cone at the end of the motor restricts air flow through the bypass. The computer can set the cone in any of 256 positions.

Although throttle cable adjustments can affect idle speed, they cannot cause a code 25 to be set.

Behind this plug is the throttle stop screw. Pry the plug out with an awl. Use a scanner to set the computer into minimum air mode. Adjust the throttle stop screw to set minimum idle. If you do not have a scanner, if your spouse decided that a couch was more important, then adjust this screw so that the throttle plate is just barely closed, then turn the screw about one-half turn in the clockwise direction.

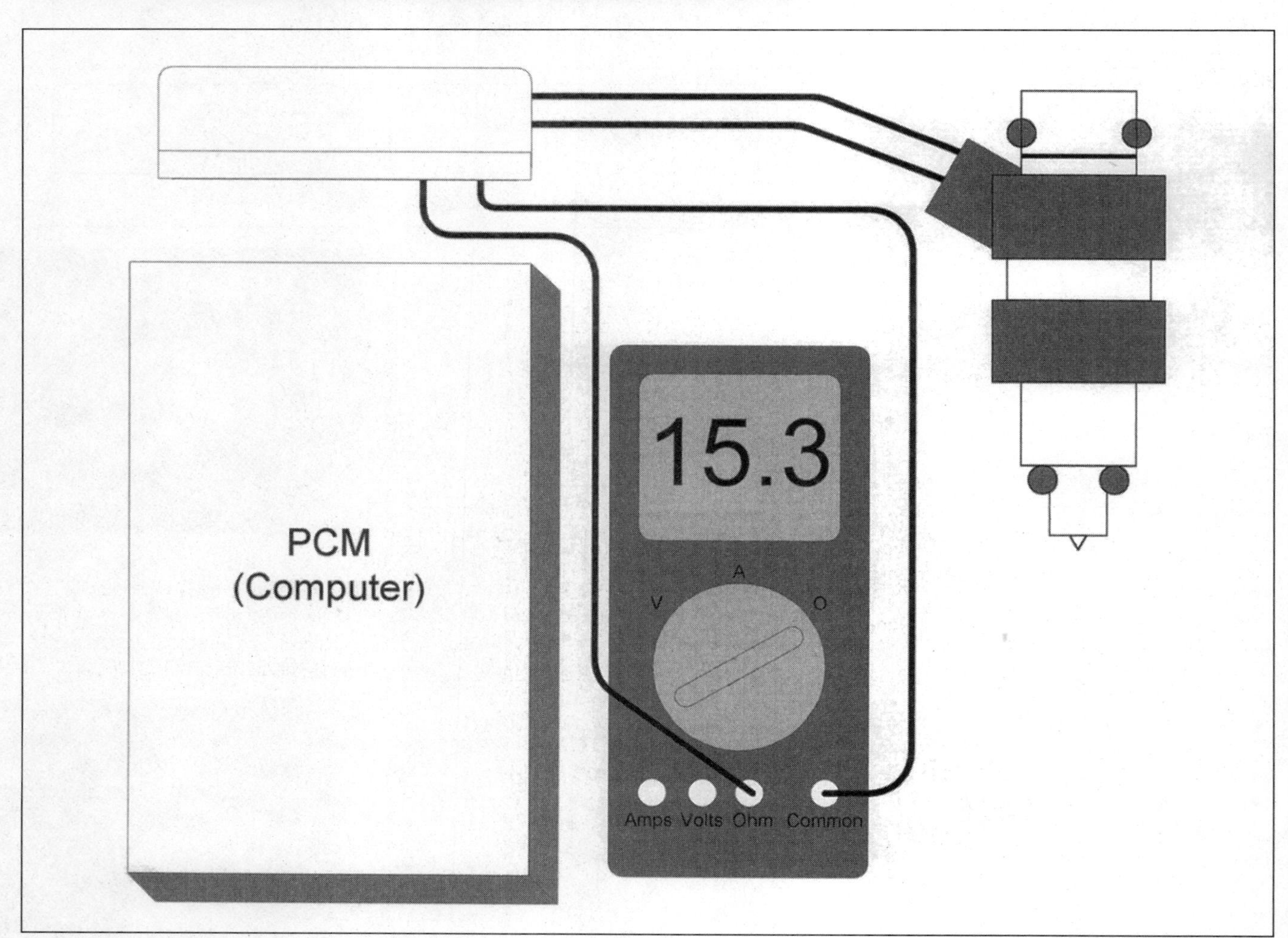

When a code 26 or 27 is received, it implies that the resistance in the injector circuit is out of range. Disconnect the harness connector at the computer. Locate the pins that supply power and ground to the injectors. Measure the resistance through the wires and injectors. The resistance should be between 4 and 20 ohms. If the resistance of the circuit is great enough to set the code, it will be way out of spec.

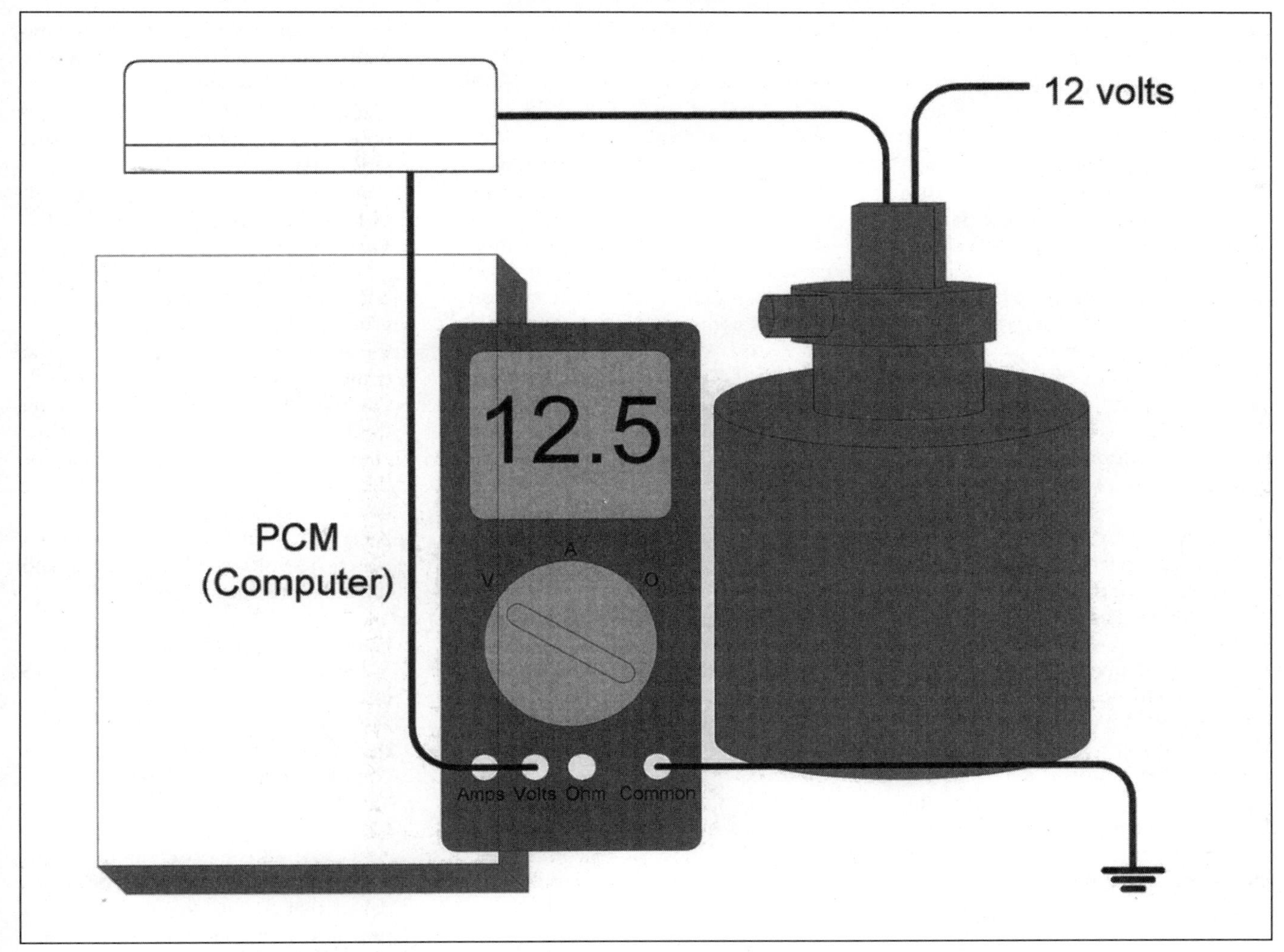

The evaporative canister holds fuel tank evaporates on the surface of its activated charcoal core. At certain times, the computer will ground the purge solenoid; this will cause a valve to open and allow the stored evaporates to be drawn into the engine and burned. Turn the key on. Check for 12 volts at the purge control valve. This is often found on a valve cover or in the right inner fender well area. Trace the control wire from the purge solenoid to the computer. With the key on and the engine not running, there should be 12 volts at the computer end of the wire. If there is not, the solenoid or the wire is bad. Leave the voltmeter connected and position it so that a co-driver can watch it while the car is being driven. Test drive for several minutes after the engine warms up. At some point, the voltage should go low. If it does not, check the connection at the computer. If the connection is good and the voltage did not drop during the test drive, replace the computer.

the computer and the injector. The resistance through these wires should be virtually zero. In fact, on most ohmmeters, it will read zero ohms.

On the four-cylinder turbo applications, this code relates to injectors number one and two.

Code 27

Code 27 relates to injectors 3 and 4 on turbo applications. On other applications, it relates to the ability of the computer to recognize. Basically, follow the instruction for code 26 but use the wire connections and terminals related to injectors 3 and 4.

Code 31

This code will set when the canister purge solenoid fails to function properly. With the ignition off, disconnect the electrical connector from the canister purge vacuum control valve solenoid. Now turn the ignition switch back on. Connect the black lead of a voltmeter to a good ground. Connect the red lead of the voltmeter to the pink or pink/black wire at the solenoid end. The voltmeter should read approximately battery voltage. If it does not, repair this supply voltage wire.

If the voltmeter reads zero, locate the other end of the pink or pink/black wire on the computer. Turn off the ignition switch. Disconnect the connector on the computer and connect the red lead of the voltmeter to the terminal end of the pink or pink/black wire. Turn the ignition on and read the voltage. If the voltmeter reads battery voltage, replace the computer. If it does not read battery voltage, test the resistance through the canister purge solenoid. The resistance should be less than 20 ohms. If greater than 20 ohms, replace the canister purge solenoid. If less than 20 ohms, and there was a correct amount of voltage at the power supply terminal, check the resistance of the canister purge solenoid control wire.

Note: Wire colors do change from one application to another. While this text does make an effort to discuss the wires in as universal a way as possible, it may be necessary to alter the aforementioned procedure to compensate for change in wire color.

Code 32

A detected failure in the operation of the exhaust gas recirculation (EGR) system will cause a code 32 to be generated. The purpose of the EGR valve is to reduce combustion temperatures. When the engine comes under a load, combustion temperatures tend to rise. If the temperature rises above 2,500 degrees F, the nitrogen constituent of the air entering the combustion chamber begins to combine with the oxygen of the air entering the combustion chamber. This burning produces NOx, or oxides of nitrogen. Oxides of nitrogen are responsible for some of the worse kinds of air pollution. When the EGR valve fails to function, the NOx emissions rise dramatically.

Now I can hear you thinking, "I do care about the environment, but right now I care more about my pocketbook, and right now I am more interested in making this month's car payment than I am in saving the environment...this month." As a starving writer of Scottish decent, I can more than relate to that kind of thought. The high combustion temperatures can, however, do severe damage to the engine.

To diagnose code 32, consider the fact that there are three primary components involved in the system related to this code. There is the EGR control solenoid, the EGR backpressure valve, and the EGR valve. Locate the EGR valve solenoid. This solenoid is located either on a bracket on one of the valve covers or on the right fender well. Of course, now that I have said that, the EGR solenoid on your vehicle will doubtless be in neither of those two places.

Turn the ignition switch on and verify that there is voltage at both terminals of the EGR solenoid. This component, like most of the others controlled by the computer, is grounded by the computer. The EGR will not be requested to work when the engine is idling, much less when it is not running. These facts mean that the voltage on both wires should be the same as battery voltage. If there is battery voltage, it means the power supply, the dark blue wire, is in good condition, and the solenoid does not have an open.

The gray/yellow wire between the solenoid and the computer is grounded by the computer to energize the solenoid. Follow this wire to the computer. Turn the ignition switch off. Disconnect the connector containing this wire from the computer. Now connect a voltmeter to the terminal for the gray/yellow wire on the computer harness and turn the ignition switch back on. If the reading is approximately 12 volts, the wire is good, and the computer must be replaced.

Note: Wire colors do change from one application to another. While this text does make an effort to discuss the wires in as universal a way as possible, it may be necessary to alter the aforementioned procedure to compensate for change in wire color.

In the early days of emission controls, the exhaust gas recirculation (EGR) valve was connected to ported vacuum on the carburetor. When the throttle valve was opened by the driver, the EGR would open all the way. Often, the fully open valve would cause stumbling or stalling. On newer Chryslers, the opening of the EGR is controlled with a vacuum solenoid.

Code 33

Code 33 is most commonly generated when the computer does not detect that the air conditioning compressor clutch control circuit is working. Oddly enough, this is an incredibly common code, especially on vehicles operated in the northern latitudes. It is also very common on many fleet-owned vehicles. The reason for it being so common is that this code will even set when there is no air conditioner installed. Knowing this, the manufacturers of many scan tools will ask, during the set-up routine for the vehicle, if the car is equipped with air conditioning. If the technician responds in the negative, the scan tool will simply ignore the code 33, will not display that the code was received, and thereby not confuse the technician with a problem that does not exist.

When a vehicle with air conditioning sets a code 33, it specifically means that the computer has not seen voltage present at the air conditioner wide open throttle (WOT) cutout relay terminal of the computer. The wire connected to this terminal is usually a dark blue wire with an orange stripe. If the vehicle is equipped with

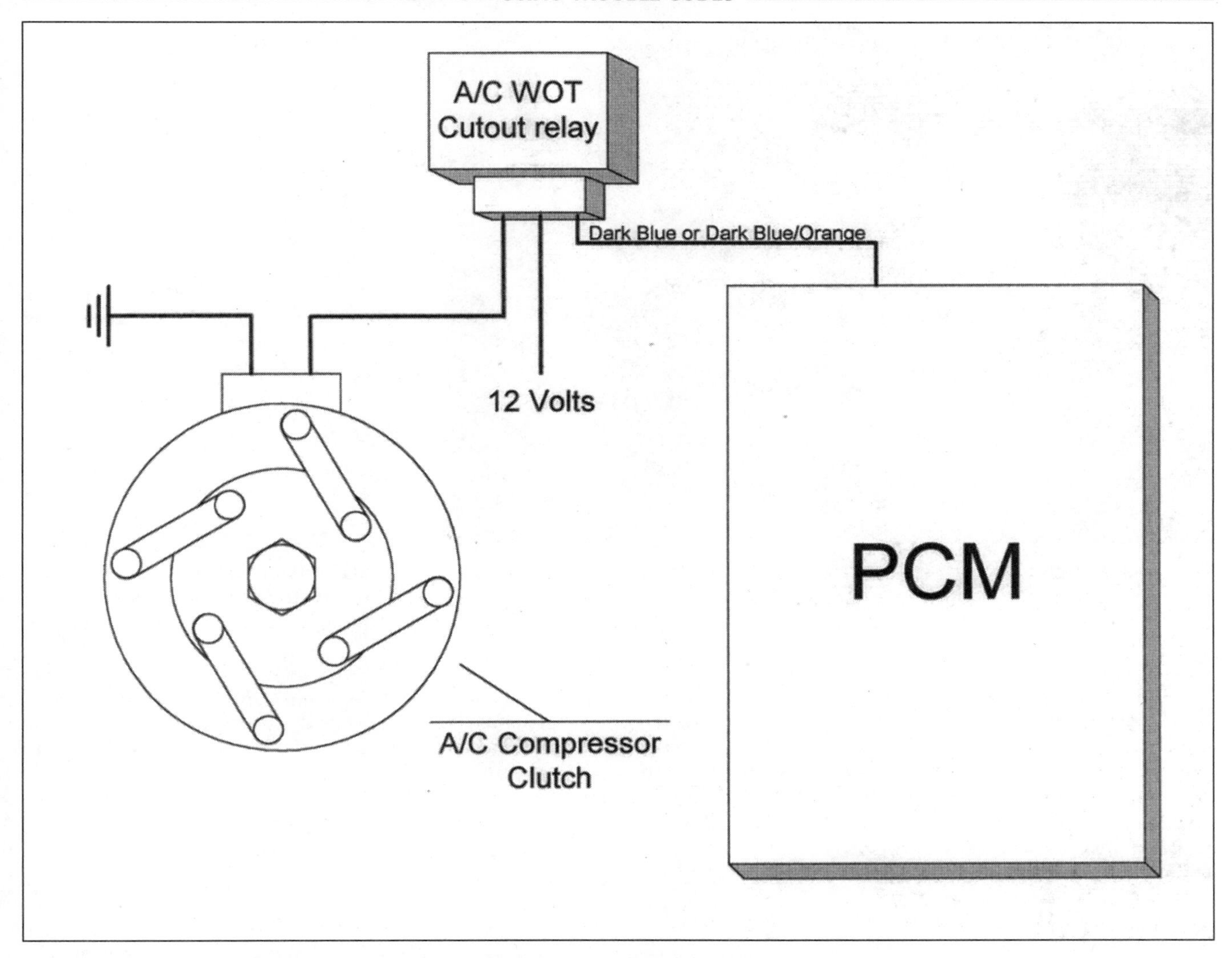

The computer controls the application of the air conditioning compressor clutch through a relay known as the A/C wide open throttle (WOT) cut-out relay. When the computer fails to see 12 volts on the dark blue or dark blue and orange wire, a code 33 is set.

air conditioning, begin by checking all fuses. If the fuses are okay, turn the ignition switch off, disconnect the computer, and connect a voltmeter to the terminal on the computer connector that corresponds to this dark blue and orange wire. Turn the ignition switch back on. Observe the voltage. If the voltage is less than 10 volts (and it really should be over 12), there is high resistance or an open in the air conditioner wide open throttle relay circuit.

Next locate the wide open throttle cutout relay. This relay is usually on the left wheelwell. If it is not, it is located somewhere else on the hood. Find the dark blue/orange wire. Check for 12 volts. If the voltage is 12 volts or more, the wire between the air conditioner wide open throttle cutout relay and the computer is damaged. Locate and repair the damage or run a replacement wire.

If the voltage to the relay end of the dark blue/orange wire is not correct, then with the ignition off, disconnect the connectors from the relay. Connect one lead of an ohmmeter to the terminal where the dark blue/orange wire had been connected and one at a time measure the resistance to the other terminals. What the exact readings are is not important. Two of the terminals should read open to the terminal where the dark blue/orange wire had been connected; one should read a low resistance. This means that the resistance should read greater than zero but not open. If there is an open between the terminal where the dark blue/orange wire had been connected and all the other terminals, replace the relay. If there is 0 ohms of resistance between the terminal where the dark blue/orange wire had been connected and one of the other terminals, replace the relay. If the resistance measurements of the relay are correct, then repair the open in the dark blue wire that runs to the A/C fuse in the fuse panel.

Note: Wire colors do change from one application to another. While this text does make an effort to discuss the wires in as universal a way as possible, it may be necessary to alter the aforementioned procedure to compensate for change in wire color.

Code 34 relates to the computer's cruise control servo. The most common reason for this code to be present is that the vehicle is not equipped with cruise control.

Code 34

On 1987 and earlier models, handle this code as described in the code 32 section. For later applications, this code means that the computer has detected a fault in the cruise control circuitry. If the vehicle is not equipped with cruise control, a code 34 will be generated. Like code 33, many scanners are programmed to ignore the presence of this code. If you get a code 34, and your vehicle is not equipped with cruise, simply ignore the code.

For vehicles equipped with cruise control, this code means that the computer is not detecting a voltage to and through the speed control servo. All Chrysler fuel-injected engines have about half of the cruise control system even if there are no cruise control switches installed. The computer for the fuel injection system is also the controller for the cruise control. All that needs to be added to a vehicle that does not have cruise control is the servo and switches. The computer grounds the windings in the servo to activate the cruise. Therefore, when there is no cruise servo or when the windings are open or the servo is disconnected, the computer will set a code 34.

First check the fuse. Now verify that there is 12+ volts on the dark blue/red wire. If there is not, repair the defect in the dark blue/red wire between the servo connector and the cruise control fuse. Consult your owner's manual or the diagram on the fuse panel cover to determine which fuse is related to the cruise control. If the fuse is good, turn the ignition switch off. Locate the tan/red wire on the computer that controls the cruise servo. Disconnect the 60-pin connector on the computer. The tan/red wire is usually connected to pin 53. Also locate the light green/red wire that is usually connected to pin 30. Disconnect the cruise servo and place a jumper wire between the tan/red and light green/red wires at the servo end. Now measure the resistance between these two wires at the computer connector end. For practical purposes, the resistance should be zero.

If the resistance is zero, measure the resistance through the servo from where the tan/red wire connects to the servo to where the dark blue/red wire connects to the servo. Repeat this process between the light green/red wire and the dark blue/red wire. In each case, the resistance should be greater than zero, but less than infinity. If these resistances measured through the servo are correct, and the wires are good, replace the computer.

Note: Wire colors do change from one application to another. While this text does make an effort to discuss the wires in as universal a way as possible, it may be necessary to alter the aforementioned procedure to compensate for change in wire color.

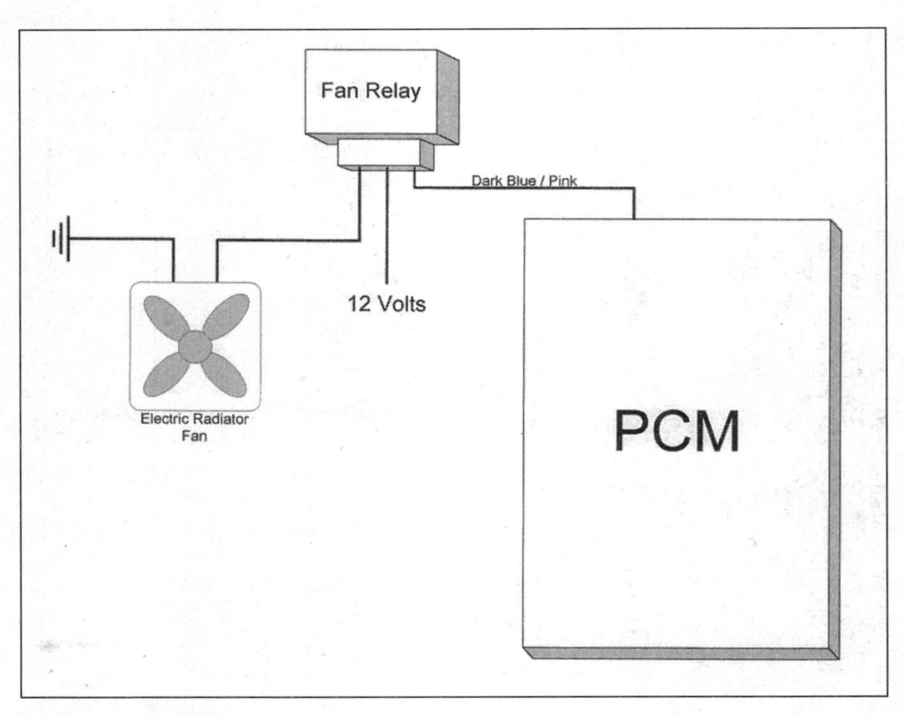

On front-wheel-drive applications, the computer uses the engine coolant temperature sensor to control the operation of the radiator fan. The fan will also come on when the computer detects a system fault and when the air conditioner compressor clutch is engaged.

Code 35 (Front-Wheel-Drive)

Front-wheel-drive applications use an electric fan to cool the radiator. On these applications, code 35 relates to the computer circuitry that controls the radiator fan. Locate the cooling fan relay. There is a dark blue/pink wire connected to this relay. With the ignition switch on, there should be battery voltage on this wire. If there is not, turn the ignition switch off, and disconnect the connector of the computer that has the dark blue/pink wire for the fan relay going into it. Now turn the ignition switch back on, and check the voltage on this wire back at the relay. If there are now 12 volts or more, replace the computer. If there are not 12 volts or more, repair the open in the dark blue/pink wire.

If, when you initially checked the voltage at the relay, you read battery voltage, then turn the ignition switch off, and disconnect the computer connector. Turn the ignition switch back on, and check the voltage on the dark blue/pink wire at the computer harness end. If there are 0 volts, repair the open or grounded wire between the fan relay and the computer. If the voltmeter reads 12 volts, it means that the power supply to the relay is good, that the relay pull-down windings are good, and that the wire from the relay to the computer is good. What is not good is the computer. Replace it.

Note: Wire colors do change from one application to another. While this text does make an effort to discuss the wires in as universal a way as possible, it may be necessary to alter the aforementioned procedure to compensate for change in wire color.

Code 35 (Rear-Wheel-Drive)

Rear-wheel-drive applications use a DC motor-powered plunger to control the throttle position for the purpose of controlling the idle speed. There is a switch located deep in the motor that closes when the driver releases the throttle to the idle position. Code 35 will set when the computer does not see the voltage on the violet wire that runs to the computer from the idle speed motor change.

Begin the test procedure by turning the ignition switch off and disconnecting the connector on the computer that contains the purple wire. Connect an ohmmeter between the cavity containing the violet wire and ground. Open and close the throttle. If the resistance to ground does not change, repair the violet wire. It my be open or shorted to ground. If the resistance to ground does change, replace the computer.

Note: Wire colors do change from one application to another. While this text does make an effort to discuss the wires in as universal a way as possible, it may be necessary to alter the aforementioned procedure to compensate for change in wire color.

Code 36

Back in the early 1980s, there were a lot of cars that would "pop" loudly on deceleration. This was often caused by the air pump control valve allowing air to be dumped into the exhaust system when there was a negative load on the engine. On carbureted engines, the throttle is, of course, closed when the engine is decelerating, yet the rpm and air flow remain high. This draws more fuel into the engine than power requirements demand and more fuel than can be burned under those engine operating conditions. When the unburned fuel hits the hot exhaust, it explodes. This is the popping sound. Many fixes were tried during the latter half of the 1970s and early 1980s. Vacuum controls and delays, timing modifications, and expletives were all tried and none of them worked. Precise control of the air pump was not possible until the fuel and emission control systems became computer-controlled. The computer gathers information about throttle position, engine load, and rpm. With this information, the computer is able to predict deceleration and divert the air pump to the atmosphere before the unburned fuel content of the exhaust gases gets high enough to explode. As a result, there is no popping.

There is a black/orange wire connected to the computer that is the carrier for code 36. Confirm that all fuses are in good condition. With the ignition switch in the off position, disconnect the connector that contains the air switching solenoid control wire. Connect the red lead of your voltmeter to the black/orange wire. Connect the black lead to ground. Turn the ignition switch to the run position. If there are 12 volts displayed on the voltmeter, check the connection between the connector and the computer. If the terminals are clean and tight, replace the computer. If there

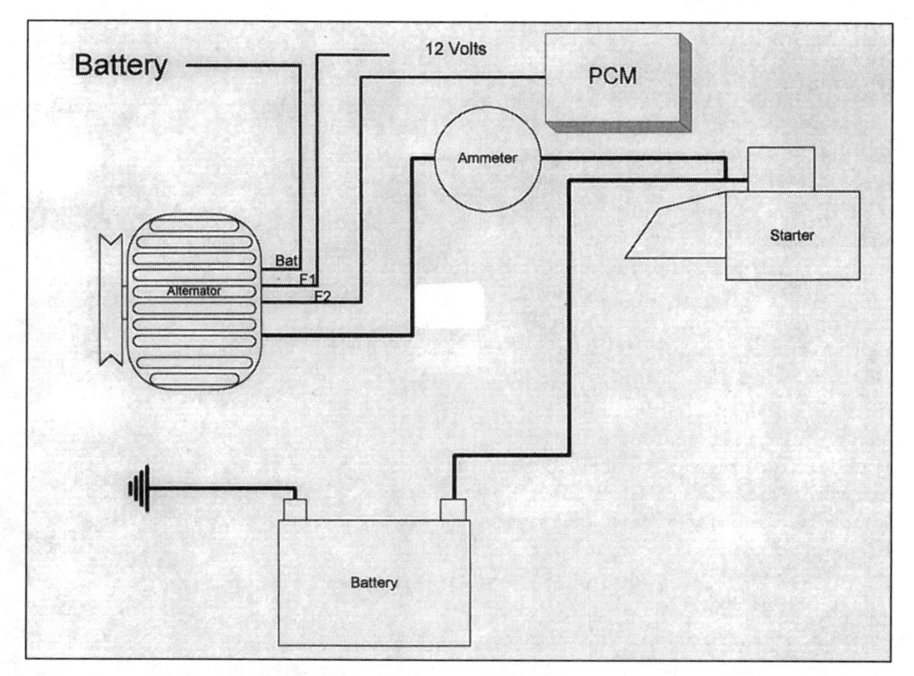

The computer grounds the F2 terminal of the alternator to energize the field. Confirm that there are 12 volts on both wires with the engine not running but the key in the on position. Now connect a digital tachometer to the F2 terminal and start the engine. The tach should read something other than zero. If it reads zero, connect the tach to the dark green wire that is connected to the F2 terminal of the computer.

are not 12 volts at this terminal, then trace the black/orange wire back to the air switching solenoid. Gently back-probe the black/orange wire on the solenoid. The voltage on this wire should be 12 volts. If it is not, check for battery voltage with the ignition switch on. Back-probe the dark blue wire on the air switching valve solenoid. There should be 12+ volts on this wire. If there is not, repair the wire. If there is, replace the solenoid.

Note: Back-probing is an illicit art. The manufacturers frown on this procedure, and so do I when it is not done with great care. Gently insert a small paper clip along the side of the wire into the back side of the connector until it contacts the metal connector end of the wire. Take your readings.

Note: Wire colors do change from one application to another. While this text does make an effort to discuss the wires in as universal a way as possible, it may be necessary to alter the aforementioned procedure to compensate for change in wire color.

Code 41

Code 41 denotes a problem with the voltage regulator. Now in the classes I teach for journey-person automotive technicians, I often make a point about this code and its troubleshooting procedure. It seems that if, indeed, the voltage regulator does end up being the cause of the code, then it will need to be replaced. It will be very expensive, but you get a free computer, a free cruise control module, and a free ignition module with it. You see, all of these components are located in the fuel injection computer. It may, at first, seem a little ridiculous to put all of these controls inside a computer. Using only one computer to operate all of the devices under the hood does make sense from the manufacturing perspective. The engine can arrive at the assembly line as a pre-wired component of the vehicle. The engine is then inserted into the engine compartment, and the computer is attached with sheet metal screws.

The most effective way to troubleshoot a code 41 is with a scanner. However, a second-best method would be to attach a frequency counter, tachometer, or dwell meter to the field control wire of the alternator. Which one is the field control wire? It is probably the terminal with the dark green wire connected to it. There are two small wires on the back of the alternator. One of these wires is field power; the other is field control. With the engine running, connect a voltmeter to each of the smaller terminals, first one then the other. One terminal should read almost the exact voltage that is found across the terminals of the battery. The other should be slightly lower. If they are both the same, there is probably an open between the alternator and the computer. Next choose a meter that can detect and confirm a pulsing DC voltage. Connect this meter to the field control wire, the one with the lower voltage. If both terminals had the same voltage, try to detect a pulse on both wires. On the frequency counter, one of the wires should read a frequency greater than zero. On the tachometer, one of the wires should read an "rpm" greater than zero. On the dwell meter (set on the four-cylinder scale), one of the wires should read a dwell greater than 0 degrees but less than 90 degrees.

If the readings are outside the range of the specifications given above, check the dark green wire to the computer for continuity. If the wire is good, replace the computer.

The repair of, or even the absence of, a code 41 does not mean that the charging system is working, it only means that the computer thinks it has the ability to control the alternator field.

Code 42

Code 42 is defined by Chrysler as indicating a problem with the auto shutdown (ASD) relay circuit. Before we dive into the troubleshooting of wiring, relay, and computer problems, let us look at one of the miscellaneous things that can cause this code. I often tell technicians in my classes that if there is no complaint about the engine stalling, simply ignore code 42. Here is why.

Imagine a 1990 Dodge Daytona five-speed turbo that is owned by an 80-year-old lady from Pasadena. She is only about 5 feet tall and can barely touch the pedals when sitting on the forward edge of the seat. She is cruising through Rancho Cucamonga when she pulls up to a stop light next to a 1994 Corvette. The guy in the Corvette revs his engine; the little old lady revs hers. The guy in the Corvette holds up his title; the lady in the Daytona holds up her title. The light turns green, the Corvette lights his tires, and the little old lady pops her clutch—a little too fast. The engine nearly dies, and a code 42 is set. Fortunately, she recovers, and the Corvette driver misses the second gear-to-third gear shift. The Daytona screams through

The black relay near the center of the photo is the auto-shutdown (ASD) relay. Code 42 means that the computer has commanded the ASD relay to shut down power to the fuel pump, the power to the ignition primary positive, and other components under the hood.

the next intersection first. The Corvette driver hands his title to the Daytona owner at the same time the police officer hands them both a ticket.

The code 42 was generated because the computer saw a sudden change in the frequency from the Hall Effects. The computer determined that this sudden change was a result of a collision, and the ASD relay was commanded by the computer to shut the fuel pump off. However, the driver of the Daytona was able to recover the speed of the engine by putting her foot through the floorboard.

Bottom line: A near stall condition can set a code 42. This situation is more common on standard transmission cars than on automatics. Now if you choose to ignore the code 42 because there are no stalling symptoms, and then the car develops stalling symptoms, then it is time to pay attention.

A legitimate code 42 may very well be the most difficult to diagnose of all the codes. In the Chrysler service manual, there is a very extensive process to determine the exact cause of this code. In the hands-on portion of my training classes, I artificially create a condition that can cause code 42. This involves moving one of the wires connected to the ASD relay to an empty position in the connector so that it no longer contacts its mating terminal on the relay. It is very entertaining to watch technicians try to work through the diagnostic charts supplied by Chrysler to find the source of this problem. A classic situation occurred in San Diego several years ago. I had a class of twelve technicians. Four of these technicians were highly skilled driveability technicians, four were skilled, experienced technicians with little driveability experience, and four were apprentices. One of these apprentices was a 40-year-old female who had been an emergency medical technician a year earlier. No one wanted to work with her. They broke up into three groups. The "experts" formed one group. The "experienced" formed another group, and the apprentices formed the third. The female was the only individual that would aggressively seek the solutions to the problems I had injected into the vehicles. The "expert" group took three hours to find the code 42 problem. The "experienced" took an hour. The apprentices, led by the female, took only 15 minutes—a record that has stood for four years.

The computer controls the ignition timing based on inputs from the Hall Effects, the MAP sensor, and the coolant sensor. When this system fails, a code 43 will be set.

There are four wires that are connected to the ASD relay. The red wire runs through a fuseable link to the battery. The dark green/black wire supplies power to the fuel pump, the ignition coil, the field of the alternator, the injector, and the oxygen sensor heater. The dark blue wire has fused ignition voltage from the computer. The dark blue/yellow wire runs to the diagnostic connector and then to the ASD control terminal on the computer.

With the ignition switch off, check for voltage on the red wire. The voltmeter should read approximately battery terminal voltage. If the voltage is low or zero, check the fuseable link. If it is okay, repair the red wire between the battery and the ASD relay.

If the red wire has the proper voltage, turn the ignition switch on and check for voltage on the dark blue wire. If there is no voltage on the dark blue wire, then trace the wire to the computer. If there is voltage at the computer, repair the wire. If there is no voltage at the computer, confirm that the computer is powered up, then replace the computer.

Turn the ignition switch off. Disconnect the connector on the ASD relay. Disconnect the terminal on the computer that contains the blue/yellow wire. With an ohmmeter, check the blue/yellow wire for continuity. If the wire has continuity, replace the computer; if it does not, repair the wire.

At this point you may have noticed that I have not said anything about the green/black wire. Although this is one of the most critical wires on the engine, and although it could cause all of the aforementioned systems to malfunction, a defect in this wire cannot cause a code 42.

Note: Wire colors do change from one application to another. While this text does make an effort to discuss the wires in as universal a way as possible, it may be necessary to alter the aforementioned procedure to compensate for change in wire color.

Code 43

Code 43 indicates a problem in the spark control circuit. This is caused by the computer recognizing an apparent problem with its control of the ignition system. There are only one or two wires involved in this circuit that might cause the problem. If these wires test good, replace the computer.

The first wire is a black/gray or black/yellow wire that runs from the ignition coil to the computer. With the ignition switch turned off, disconnect the connector on the computer that contains the aforementioned wire. Use an ohmmeter to check the continuity of this wire. Please note that a voltage test will not be possible since the coil is not powered until the engine is cranking with all wires connected to the computer. If the wire is open or shorted to power or to ground, repair the wire.

Now it gets complicated. Remember from earlier in this book that Chrysler has used three basic computer configurations. There was the two-module setup used on the 1984-1987 models. Then there was the single-module engine controller used from 1988 into the early 1990s. Although it had only one module, it had two connectors on that module. One had 14 pins, the other 60 pins. Then along came the single-board engine controller; it has only one 60-pin connector.

If you have a vehicle with a dual-module setup, find a yellow wire that runs from the logic module to the power module. If this wire is in good condition, replace the power module. If the code persists, replace the logic module. I am not a believer in being a "parts changer," but this is one diagnosis where parts changing may be the bottom line.

If you have a single-board engine controller, check the continuity on the yellow wire that runs from pin 13 of the 14-pin connector to pin 34 of the 60-pin connector. If the wire has continuity, replace the computer.

Note: Wire colors do change from one application to another. While this text does make an effort to discuss the wires in as universal a way as possible, it may be necessary to alter the aforementioned procedure to compensate for change in wire color.

Code 44 (1990 and Earlier)

This code is one of my favorites to get technicians thinking like a Chrysler technician. Code 44 means that the computer has detected a fault in the battery temperature sensing circuit. Battery temperature sensing? Remember that the computer is also the voltage regulator for the alternator. Unlike most things, when you cool a battery, its resistance rises. This means that on those crisp Fairbanks mornings when the temperature of the engine and the battery was a balmy -40 degrees F, the resistance of the battery was extremely high. To get the battery to take a charge, it was necessary to increase the alternator output voltage. This is a common practice in the industry. Since the computer, or on the two-module systems the power module, is under the hood, it is exposed to the same ambient conditions as the battery. Chrysler places the battery temperature sensor inside the computer. When the temperature is low, it can tell the voltage regulator circuitry of the computer to kick out an extra volt or so. The only cure for a code 44 is to replace the battery temperature sensor. When you replace the battery temperature sensor, it will be real expensive, but you will get a free computer, a free cruise control module, and a free ignition module with it. Replace the power module (on 1984-1987 cars) or the computer (on 1988-1990 cars) if you get this code.

Code 44. The computer does many jobs around the engine compartment. On 1990 and earlier models, one of these jobs is performed by a sensor located inside the computer. This sensor is called a battery temperature sensor. Since the battery is exposed to the same approximate temperatures as the computer, it can be assumed that their temperatures are about the same. When the battery temperature sensor detects a temperature around 20 degrees F, the alternator output voltage will be increased. At these lower temperatures, the internal resistance of the battery is quite high. The higher output voltage overcomes this internal resistance so the battery will recharge.

Code 44 (1991 and Later)

This code still relates to the charging system on later-model applications. However, the meaning has changed a little. Instead of being specifically directed at the battery temperature sensor, it reports a general charging system problem. With the ignition switch off, connect a voltmeter to the battery terminal (the big wire) of the alternator. Are there 12 volts? If not, repair the wire between the battery and the alternator. If there are, start the engine and check for voltage and pulses on the field terminals as described in the code 41 description. If necessary, repair as described under code 41. If these items test good, turn the ignition switch off. Connect an AC voltmeter across the terminals of the battery and take a reading with the engine not running. The voltmeter will read either 0 volts or about 18 volts. Note the reading. Now start the engine and take a reading on the battery terminal (the big wire) of the alternator. The reading should be almost identical to the reading that was taken at the battery. If the reading is significantly different, the diodes in the alternator are defective, and the alternator must either be rebuilt or replaced. If all of the above tests good, replace the computer.

Code 45

Code 45 means that the MAP sensor has detected an over-boost condition in the intake manifold. This is not generally related to electrical or electronic problems but rather to the mechanical operation of the turbocharger wastegate. The wastegate

On late-model applications, code 44 is generated when the alternator output is incorrect. Keep in mind that the code is a result of an improper voltage being received by the computer. Before condemning the charging system, check all battery connections and terminals. Many applications use junction blocks for distributing power and to act as a common ground. Be sure to check these for tightness and cleanliness.

This is a positive battery distribution point. With the ignition switch off, remove all the cables and clean the terminal ends.

should bypass exhaust gases around the turbine when maximum boost is reached. Verify that there is no code 36. Ensure that the wastegate is properly installed and that all hoses to it are connected properly to the exhaust and intake system. A restricted exhaust might also result in a code 45.

Codes 46 and 47 (Refer to Code 44)
Code 51 (1984-1985)

This code means that there is a problem with the oxygen feedback circuit or closed loop fuel system. Connect a voltmeter to the oxygen sensor where it connects to the main wire going to the computer wiring harness. Depending on the year and model, this connector may have one, three, or four wires running through it. The wire that you need to connect your voltmeter to is either a black or black/green wire. With the engine cold and the oxygen sensor cold, turn the ignition switch to the run position. Observe the voltmeter immediately. The voltmeter should read about 0.45 volts, maybe a little less. If the voltage read begins to drop almost immediately, even before the engine is started, that is okay. If the voltmeter reading does not begin to drop, that is okay too. If there is no voltage at the oxygen sensor wire, turn the ignition switch off, wait two minutes, and start the engine. Again observe the voltage immediately. If there is still no voltage, inspect the wire that runs from this connector to the computer. Look for opens and grounds. If the wire is good, replace the computer.

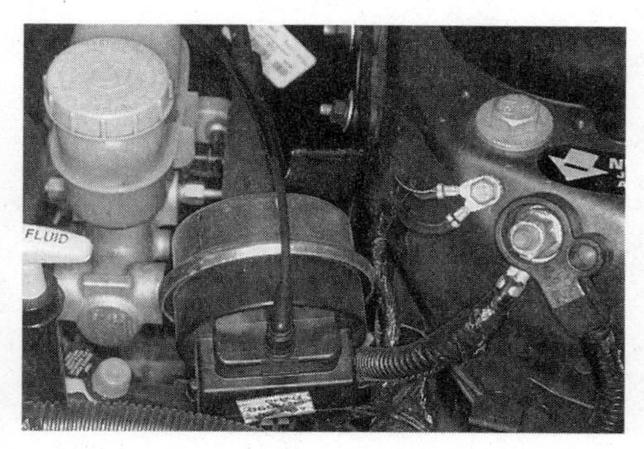

Do not forget to clean the negative cables as well.

Start the engine and allow the engine to come to operating temperature. Please note that although the voltmeter is connected to the oxygen sensor wire, the oxygen sensor wire must still be connected to the computer. When the engine reaches operating temperature, begin to observe the voltmeter. If the voltmeter reading is moving rapidly between nearly zero and about 1 volt, then the problem that set the code 51 no longer exists. Clear the code and put the vehicle back into service. If the code returns, replace the oxygen sensor. If replacing the oxygen sensor does not permanently resolve the code, replace the computer.

Note: Wire colors do change from one application to another. While this text does make an effort to discuss the wires in as universal a way as possible, it

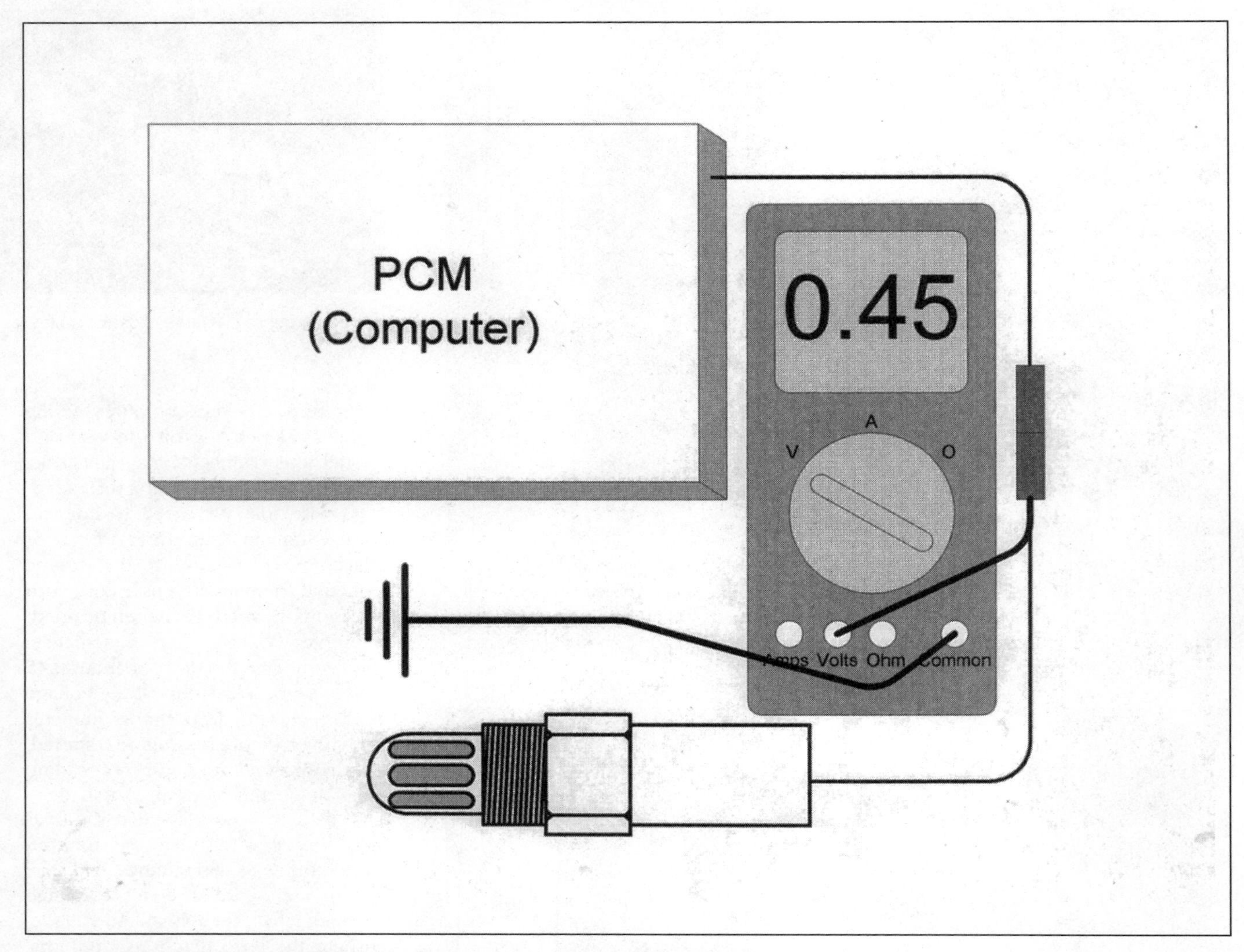

Code 51 (1984-1985). The computer places 0.45 volts on the oxygen sensor signal wire when there is no current from the oxygen sensor. Once the oxygen sensor begins to produce a signal, the voltage will usually first drop, then rise, and then begin to toggle. If it does not do all of the above, inspect the wires, then replace the oxygen sensor.

may be necessary to alter the aforementioned procedure to compensate for change in wire color.

Code 51 (1986-1990)

For these models, this code refers to a problem with stand-by memory. Verify that the following are within parameters. The coolant temperature sensor signal to the computer should read between 0.5 volts and 4.5 volts. At -40 degrees F, the reading should be nearly 5.0 volts. As the temperature of the engine approaches 120 degrees F, the voltage should be near 1.2 volts. Shortly thereafter, the voltage should rise to 3.7 and begin to fall again. When the engine is thoroughly warmed up, the voltage should be less than 1 volt. Shut the engine off and turn the ignition switch to the run position. The TPS (throttle position sensor) should read about 0.8 to 1.2 volts when the throttle is closed and over 3.5 volts when the throttle is depressed to the floor. Connect the voltmeter to the MAP sensor. Connect a hand-held vacuum pump to the MAP sensor vacuum port. Apply 5 inches hg vacuum to the MAP sensor and take a voltage reading. Now apply 20 inches hg vacuum to the MAP sensor and take a voltage reading. Subtract the second voltage reading from the first. The difference should be 1.26 ±0.13. Now connect a voltmeter to the charge temperature sensor. Assuming you are not in Fairbanks during the winter or Death Valley during the summer, the voltage should be between 1.0 and 3.0 volts. If any of these readings are outside the listed specifications, repair the circuit as necessary. If all are within spec, clear the code and drive the vehicle. After an extensive test drive, check for the code 51 again. If the code has returned, check the voltage listed above again. If they are still correct, replace the logic module, SMEC, or SBEC (the computer).

Code 51 (1991-Present)

Code 51 on the 1991-and-later applications indicates that the oxygen sensor is indicating lean for too long at a time. If the code 51 is received during the test, it indicates the EGO sensor output

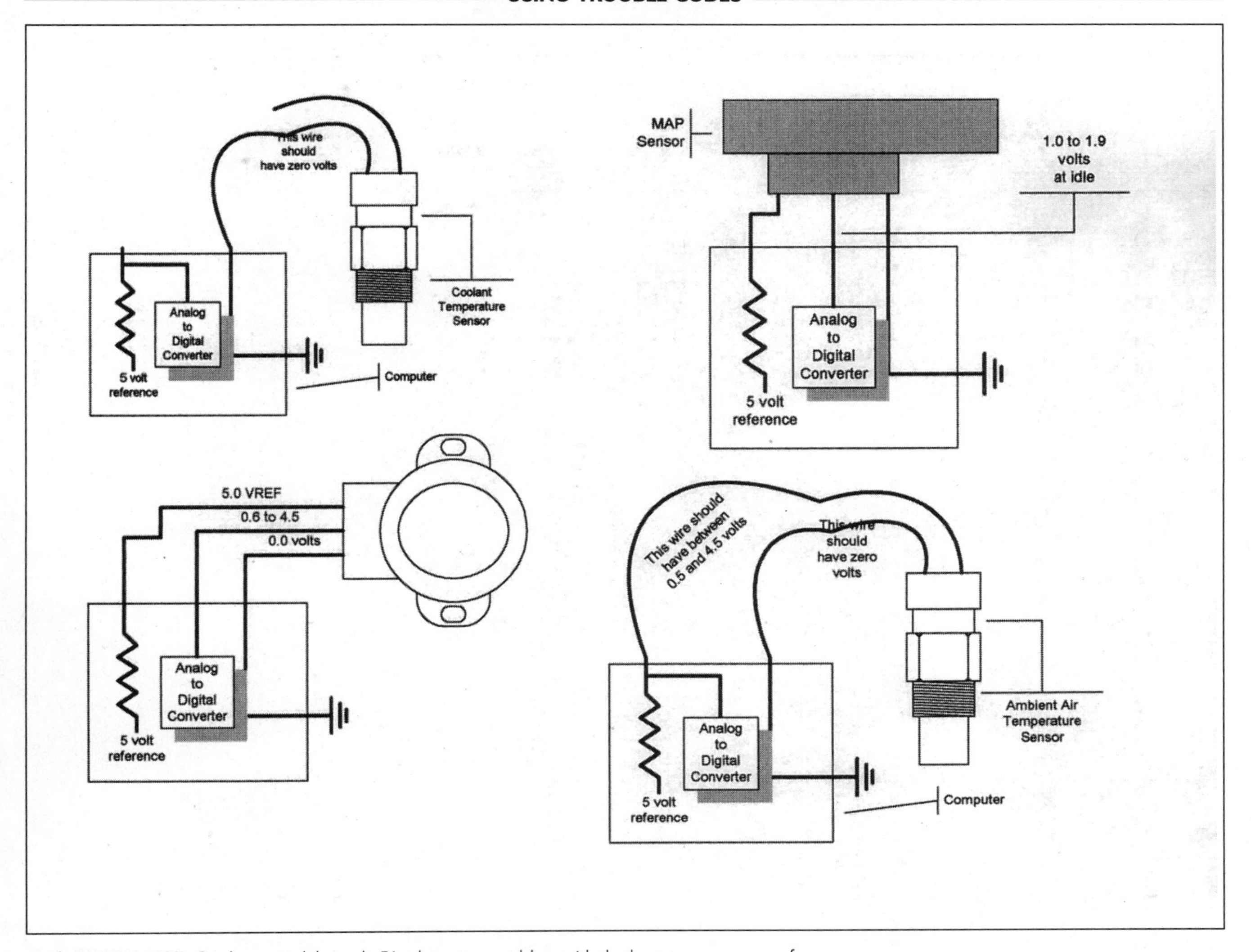

Code 51 (1986-1990). On these models, code 51 relates to a problem with the long-term memory of the computer. Before condemning the computer, be sure to check the coolant sensor, the MAP sensor, the TPS, and the air temperature sensor for proper operation.

voltage was stuck low for an extended time during the last 40 operations of the engine. If the code 51 is received during the test, it means the EGO voltage remained low throughout the test.

Begin troubleshooting the code 51 by connecting a high-impedance voltmeter in parallel to the EGO sensor. Leave the EGO sensor connected. Start the engine and run at 2,000 rpm for two minutes. At the end of the two minutes, observe the voltmeter. The voltage should be switching from below 0.450 volts to above 0.450 volts several times during each five-second interval. If the voltage remains low, disconnect the hoses between the air cleaner and the throttle assembly. Place the end of an unlit propane torch in the intake and open the propane control valve. If the voltage does not increase to above the 0.450 voltage threshold, check the distributor cap, distributor rotor, spark plug wires, ignition coil, spark plugs, EGR system, fuel pressure (should be 30-40 psi), and intake leaks (also called vacuum leaks). If there are no problems with any of these, replace the EGO sensor.

Codes 52, 53, and 54 (1984)

For 1984 and earlier applications, these codes refer to a problem in the logic module. If you receive this code, simply replace the logic module.

Code 52 (1985 and Later)

The oxygen sensor circuit is indicating a chronic rich condition. If the code 52 is received during the test, it indicates the oxygen sensor or sensor output voltage was stuck high for an extended time during the last 20 to 40 operations of the engine. Begin troubleshooting the code 52 by connecting a high-impedance voltmeter in parallel to the oxygen sensor. Leave the oxygen sensor connected. Start the engine and run at 2,000 rpm for two minutes. At the end of the two minutes, observe the voltmeter. The voltage should be switching from above 0.450 volts to below 0.450 volts several times during each five-second interval. If the voltage remains high, disconnect the hoses between the air cleaner and the throttle assembly. Create a vacuum leak. If the voltage does

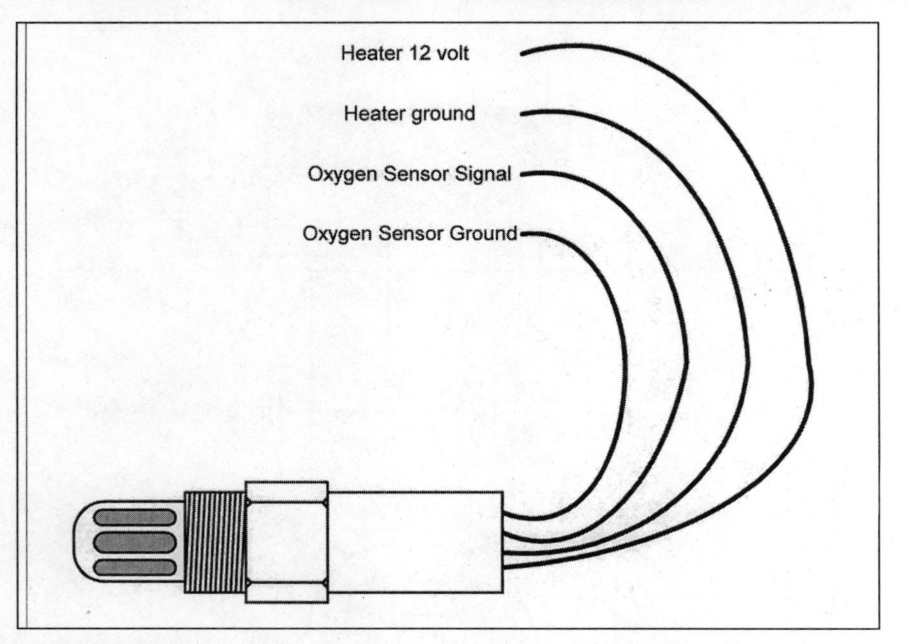

On 1991 applications, code 51 means that the oxygen sensor voltage has been stuck low for an extended time. Review the history and operation of the vehicle for events that may have led to contamination of the oxygen sensor.

not decrease to below the 0.450 voltage threshold, check for a saturated evaporative canister, a defective canister purge valve, a defective PCV valve, contaminated engine oil, and high fuel pressure (30-40 psi).

Code 53 (1985 and Later)

There is an internal logic module problem. Confirm that there is proper power and a proper ground to the logic module, SMEC, or SBEC. If not, replace the logic module, SMEC, or SBEC (computer).

Code 54 (1985 and Later)

This code indicates that there is no fuel sync signal to the computer. The fuel sync pickup is located on the bottom of the ignition Hall Effects pickup. There are three wires connected to this pickup. One is orange; this is the 8-volt power supply for the sensor. The second is tan/yellow; this is the distributor sync signal. The third is black/light blue; this is the ground. Turn off the ignition switch. Disconnect the sensor from vehicle wiring harness. Turn the ignition switch back on. Measure the voltage between the orange wire and the black/light blue wire. The reading should be about 8 volts. If it is not, find where this wire is connected to the computer and measure the voltage there. If the voltage is now 8 volts, repair the wire. If the voltage is still significantly less than 8 volts, inspect the connector. If the connection is good, replace the computer. Now measure the voltage between the tan/yellow wire and the black/light blue wire. The reading should be about 8 volts. If it is not, find where this wire is connected to the computer and measure the voltage there. If the voltage is now 8 volts, repair the wire. If the voltage is still significantly less than 8 volts, inspect the connector. If the connection is good, replace the computer.

Note: Wire colors do change from one application to another. While this text does make an effort to discuss the wires in as universal a way as possible, it may be necessary to alter the aforementioned procedure to compensate for change in wire color.

Code 55

End of message. This code is received when codes are pulled through the Check Engine light or on some scanners. Code 55 simply means that no other codes will follow. Codes related to all detected failures have already been given.

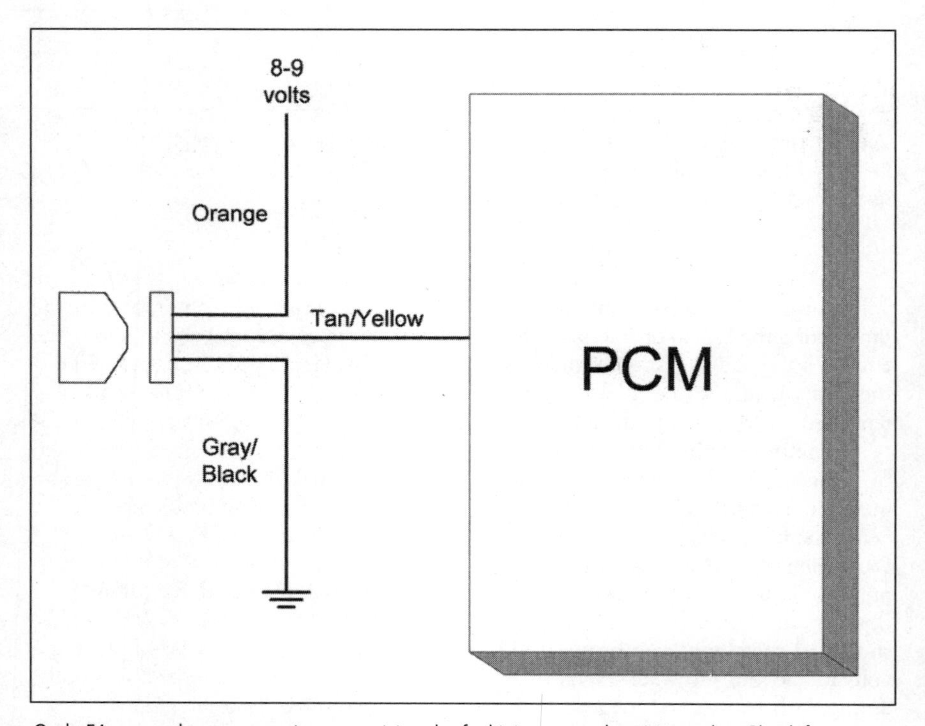

Code 54 means the computer is not receiving the fuel injector synchronizing pulse. Check for voltage on the orange wire and a ground on the gray and black wire. Connect a voltmeter to the tan and yellow wire. There should be a voltage that will go from 0 to about 9 volts and then about 9 to 0 volts as the distributor is rotated.

Code 61

This code relates to a perceived open or shorted condition in the barometric read solenoid circuit. You will find this solenoid located near the MAP sensor, or on the 1984 model, on the right front strut tower. There are two wires going to the baro read solenoid. One of the two wires is dark blue, the other is light blue. With the engine running, both wires should have 12 volts. If only the dark blue wire has 12 volts, but the engine seems to idle, replace the barometric read solenoid. If the dark blue wire has 12 volts, but the light blue wire has zero and the engine runs poorly, then repair the grounded light blue wire. If the light blue wire is in good condition, replace the computer.

If both wires have 12 volts at the baro read solenoid, then check the voltage on the light blue wire at the computer. If there is voltage, inspect the connection. If the connection is good, replace the computer.

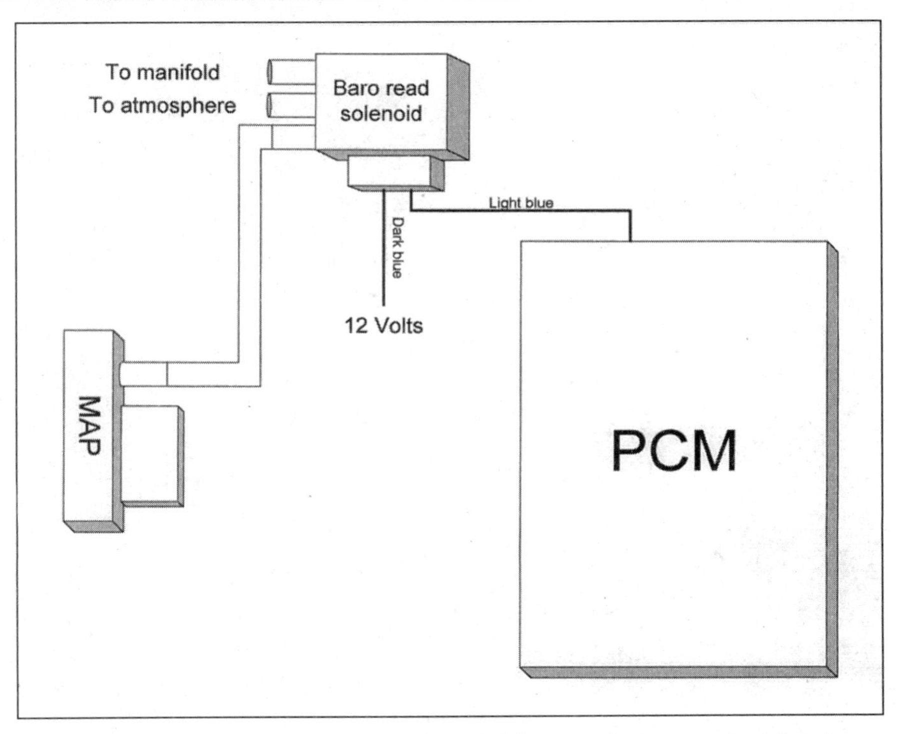

The typical Chrysler fuel injection system uses the speed density method to measure air flow. For this to be effective, the computer has to know both manifold and barometric pressure. Since there is only one pressure sensor, it needs to be able to do both duties. When the key is rotated through the on position heading toward start, the computer samples the pressure in the manifold, which at that point is barometric. Once the engine is started, the pressure sensor now measures manifold pressure. On the turbo models, the baro read solenoid occasionally switches so that the computer can be updated on the current barometric pressure.

Code 62

Defective computer. EMR (Emission Maintenance Reminder) was unable to update and reset.

Code 63

Defective computer. One of the memory address locations in the computer is defective.

Code 66

Fuel injection computer did not receive information from the transmission computer module (TCM). Refer this condition to the dealer.

Code 66 (BCM)

Fuel injection computer did not receive information from the body and chassis computer module. Refer this condition to the dealer.

On-Board Diagnostics 2 (OBD II)

Cars after 1995 are equipped with the on-board diagnostics 2 (OBD II) system. This system has been designed to assist the technician in the maintenance and repair of the fuel injection system and the emission control system. The system must also be capable of determining when the normal events of engine operation occur. The system not only looks for blatant defects in the system but also looks for system readiness problems. Under the old system, a code would be generated when the computer recognized an electronic failure in one of the sensor circuits, say the coolant sensor. Under OBD II, sensor circuits are monitored, but the system must also be able to recognize that the engine has warmed up. Taking the vehicle that has not been properly warmed up to an emission testing station would result in an emission test failure. The following is an excerpt from Federal Register, August 6, 1996 (Volume 61, Number 152) [Rules and Regulations] [Page 40939-40948]:

1. Summary of Proposal. The proposal required that all vehicles subject to an I/M test requirement undergo an OBD test beginning January 1, 1998. The proposal also stated that any vehicle which failed the OBD portion of the I/M test would fail the I/M test as of January 1, 1998. One of the possible reasons for failing the OBD test would be if all the vehicle's readiness codes were not cleared when it arrived at the test station. The readiness code status provides an indication of whether or not a specific monitor has been exercised. A code is set when the monitor has not yet had a sufficient chance to make an accurate evaluation of the component's operation. The readiness code is cleared when an accurate determination has been made, thus indicating I/M readiness.

2. Summary of Comments. On September 26, 1995, several vehicle manufacturers met with EPA to discuss the OBD rule. At this meeting and again in written comments, manufacturers expressed the concern that vehicles would be rejected from testing because all the OBD readiness codes for the vehicle would not be cleared when the vehicle arrived at the test station. In particular, the manufacturers were concerned that extreme cold weather or high altitude might prevent certain readiness codes from clearing. Since that time, three man-

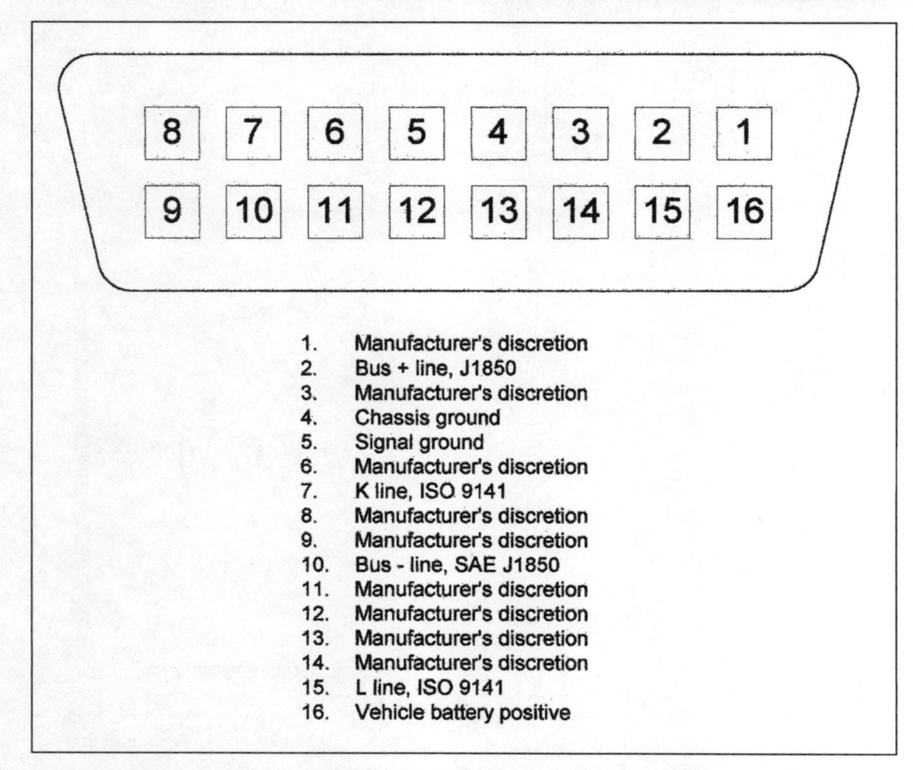

Beginning in 1994, the federal government began a mandate of standardized on-board diagnostics. From the 1996 model year forward, all applications will be equipped with the OBD-2 diagnostic connector.

The OBD-2 connector is located under the dash near the centerline of the vehicle.

ufacturers have notified EPA that there were problems with the design of the OBD readiness codes in a portion of the 1996 model year fleet and that it was likely that all of the codes would not be cleared when these vehicles arrived at the test station even though the vehicle was functioning normally. Some commenters also noted that OBD system checks should be incorporated in a manner that encourages public support and acceptance of OBD systems, especially during the early stages of implementation when technology for OBD systems is still relatively new. To deal with these issues, stakeholders suggested that a data collection period on the OBD system would be prudent. This would give EPA, the states, and the manufacturers time to assess the effectiveness of the OBD tests, identify any problems, and implement refinements.

3. Response to Comments. EPA agrees with commentors that because the OBD technology is new, a period of study is warranted. Therefore, although this action makes OBD testing mandatory for most I/M programs as of January 1, 1998, for the first two years of the program, until December 31, 1999, vehicles that fail the OBD test will not automatically fail the I/M test or be required to obtain repairs. From January 1, 1998 to December 31, 1999, vehicles that fail the OBD test can still pass the I/M test provided they undergo and pass the tail-pipe emission test, and, where applicable, the evaporative system tests. This will give EPA, the states, and vehicle manufacturers two years to collect data on OBD test results and the interaction between OBD test failures and exhaust and evaporative test results. This test period should allow for the resolution of any vehicle software problems to ensure that vehicle owners will not be turned away from the test center solely because of the way in which their vehicle's readiness codes were programmed. In addition, this two-year period will allow time to correct any other unforeseen problems that may arise with readiness and diagnostic trouble codes or any other element of OBD testing. By providing this test-only period, EPA hopes to identify and solve potential problems so that consumers will face the least amount of inconvenience possible. EPA does not believe there will be any lost emission reductions as a result of this two-year data collection period because most vehicles will still have to undergo tailpipe emission and, where applicable, evaporative tests. Furthermore, since OBD testing is only required on 1996 and newer vehicles, these vehicles will still be new and "clean" in 1998 and 1999. Because of this, EPA expects that very few of these vehicles will fail the I/M test. EPA considered providing more detailed guidance on what the vehicle operator should be told (beginning in 2000) in the event their vehicle is rejected from testing because all of its readiness codes are not cleared. The proposed language of Sec. 85.2223(a)(3) stated that the operator should be told to return after driving the vehicle "long enough" to allow the readiness codes to clear. Because time is not the only condition which will affect readiness code status, EPA changed this language (now in Sec. 85.2222(c)) to provide that the operator be told to return after driving the vehicle under the conditions necessary for it to provide an accurate readiness determination. At this time, EPA does not feel it is appropriate to specify in the regulation what the vehicle operator should be told and instead believes it is best left to the states to devise a solution that meets local program needs. As a

result of the general language in this portion of the regulation, it is imperative that I/M inspectors obtain education about OBD so they can assess each individual operator's situation and provide advice on what should be done to ensure that the vehicle is ready when it returns to the test station. By way of example, EPA is including the following scenarios. First, evaporative system leak detection monitors generally require ambient temperatures above 40 degrees Fahrenheit, and an overnight soak or extended period of non-operation, prior to exercising the monitor. In a situation where the evaporative system readiness code is not cleared, an operator should be told to return after starting their vehicle in warmer ambient temperature conditions with a near full tank of gasoline. Second, continued low-speed operation could provide little opportunity for exercising the exhaust gas recirculation (EGR) monitor. In a situation where the EGR readiness code has not cleared, an operator should be told to return after driving at higher speeds on the highway so that EGR would occur and the EGR monitor could be exercised.

B. Verifying Codes at Test Station

1. Summary of Proposal. Under the proposal any vehicle whose malfunction indicator light (MIL) is commanded to be illuminated and who has certain diagnostic trouble codes (DTCs) present fails the OBD test.

2. Summary of Comments. One commentor urged EPA to establish a procedure to determine at the test center if a DTC could be false.

3. Response to Comments. Currently, the technology is not available to determine if a DTC is false at the test center. EPA believes that the two-year test period discussed above in section V.A will allow for development and refinement of OBD systems so that false failures will be less likely.

Now upon first read, it looks like they may be doing an excellent job of precluding the do-it-yourselfer from the auto repair loop. There are probably those among you who are sure there is a conspiracy. I do not see it this way. I feel that there will just be a need for the do-it-yourselfer to be better informed.

Powertrain Diagnostic Trouble Codes for OBD II

P01XX Fuel and Air Metering

P0100	Mass or Volume Air Flow Circuit Malfunction
P0101	Mass or Volume Air Flow Circuit Range/Performance Problem
P0102	Mass or Volume Air Flow Circuit Low Input
P0103	Mass or Volume Air Flow Circuit High Input
P0104	Mass or Volume Air Flow Circuit Intermittent
P0105	Manifold Absolute Pressure/Barometric Pressure Circuit Malfunction
P0106	Manifold Absolute Pressure/Barometric Pressure Circuit Range/Performance Problem
P0107	Manifold Absolute Pressure/Barometric Pressure Circuit Low Input
P0108	Manifold Absolute Pressure/Barometric Pressure Circuit High Input
P0109	Manifold Absolute Pressure/Barometric Pressure Circuit Intermittent
P0110	Intake Air Temperature Circuit Malfunction
P0111	Intake Air Temperature Circuit Range/Performance Problem
P0112	Intake Air Temperature Circuit Low Input
P0113	Intake Air Temperature Circuit High Input
P0114	Intake Air Temperature Circuit Intermittent
P0115	Engine Coolant Temperature Circuit Malfunction
P0116	Engine Coolant Temperature Circuit Range/Performance Problem
P0117	Engine Coolant Temperature Circuit Low Input
P0118	Engine Coolant Temperature Circuit High Input
P0119	Engine Coolant Temperature Circuit Intermittent
P0120	Throttle/Pedal Position Sensor/Switch A Circuit Malfunction
P0121	Throttle/Pedal Position Sensor/Switch A Circuit Range/Performance Problem
P0122	Throttle/Pedal Position Sensor/Switch A Circuit Low Input
P0123	Throttle/Pedal Position Sensor/Switch A Circuit High Input
P0124	Throttle/Pedal Position Sensor/Switch A Circuit Intermittent
P0125	Insufficient Coolant Temperature for Closed Loop Control
P0126	Insufficient Coolant Temperature for Stable Operation
P0130	O_2 Sensor Circuit Malfunction (Bank 1 Sensor 1)
P0131	O_2 Sensor Circuit Low Voltage (Bank 1 Sensor 1)
P0132	O_2 Sensor Circuit High Voltage (Bank 1 Sensor 1)
P0133	O_2 Sensor Circuit Slow Response (Bank 1 Sensor 1)

P0134 O_2 Sensor Circuit No Activity Detected (Bank 1 Sensor 1)
P0135 O_2 Sensor Heater Circuit Malfunction
P0136 O_2 Sensor Circuit Malfunction (Bank 1 Sensor 2)
P0137 O_2 Sensor Circuit Low Voltage (Bank 1 Sensor 2)
P0138 O_2 Sensor Circuit High Voltage (Bank 1 Sensor 2)
P0139 O_2 Sensor Circuit Slow Response (Bank 1 Sensor 2)
P0140 O_2 Sensor Circuit No Activity Detected (Bank 1 Sensor 2)
P0141 O_2 Heater Circuit Malfunction (Bank 1 Sensor 2)
P0142 O_2 Sensor Circuit Malfunction (Bank 1 Sensor 3)
P0143 O_2 Sensor Circuit Low Voltage (Bank 1 Sensor 3)
P0144 O_2 Sensor Circuit High Voltage (Bank 1 Sensor 3)
P0145 O_2 Sensor Circuit Slow Response (Bank 1 Sensor 3)
P0146 O_2 Sensor Circuit No Activity Detected (Bank 1 Sensor 3)
P0147 O_2 Heater Circuit Malfunction (Bank 1 Sensor 3)
P0150 O_2 Sensor Circuit Malfunction (Bank 2 Sensor 1)
P0151 O_2 Sensor Circuit Low Voltage (Bank 2 Sensor 1)
P0152 O_2 Sensor Circuit High Voltage (Bank 2 Sensor 1)
P0153 O_2 Sensor Circuit Slow Response (Bank 2 Sensor 1)
P0154 O_2 Sensor Circuit No Activity Detected (Bank 2 Sensor 1)
P0155 O_2 Heater Circuit Malfunction (Bank 2 Sensor 1)
P0156 O_2 Sensor Circuit Malfunction (Bank 2 Sensor 2)
P0157 O_2 Sensor Circuit Low Voltage (Bank 2 Sensor 2)
P0158 O_2 Sensor Circuit High Voltage (Bank 2 Sensor 2)
P0159 O_2 Sensor Circuit Slow Response (Bank 2 Sensor 2)
P0160 O_2 Sensor Circuit No Activity Detected (Bank 2 Sensor 2)
P0161 O_2 Heater Circuit Malfunction (Bank 2 Sensor 2)
P0162 O_2 Sensor Circuit Malfunction (Bank 2 Sensor 3)
P0163 O_2 Sensor Circuit Low Voltage (Bank 2 Sensor 3)
P0164 O_2 Sensor Circuit High Voltage (Bank 2 Sensor 3)
P0165 O_2 Sensor Circuit Slow Response (Bank 2 Sensor 3)
P0166 O_2 Sensor Circuit No Activity Detected (Bank 2 Sensor 3)
P0167 O_2 Heater Circuit Malfunction (Bank 2 Sensor 3)
P0170 Fuel Trim Malfunction (Bank 1)
P0171 System Too Lean (Bank 1)
P0172 System Too Rich (Bank 1)
P0173 Fuel Trim Malfunction (Bank 2)
P0174 System Too Lean (Bank 2)
P0175 System Too Rich (Bank 2)
P0176 Fuel Composition Sensor Circuit Malfunction
P0177 Fuel Composition Sensor Circuit Range/Performance
P0178 Fuel Composition Sensor Circuit Low Input
P0179 Fuel Composition Sensor Circuit High Input
P0180 Fuel Temperature Sensor A Circuit Malfunction
P0181 Fuel Temperature Sensor A Circuit Range/ Performance
P0182 Fuel Temperature Sensor A Circuit Low Input
P0183 Fuel Temperature Sensor A Circuit High Input
P0184 Fuel Temperature Sensor A Circuit Intermittent
P0185 Fuel Temperature Sensor B Circuit Malfunction
P0186 Fuel Temperature Sensor B Circuit Range/ Performance
P0187 Fuel Temperature Sensor B Circuit Low Input
P0188 Fuel Temperature Sensor B Circuit High Input
P0189 Fuel Temperature Sensor B Circuit Intermittent
P0190 Fuel Rail Pressure Sensor Circuit Malfunction
P0191 Fuel Rail Pressure Sensor Circuit Range/ Performance
P0192 Fuel Rail Pressure Sensor Circuit Low Input
P0193 Fuel Rail Pressure Sensor Circuit High Input
P0194 Fuel Rail Pressure Sensor Circuit Intermittent
P0195 Engine Oil Temperature Sensor Malfunction
P0196 Engine Oil Temperature Sensor Range/Performance
P0197 Engine Oil Temperature Sensor Low
P0198 Engine Oil Temperature Sensor High
P0199 Engine Oil Temperature Sensor Intermittent

P02XX Fuel And Air Metering

P0200 Injector Circuit Malfunction
P0201 Injector Circuit Malfunction - Cylinder 1
P0202 Injector Circuit Malfunction - Cylinder 2
P0203 Injector Circuit Malfunction - Cylinder 3
P0204 Injector Circuit Malfunction - Cylinder 4
P0205 Injector Circuit Malfunction - Cylinder 5
P0206 Injector Circuit Malfunction - Cylinder 6
P0207 Injector Circuit Malfunction - Cylinder 7
P0208 Injector Circuit Malfunction - Cylinder 8
P0209 Injector Circuit Malfunction - Cylinder 9
P0210 Injector Circuit Malfunction - Cylinder 10
P0211 Injector Circuit Malfunction - Cylinder 11
P0212 Injector Circuit Malfunction - Cylinder 12
P0213 Cold Start Injector 1 Malfunction
P0214 Cold Start Injector 2 Malfunction
P0215 Engine Shutoff Solenoid Malfunction
P0216 Injection Timing Control Circuit Malfunction
P0217 Engine Over Temperature Condition
Transmission Over Temperature Condition
P0219 Engine Overspeed Condition
P0220 Throttle/Pedal Position Sensor/Switch B Circuit Malfunction
P0221 Throttle/Pedal Position Sensor/Switch B Circuit Range/ Performance Problem
P0222 Throttle/Pedal Position Sensor/Switch B Circuit Low Input
P0223 Throttle/Pedal Position Sensor/Switch B Circuit High Input
P0224 Throttle/Pedal Position Sensor/Switch B Circuit Intermittent
P0225 Throttle/Pedal Position Sensor/Switch C Circuit Malfunction
P0226 Throttle/Pedal Position Sensor/Switch C Circuit Range/ Performance Problem
P0227 Throttle/Pedal Position Sensor/Switch C Circuit Low Input
P0228 Throttle/Pedal Position Sensor/Switch C Circuit High Input
P0229 Throttle/Pedal Position Sensor/Switch C Circuit Intermittent
P0230 Fuel Pump Primary Circuit Malfunction
P0231 Fuel Pump Secondary Circuit Low
P0232 Fuel Pump Secondary Circuit High
P0233 Fuel Pump Secondary Circuit Intermittent
P0235 Turbocharger Boost Sensor A Circuit Malfunction
P0236 Turbocharger Boost Sensor A Circuit Range/ Performance
P0237 Turbocharger Boost Sensor A Circuit Low
P0238 Turbocharger Boost Sensor A Circuit High
P0239 Turbocharger Boost Sensor B Malfunction
P0240 Turbocharger Boost Sensor B Circuit Range/ Performance
P0241 Turbocharger Boost Sensor B Circuit Low

P0242 Turbocharger Boost Sensor B Circuit High
P0243 Turbocharger Wastegate Solenoid A Malfunction
P0244 Turbocharger Wastegate Solenoid A Range/Performance
P0245 Turbocharger Wastegate Solenoid A Low
P0246 Turbocharger Wastegate Solenoid A High
P0247 Turbocharger Wastegate Solenoid B Malfunction
P0248 Turbocharger Wastegate Solenoid B Range/Performance
P0249 Turbocharger Wastegate Solenoid B Low
P0250 Turbocharger Wastegate Solenoid B High
P0251 Injection Pump A Rotor/Cam Malfunction
P0252 Injection Pump A Rotor/Cam Range/Performance
P0253 Injection Pump A Rotor/Cam Low
P0254 Injection Pump A Rotor/Cam High
P0255 Injection Pump A Rotor/Cam Intermittent
P0256 Injection Pump B Rotor/Cam Malfunction
P0257 Injection Pump B Rotor/Cam Range/Performance
P0258 Injection Pump B Rotor/Cam Low
P0259 Injection Pump B Rotor/Cam High
P0260 Injection Pump B Rotor/Cam Intermittent
P0261 Cylinder 1 Injector Circuit Low
P0262 Cylinder 1 Injector Circuit High
P0263 Cylinder 1 Contribution/Balance Fault
P0264 Cylinder 2 Injector Circuit Low
P0265 Cylinder 2 Injector Circuit High
P0266 Cylinder 2 Contribution/Balance Fault
P0267 Cylinder 3 Injector Circuit Low
P0268 Cylinder 3 Injector Circuit High
P0269 Cylinder 3 Contribution/Balance Fault
P0270 Cylinder 4 Injector Circuit Low
P0271 Cylinder 4 Injector Circuit High
P0272 Cylinder 4 Contribution/Balance Fault
P0273 Cylinder 5 Injector Circuit Low
P0274 Cylinder 5 Injector Circuit High
P0275 Cylinder 5 Contribution/Balance Fault
P0276 Cylinder 6 Injector Circuit Low
P0277 Cylinder 6 Injector Circuit High
P0278 Cylinder 6 Contribution/Balance Fault
P0279 Cylinder 7 Injector Circuit Low
P0280 Cylinder 7 Injector Circuit High
P0281 Cylinder 7 Contribution/Balance Fault
P0282 Cylinder 8 Injector Circuit Low
P0283 Cylinder 8 Injector Circuit High
P0284 Cylinder 8 Contribution/Balance Fault
P0285 Cylinder 9 Injector Circuit Low
P0286 Cylinder 9 Injector Circuit High
P0287 Cylinder 9 Contribution/Balance Fault
P0288 Cylinder 10 Injector Circuit Low
P0289 Cylinder 10 Injector Circuit High
P0290 Cylinder 10 Contribution/Balance Fault
P0291 Cylinder 11 Injector Circuit Low
P0292 Cylinder 11 Injector Circuit High
P0293 Cylinder 11 Contribution/Balance Fault
P0294 Cylinder 12 Injector Circuit Low
P0295 Cylinder 12 Injector Circuit High
P0296 Cylinder 12 Contribution/Range Fault

P03XX Ignition System or Misfire

P0300 Random/Multiple Cylinder Misfire Detected
P0301 Cylinder 1 Misfire Detected
P0302 Cylinder 2 Misfire Detected
P0303 Cylinder 3 Misfire Detected
P0304 Cylinder 4 Misfire Detected
P0305 Cylinder 5 Misfire Detected
P0306 Cylinder 6 Misfire Detected
P0307 Cylinder 7 Misfire Detected
P0308 Cylinder 8 Misfire Detected
P0309 Cylinder 9 Misfire Detected
P0310 Cylinder 10 Misfire Detected
P0311 Cylinder 11 Misfire Detected
P0312 Cylinder 12 Misfire Detected
P0320 Ignition/Distributor Engine Speed Input Circuit Malfunction
P0321 Ignition/Distributor Engine Speed Input Circuit Range/Performance
P0322 Ignition/Distributor Engine Speed Input Circuit No Signal
P0323 Ignition/Distributor Engine Speed Input Circuit Intermittent
P0325 Knock Sensor 1 Circuit Malfunction (Bank 1 or Single Sensor 3)
P0326 Knock Sensor 1 Circuit Range/Performance (Bank 1 or Single Sensor 3)
P0327 Knock Sensor 1 Circuit Low Input (Bank 1 or Single Sensor 3)
P0328 Knock Sensor 1 Circuit High Input (Bank 1 or Single Sensor 3)
P0329 Knock Sensor 1 Circuit Intermittent (Bank 1 or Single Sensor 3)
P0330 Knock Sensor 2 Circuit Malfunction (Bank 2)
P0331 Knock Sensor 2 Circuit Range/Performance (Bank 2)
P0332 Knock Sensor 2 Circuit Low Input (Bank 2)
P0333 Knock Sensor 2 Circuit High Input (Bank 2)
P0334 Knock Sensor 2 Circuit Intermittent (Bank 2)
P0335 Crankshaft Position Sensor A Circuit Malfunction
P0336 Crankshaft Position Sensor A Circuit Range/Performance
P0337 Crankshaft Position Sensor A Circuit Low Input
P0338 Crankshaft Position Sensor A Circuit High Input
P0339 Crankshaft Position Sensor A Circuit Intermittent
P0340 Camshaft Position Sensor Circuit Malfunction
P0341 Camshaft Position Sensor Circuit Range/Performance
P0342 Camshaft Position Sensor Circuit Low Input
P0343 Camshaft Position Sensor Circuit High Input
P0344 Camshaft Position Sensor Circuit Intermittent
P0350 Ignition Coil Primary/Secondary Circuit Malfunction
P0351 Ignition Coil A Primary/Secondary Circuit Malfunction
P0352 Ignition Coil B Primary/Secondary Circuit Malfunction
P0353 Ignition Coil C Primary/Secondary Circuit Malfunction
P0354 Ignition Coil D Primary/Secondary Circuit Malfunction
P0355 Ignition Coil E Primary/Secondary Circuit Malfunction
P0356 Ignition Coil F Primary/Secondary Circuit Malfunction
P0357 Ignition Coil G Primary/Secondary Circuit Malfunction
P0358 Ignition Coil H Primary/Secondary Circuit Malfunction
P0359 Ignition Coil I Primary/Secondary Circuit Malfunction
P0360 Ignition Coil J Primary/Secondary Circuit Malfunction
P0361 Ignition Coil K Primary/Secondary Circuit Malfunction
P0362 Ignition Coil L Primary/Secondary Circuit Malfunction

P0370 Timing Reference High Resolution Signal A Malfunction
P0371 Timing Reference High Resolution Signal A Too Many Pulses
P0372 Timing Reference High Resolution Signal A Too Few Pulses
P0373 Timing Reference High Resolution Signal A ntermittent/ Erratic Pulses
P0374 Timing Reference High Resolution Signal A No Pulses
P0375 Timing Reference High Resolution Signal B Malfunction
P0376 Timing Reference High Resolution Signal B Too Many Pulses
P0377 Timing Reference High Resolution Signal B Too Few Pulses
P0378 Timing Reference High Resolution Signal B Intermittent/ Erratic Pulses
P0379 Timing Reference High Resolution Signal B No Pulses
P0380 Glow Plug/Heater Circuit Malfunction
P0381 Glow Plug/Heater Indicator Circuit Malfunction
P0385 Crankshaft Position Sensor B Circuit Malfunction
P0386 Crankshaft Position Sensor B Circuit Range/ Performance
P0387 Crankshaft Position Sensor B Circuit Low Input
P0388 Crankshaft Position Sensor B Circuit High Input
P0389 Crankshaft Position Sensor B Circuit Intermittent

P04XX Auxiliary Emission Controls

P0400 Exhaust Gas Recirculation Flow Malfunction
P0401 Exhaust Gas Recirculation Flow Insufficient Detected
P0402 Exhaust Gas Recirculation Flow Excessive Detected
P0403 Exhaust Gas Recirculation Circuit Malfunction
P0404 Exhaust Gas Recirculation Circuit Range/ Performance
P0405 Exhaust Gas Recirculation Sensor A Circuit Low
P0406 Exhaust Gas Recirculation Sensor A Circuit High
P0407 Exhaust Gas Recirculation Sensor B Circuit Low
P0408 Exhaust Gas Recirculation Sensor B Circuit High
P0410 Secondary Air Injection System Malfunction
P0411 Secondary Air Injection System Incorrect Flow Detected
P0412 Secondary Air Injection System Switching Valve A Circuit Malfunction
P0413 Secondary Air Injection System Switching Valve A Circuit Open
P0414 Secondary Air Injection System Switching Valve A Circuit Shorted
P0415 Secondary Air Injection System Switching Valve B Circuit Malfunction
P0416 Secondary Air Injection System Switching Valve B Circuit Open
P0417 Secondary Air Injection System Switching Valve B Circuit Shorted
P0420 Catalyst System Efficiency Below Threshold (Bank 14)
P0421 Warm Up Catalyst Efficiency Below Threshold (Bank 14)
P0422 Main Catalyst Efficiency Below Threshold (Bank 14)
P0423 Heated Catalyst Efficiency Below Threshold (Bank 14)
P0424 Heated Catalyst Temperature Below Threshold (Bank 14)
P0430 Catalyst System Efficiency Below Threshold (Bank 2)
P0431 Warm Up Catalyst Efficiency Below Threshold (Bank 2)
P0432 Main Catalyst Efficiency Below Threshold (Bank 2)
P0433 Heated Catalyst Efficiency Below Threshold (Bank 2)
P0434 Heated Catalyst Temperature Below Threshold (Bank 2)
P0440 Evaporative Emission Control System Malfunction
P0441 Evaporative Emission Control System Incorrect Purge Flow
P0442 Evaporative Emission Control System Leak Detected (small leak)
P0443 Evaporative Emission Control System Purge Control Valve Circuit Malfunction
P0444 Evaporative Emission Control System Purge Control Valve Circuit Open
P0445 Evaporative Emission Control System Purge Control Valve Circuit Shorted
P0450 Evaporative Emission Control System Pressure Sensor Malfunction
P0451 Evaporative Emission Control System Pressure Sensor Range/Performance
P0452 Evaporative Emission Control System Pressure Sensor Low Input
P0453 Evaporative Emission Control System Pressure Sensor High Input
P0454 Evaporative Emission Control System Pressure Sensor Intermittent
P0455 Evaporative Emission Control System Leak Detected (gross leak)
P0460 Fuel Level Sensor Circuit Malfunction
P0461 Fuel Level Sensor Circuit Range/Performance
P0462 Fuel Level Sensor Circuit Low Input
P0463 Fuel Level Sensor Circuit High Input
P0464 Fuel Level Sensor Circuit Intermittent
P0465 Purge Flow Sensor Circuit Malfunction
P0466 Purge Flow Sensor Circuit Range/Performance
P0467 Purge Flow Sensor Circuit Low Input
P0468 Purge Flow Sensor Circuit High Input
P0469 Purge Flow Sensor Circuit Intermittent
P0470 Exhaust Pressure Sensor Malfunction
P0471 Exhaust Pressure Sensor Range/Performance
P0472 Exhaust Pressure Sensor Low Input
P0473 Exhaust Pressure Sensor High Input
P0474 Exhaust Pressure Sensor Intermittent
P0475 Exhaust Pressure Control Valve Malfunction
P0476 Exhaust Pressure Control Valve Range/Performance
P0477 Exhaust Pressure Control Valve Low Input
P0478 Exhaust Pressure Control Valve High Input
P0479 Exhaust Pressure Control Valve Intermittent

P05XX Vehicle Speed, Idle Control, and Auxiliary Inputs

P0500 Vehicle Speed Sensor Malfunction
P0501 Vehicle Speed Sensor Range/Performance
P0502 Vehicle Speed Sensor Low Input
P0503 Vehicle Speed Sensor Intermittent/Erratic/High
P0505 Idle Control System Malfunction
P0506 Idle Control System RPM Lower Than Expected
P0507 Idle Control System RPM Higher Than Expected
P0510 Closed Throttle Position Switch Malfunction
P0530 A/C Refrigerant Pressure Sensor Circuit Malfunction
P0531 A/C Refrigerant Pressure Sensor Circuit Range/Performance
P0532 A/C Refrigerant Pressure Sensor Circuit Low Input

P0533 A/C Refrigerant Pressure Sensor Circuit High Input
P0534 Air Conditioner Refrigerant Charge Loss
P0550 Power Steering Pressure Sensor Circuit Malfunction
P0551 Power Steering Pressure Sensor Circuit Range/Performance
P0552 Power Steering Pressure Sensor Circuit Low Input
P0553 Power Steering Pressure Sensor Circuit High Input
P0554 Power Steering Pressure Sensor Circuit Intermittent
P0560 System Voltage Malfunction
P0561 System Voltage Unstable
P0562 System Voltage Low
P0563 System Voltage High
P0565 Cruise Control On Signal Malfunction
P0566 Cruise Control Off Signal Malfunction
P0567 Cruise Control Resume Signal Malfunction
P0568 Cruise Control Set Signal Malfunction
P0569 Cruise Control Coast Signal Malfunction
P0570 Cruise Control Accel Signal Malfunction
P0571 Cruise Control/Brake Switch A Circuit Malfunction
P0572 Cruise Control/Brake Switch A Circuit Low
P0573 Cruise Control/Brake Switch A Circuit High
P0574 P0574 Through P0580 Reserved For Cruise Codes

P06XX Computer and Auxiliary Outputs

P0600 Serial Communication Link Malfunction
P0601 Internal Control Module Memory Check Sum Error
P0602 Control Module Programming Error
P0603 Internal Control Module Keep Alive Memory (KAM) Error
P0604 Internal Control Module Random Access Memory (RAM) Error
P0605 Internal Control Module Read Only Memory (ROM) Error
P0606 PCM Processor Fault

P07XX Transmission

P0703 Brake Switch Input Failure
P0705 Transmission Range Sensor Circuit Malfunction (PRNDL Input)
P0706 Transmission Range Sensor Circuit Range/Performance
P0707 Transmission Range Sensor Circuit Low Input
P0708 Transmission Range Sensor Circuit High Input
P0710 Transmission Fluid Temperature Sensor Circuit Malfunction
P0711 Transmission Fluid Temperature Sensor Circuit Range/Performance
P0712 Transmission Fluid Temperature Sensor Circuit Low Input
P0713 Transmission Fluid Temperature Sensor Circuit High Input
P0715 Input/Turbine Speed Sensor Circuit Malfunction
P0716 Input/Turbine Speed Sensor Circuit Range/Performance
P0717 Input/Turbine Speed Sensor Circuit No Signal
P0720 Output Speed Sensor Circuit Malfunction
P0721 Output Speed Sensor Circuit Range/Performance
P0722 Output Speed Sensor Circuit No Signal
P0725 Engine Speed Input Circuit Malfunction
P0726 Engine Speed Input Circuit Range/Performance
P0727 Engine Speed Input Circuit No Signal
P0730 Incorrect Gear Ratio
P0731 Gear 1 Incorrect Ratio
P0732 Gear 2 Incorrect Ratio
P0733 Gear 3 Incorrect Ratio
P0734 Gear 4 Incorrect Ratio
P0735 Gear 5 Incorrect Ratio
P0736 Reverse Incorrect Ratio
P0740 Torque Converter Clutch System Malfunction
P0741 Torque Converter Clutch System Performance or Stuck Off
P0742 Torque Converter Clutch System Stuck On
P0743 Torque Converter Clutch System Electrical
P0745 Pressure Control Solenoid Malfunction
P0746 Pressure Control Solenoid Performance Or Stuck Off
P0747 Pressure Control Solenoid Stuck On
P0748 Pressure Control Solenoid Electrical
P0750 Shift Solenoid "A" Malfunction
P0751 Shift Solenoid "A" Performance Or Stuck Off
P0752 Shift Solenoid "A" Stuck On
P0753 Shift Solenoid "A" Electrical
P0755 Shift Solenoid "B" Malfunction
P0756 Shift Solenoid "B" Performance Or Stuck Off
P0757 Shift Solenoid "B" Stuck On
P0758 Shift Solenoid "B" Electrical
P0760 Shift Solenoid "C" Malfunction
P0761 Shift Solenoid "C" Performance Or Stuck Off
P0762 Shift Solenoid "C" Stuck On
P0763 Shift Solenoid "C" Electrical
P0765 Shift Solenoid "D" Malfunction
P0766 Shift Solenoid "D" Performance Or Stuck Off
P0767 Shift Solenoid "D" Stuck On
P0768 Shift Solenoid "D" Electrical
P0770 Shift Solenoid "E" Malfunction
P0771 Shift Solenoid "E" Performance Or Stuck Off
P0772 Shift Solenoid "E" Stuck On
P0773 Shift Solenoid "E" Electrical

CHAPTER EIGHT

SERIAL DATA STREAM

Data Fields

Serial data is one of the handiest tools a technician has at his/her disposal. Unfortunately, some rather expensive hardware is required to read the serial data stream. A scanner is used to read the data stream. A scanner connects to the diagnostic connector located under the hood and translates computer codes from the engine controller into digital information about what the computer is seeing, thinking, and doing.

Scanners come in all sizes, price ranges, and user friendliness. Scanners are incorporated into engine analyzers costing tens of thousands of dollars and in hand-held units such as those marketed by Snap-On, OTC, and others at between $600 and $2,000

Although a scanner is not a necessary tool for the average enthusiast, it is definitely a very valuable tool. Its real value comes from its use in conjunction with flow charts for diagnosing problems related to trouble codes.

For those who have an IBM PC that operates at 33 megahertz or less, there is another solution. Keep in mind that when I began writing books on fuel injection, I bought an 8086 computer that operated at 4.7 megahertz. Today, I am using a 90 megahertz unit that is rapidly becoming more obsolete with every stroke of the keyboard. That 8086 could be found at a garage sale today for around $50. The alternative solution that I am talking about is called DIACOM by Rinda Technologies. This product ran great on that 8086 I had, but will not even start on my Pentium. The trick in getting your hands on DIACOM is getting in touch with Rinda Technologies at 5112 North Elston Ave. Chicago, Illinois 60630 (312) 736-6633.

There are two types of data located in the serial data stream for Chrysler: values and voltage. Values is information that emulates tools like pressure gauges and thermometers. The problem with the values data fields is that in some cases the information in these data fields is the limp-in values rather than the true values of the sensors. Voltages

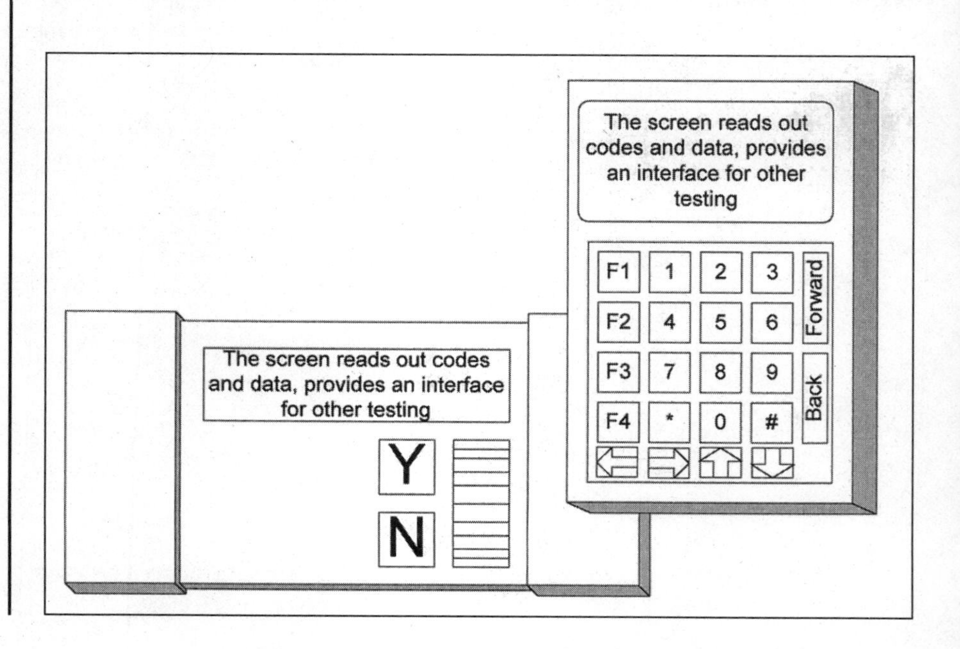

Scanners are available from several manufacturers. Most cost between $1,000 and $2,000. Do a little research, make a few calls, visit a few pawn shops; you can probably find a deal.

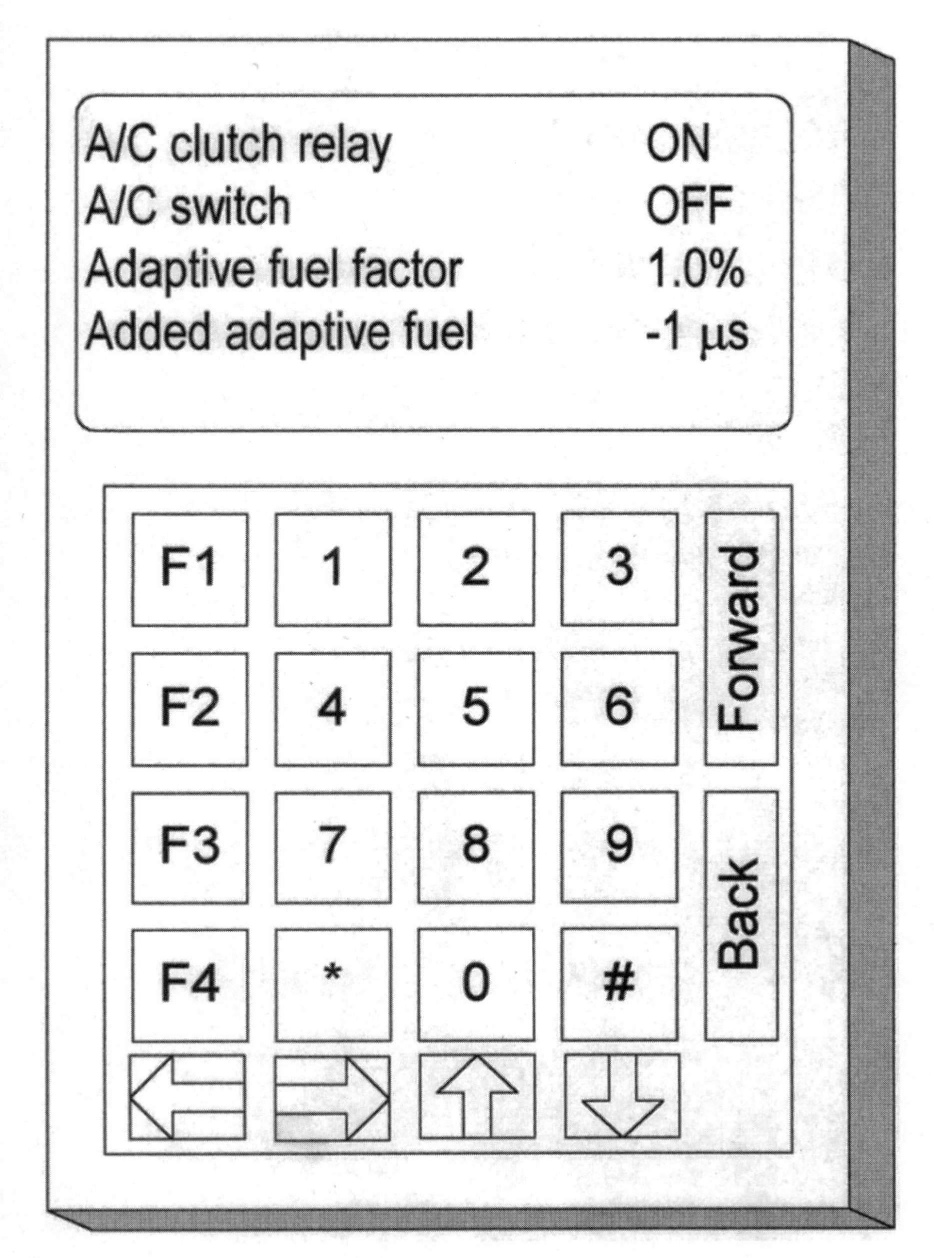

The A/C clutch relay is on; this indicates that the compressor clutch is engaged. What seems incongruent is that the A/C switch is turned off. The air conditioner compressor clutch will be on when the defroster is operating. The adaptive fuel factor indicates that there is a very slight short-term correction toward rich for air/fuel ratio. The added adaptive fuel shows that there is a slight long-term correction toward lean for air fuel ratio.

will be discussed later. For data fields where there are both values and voltage readings to be used, favor the accuracy of the voltage readings.

Values

A/C Clutch Relay

This data field informs the technician whether or not the computer is attempting to energize the A/C compressor clutch relay. If the computer is attempting to turn on the relay, this field will read on. If the computer is not attempting to do so, the scanner will read off.

A/C Switch

This data field reports to the scanner what the computer believes to be the state of the A/C on-off switch on the air conditioner control panel or climate control panel. When the technician has the switch on, that is what this data field will report. When the technician has the switch off, this data field will report off.

Adaptive Fuel Factor (AFF)

This is the short-term adjustment for air/fuel ratio problems made by the computer in response to oxygen sensor readings. This is one of the key pieces of serial data information. The data field is in units of percentage increase and decrease for injector on-time. With the engine warmed up and in closed loop, request that the scanner display the AFF function. Minus 10 percent to plus 10 percent is a normal reading. If the number displayed is greater than +10 percent, then it indicates the fuel injection system is responding to a request for more fuel. The oxygen sensor must be sensing a high oxygen content in the exhaust, which it interprets as a lean condition. The computer assumes the engine is running lean and enriches the mixture. Keep in mind that a lean air/fuel ratio is only one of the possible causes of a high oxygen content in the exhaust.

If the AFF number is less than -10 percent, it indicates that the engine is receiving extra fuel from somewhere. The AFF is therefore compensating by decreasing the amount of fuel that the injectors are delivering to the engine.

The AFF changes in direct response to changes in exhaust oxygen content. Whenever the engine is in open loop operation, the AFF number will be fixed at 0 percent. Once the engine enters closed loop operation, the AFF will be constantly changing to correct minor errors in air/fuel ratio.

If the AFF number is high, look for a source of extra oxygen in the exhaust (such as vacuum leak, restricted injector(s), cracked exhaust manifold, defective air pump upstream/downstream switching valve, mechanical engine problems, low fuel pressure, and defect in the ignition, including plugs, cap, rotor, and plug wires).

If the AFF number is low, look for a source of extra fuel (such as defective evaporative canister purge system, leaking injector, contaminated crankcase, and high fuel pressure).

Added Adaptive Fuel Time (AAF)

This data field reports the computer's long-term compensation for air/fuel ratio problems. The added adaptive fuel time is a long-term version of the AFF that is stored in a memory. The compensating that it does is for longer-term trends in air/fuel ratio. Because the AAF is stored in a memory, it even has the ability to compensate for air/fuel ratio problems while the engine is in open loop, providing the problem was detected the last time that the engine ran in closed loop. Just like the AFF, a higher number indicates that the computer is adding more fuel to compensate for a perceived lean condition. A lower number indicates the computer is compensating for a perceived rich condition. Any abnormal reading in the AAF indicates the perceived need to adjust air/fuel ratio has been there for a minimum of several seconds. Several seconds may not seem like a very "long-term" situation, but do not forget we are dealing with a computer. The units of measurement in this data field is microseconds. A reading of +10mS indicates that the injector on-time has been increased by 0.00001 seconds or 1/100 of a millisecond. A minuscule amount. Do not worry too much about this data field unless it is out of range at a level approaching 1,000.

Always keep in mind when troubleshooting that rich and lean on any modern oxygen feedback fuel injection system is determined by the oxygen content of the exhaust gases, not by fuel entering the engine.

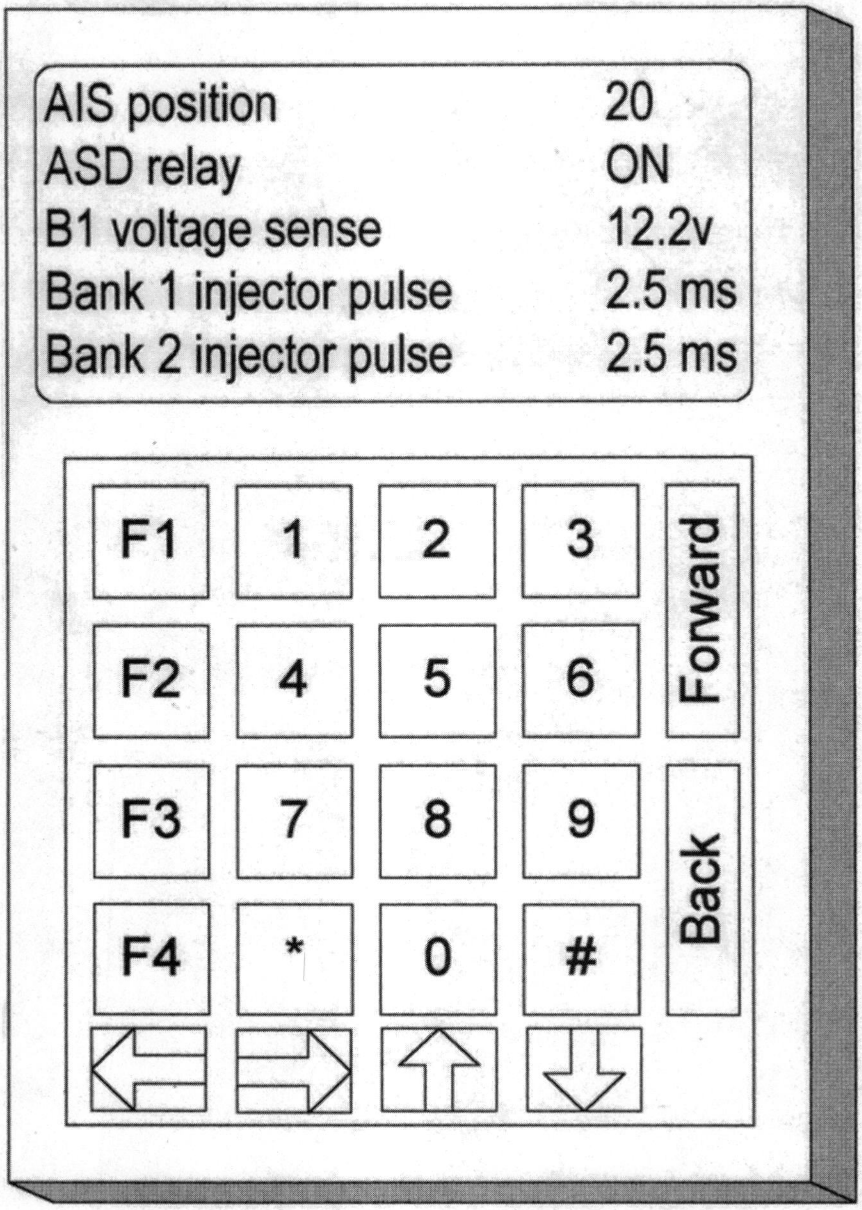

The AIS stepper motor is in a normal position for a warm engine idling with minimal load. The ASD relay is indicating on, as it should when the engine is running. B1 voltage sense is the main power supply wire for the engine control actuators. The 12.2 volts is a little below alternator output voltage, but not low enough to be of concern. In fact, this is a fairly normal reading. The injector pulse width is in the middle of normal.

Air Switch Solenoid

When the air switch solenoid data field says ON, it means that the computer is attempting to make the air pump deliver air upstream of the oxygen sensor. This is done when the computer is attempting to keep the oxygen sensor hot. On many Chrysler applications, especially light-duty trucks, air is pumped upstream when the engine is idling. Therefore, these applications will have 0 volts being delivered from the oxygen sensor to the computer when the engine is idling. When the computer senses that the engine is running under conditions where upstream air is no longer needed to keep the oxygen sensor warm, the computer will signal the air switch solenoid to deliver the air downstream of the oxygen sensor and this data field will report OFF.

AIS Motor Position

This is the position of the idle air control motor. There are 256 positions that this motor could be in. The lowest number is zero; the highest is 255. The lower the number is the slower the computer is trying to make the engine run. Seven to 30 are typical at idle for most cars when the engine is fully warmed up and the air conditioner compressor and other loads are turned off. If the number is at or close to zero, this could indicate a vacuum leak or incorrectly adjusted minimum air.

Auto Shutdown Relay

This data field reports the state of the auto shutdown relay. The scanner should indicate that this relay is on when the engine is being started and when the engine is running. Key on, engine off, this field should indicate that the ASD is off.

B1 Voltage Sense

This is the voltage delivered by the auto shutdown relay via the green/black wire to the injectors, the ignition coil, the alternator field, etc. This data field reports the voltage sampled by the computer on this wire. The reading should be at or near alternator charging voltage.

Bank 1 Injector Pulse Width

This is the injector on-time measured in thousandths of a second. At an idle, in fact at any sustained speed, the injector on-time for throttle body applications should be about 1.0 milliseconds. On multipoint injection systems that are gang- or group-fired, the injector on-time should be about 1.0 to 3.0 milliseconds. On sequential injection applications, the injector on-time should be 2.0 to 6 milliseconds.

Bank 2 Injector Pulse Width

Same as bank 1, except for bank 2.

Barometric Pressure

This is the value reading for barometric pressure. Barometric pressure is largely a matter of altitude. If you live where you can walk out your back door and go surfing, then expect the barometric pressure reading to be about 30 inches of mercury or about 101 kPa. On the other hand, if you live where you can walk out your back door and go snow skiing in a northern hemisphere in July, then expect this to read about 25 inches of mercury or about 80 kPa. Same as barometric pressure.

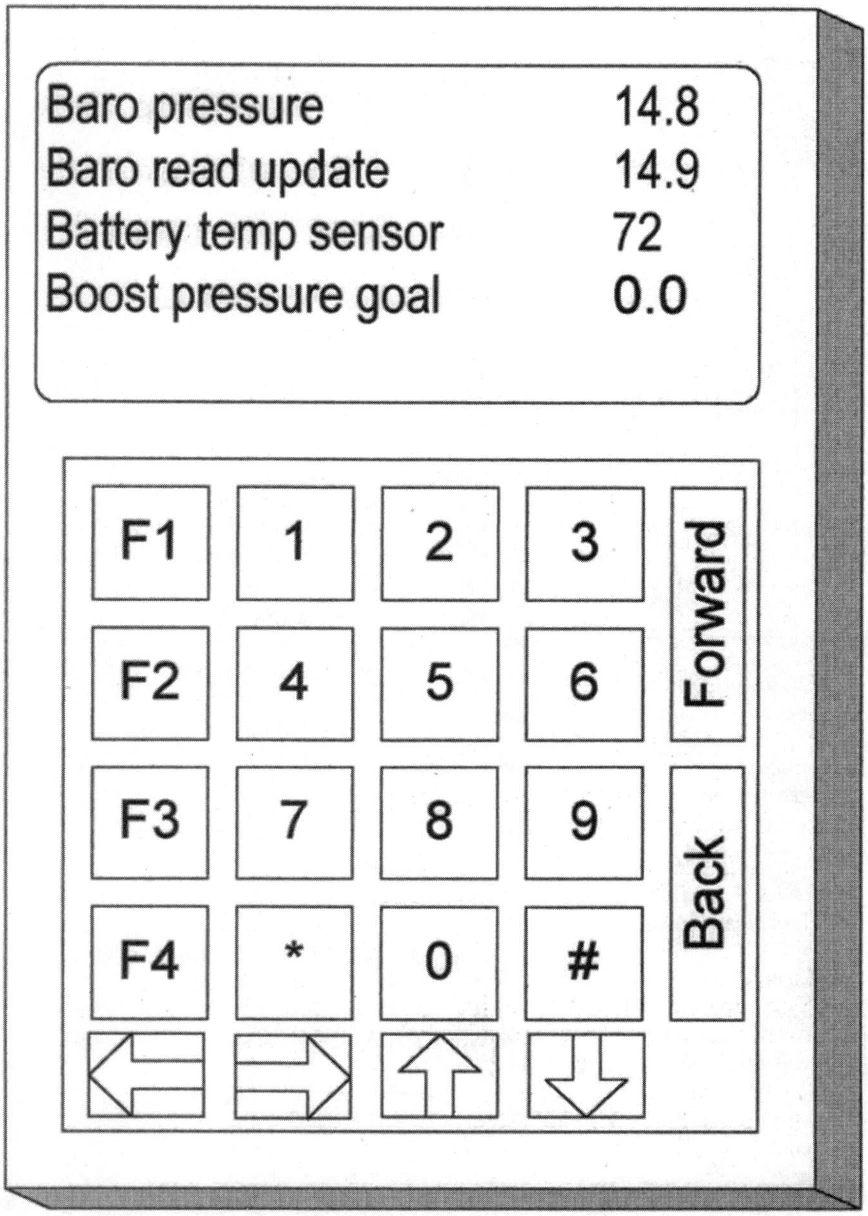

The computer is interpreting the reading from the barometric pressure sensor to be 14.8 psi. Normal barometric pressure is about 14.7 psi. Today, the atmospheric pressure must be slightly higher than normal. The baro read update data field is indicating 14.9 psi. Although it would seem logical that these two data fields be exactly the same, they often are not. The battery temperature sensor is indicating that the temperature of the air passing through the computer is 72 degrees. The computer is not attempting to let the turbocharger create boost.

Battery Temperature Sensor

This is the temperature sensed by the battery temperature sensor located in the computer. If the engine has yet to be started today, or if the engine has run only a short time, the reading should be almost the same as ambient temperature. If the engine has been running for a while, particularly at idle, the temperature displayed in this data field may be higher than ambient.

Boost Pressure Goal

This data field shows how much the computer is attempting to increase the intake manifold pressure above barometric. Boost pressure is created by the turbocharger.

Brake Switch

This data field reports the status of the brake pedal switch. When the scanner displays the word "LOW," it indicates that the brake pedal has been

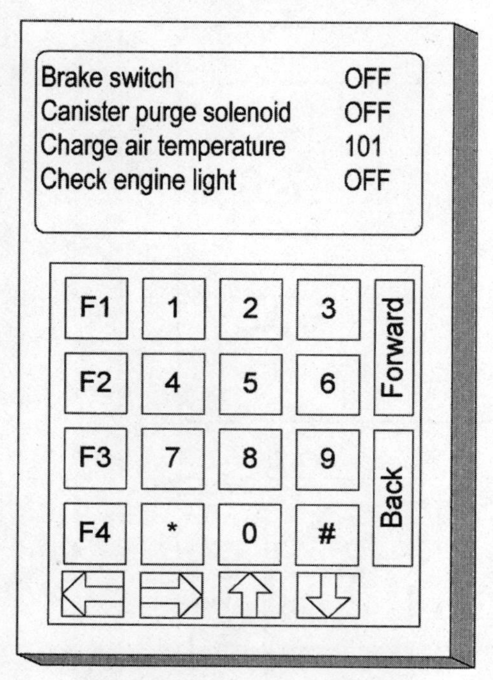

While the scanner was reading these data fields, the brakes were not applied. The evaporative canister is not currently being purged. The temperature of the air in the intake manifold is 101 degrees. There are no faults currently keeping the Check Engine light activated.

depressed and therefore the brake switch is closed.

Canister Purge Solenoid

This indicates the status of the evaporative canister purge solenoid. The data field should only indicate "ON" when the engine is warm or the vehicle is being driven down the road.

Charge Air Temperature

The charge air temperature is the temperature of the air in the intake manifold. The reading should be about the same as ambient when the engine is cold. As the engine runs and warms, the reading should rise above ambient but should remain within about 50 degrees of ambient. When the intake manifold pressure is boosted well above barometric, the temperature of the charge air can easily rise above the boiling temperature for water.

Note: The 1988-1989 turbo models do not have a charge temperature sensor. This data field is therefore calculated by the computer based on coolant temperature.

Check Engine Light

If this data field says "ON," then the Check Engine light should be illuminated.

Closed Throttle Switch

When this data field says "ON," it means that the computer believes that the throttle is closed.

Coolant Temperature

This data field shows the coolant temperature that the computer is using as a basis for fuel metering and emission control calculations. Observe this reading for several seconds before passing judgment on how acceptable the reading is. When there is a defect in the coolant temperature sensing circuit, the computer will display a believable but steady reading. For example, a reading of 104 degrees F could mean that the engine temperature is passing through 104 degrees on its way to operating temperature, or it could mean that the computer is reporting a default temperature to the scanner.

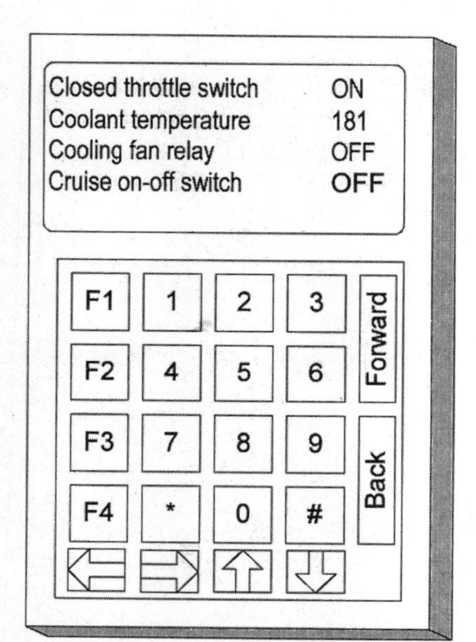

The data field that indicates the closed throttle switch is on indicates that the throttle is closed. For the most part, this is a data field that is only available on carbureted applications. The coolant temperature is 181 degrees F. Many scanners are set up to read in Celsius but can usually be adjusted to read Fahrenheit. The radiator cooling is not being requested to turn on by the computer, and the operator has the cruise control turned off.

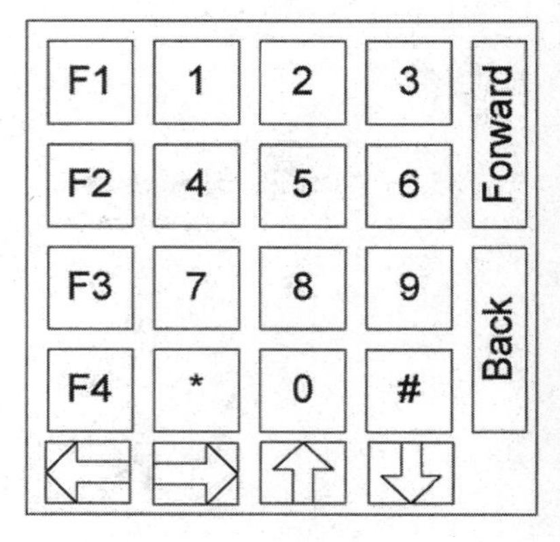

The vehicle operator is not signaling the computer to resume the pre-set cruise control speed. The Set switch is not depressed, and no target speed has yet been set. The cruise control vacuum solenoid is off. Basically, these are the readings one would expect when the vehicle is sitting still. The technician can, however, engage the switches at this time to test their operation.

Cooling Fan Relay

The scanner will be told to report "ON" when the computer is attempting to make the engine cooling fan turn on.

Cruise On/Off Switch

This shows the status of the cruise control on/off switch.

Cruise Resume Switch

This shows the status of the cruise control resume switch.

Cruise Set Switch

This shows the status of the cruise control set speed switch.

Cruise Target Speed

The cruise target speed is the speed that the driver of the vehicle has set as the desired speed for the vehicle. It is the speed at which the computer is attempting to make the vehicle cruise.

Cruise Vacuum Solenoid

"ON" in this data field means that the computer is commanding the cruise control servo to accelerate the vehicle.

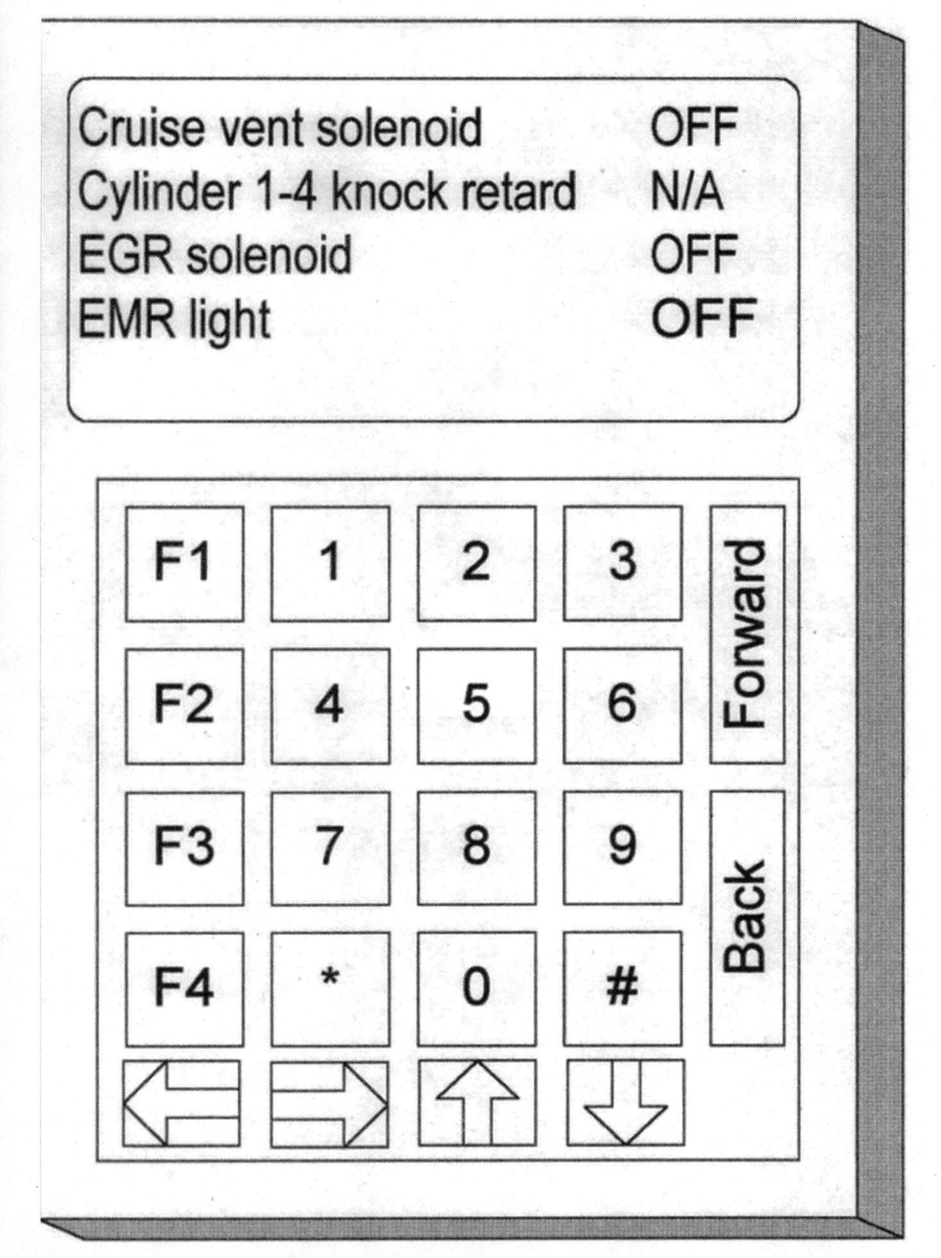

The cruise control vent solenoid data field should indicate on only when the cruise servo is being told by the computer to decelerate the engine. This application was not equipped with a knock sensor, which is why the 1-4 knock retard data field indicates non-applicable. The solenoid-operated vacuum control valve used to operate the EGR valve is off, not allowing the EGR to be opened. The EMR light is the Emission Maintenance Reminder light. This light will come on when the engine operation time that the computer feels equates to about 50,000. The light must be turned out with a scanner.

Cruise Vent Solenoid

"ON" in this data field means that the computer is commanding the cruise control servo to decelerate the vehicle.

Cylinder 1-4 Knock Retard

When the knock sensor detects a detonation, the computer responds by retarding the ignition timing. The computer retards the timing in 5-degree increments until the detonation stops. Once the detonation has stopped, the computer will advance the timing in 2-degree increments until the normal calculated timing for the current rpm and load conditions is reached. This data field displays the amount of retard that the computer has initiated in response to the detonation sensor.

EGR Solenoid

The EGR valve has been considered a nuisance and power thief since its introduction in the early 1970s. It is true that the EGR valve renders approximately 7 percent of the combustion chamber volume useless for the creation of power and therefore should be eliminated for track use where power, and not emissions, is the prime consideration. The average car owner/operator will not feel a significant increase in performance or economy when the EGR valve is eliminated. In fact, he/she may sense a loss of power. The EGR valve reduces nitrogen oxide emissions by using inert recirculated exhaust to cool the combustion process. This cooling reduces the tendency for the engine to rob power. This data field reports the computer's control of the EGR control solenoid. When the solenoid is energized to open the EGR valve, the field will say "ON."

EMR Light

The computer on some Chrysler applications is designed to turn on an Emission Maintenance Reminder light after the vehicle has been driven a certain number of miles. This mileage varies somewhat by application and year model.

Engine Speed

This is what the computer believes to be the current engine speed.

Idle Speed

This is the speed at which the computer will attempt to make the engine run.

Injector Pulse Width

This is the injector on-time measured in thousandths of a second. At an idle, in fact at any sustained speed, the injector on-time for throttle body applications should be about 1.0 milliseconds. On multipoint injection systems that are gang- or group-fired, the injector on-time should be about 1.0 to 3.0 milliseconds. On sequential injection applications, the injector on-time should be 2.0 to 6 milliseconds.

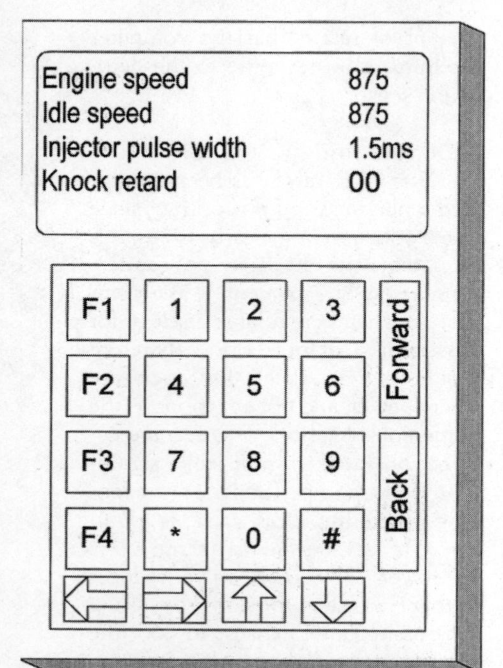

The four data fields shown here include the current engine speed of 875 rpm. In this case, that is exactly the engine speed at which the computer is attempting to make the engine idle. The data field, called on this scanner simply Idle Speed, is the speed at which the computer will attempt to make the engine idle when the throttle position sensor indicates the throttle is closed. The current injector pulse width is 1.5 milliseconds. The computer is not retarding the ignition timing based on knock sensor inputs.

Knock Retard

See cylinder 1-4 knock retard.

Knock Sensor

Same as knock retard.

Manifold Absolute Pressure

This is the reading of manifold pressure displayed in either inches of mercury pressure or kiloPascals. This, like the coolant temperature data field, is one of those that can be misleading. Imagine that you start the engine and allow the engine to idle. You would expect the manifold pressure to read about 12 inches of mercury. When the throttle is snapped, you would expect the pressure to rise to nearly 30 inches of mercury then fall to well below 12 when the throttle is allowed to snap shut. Eventually, the manifold pressure would level off at 12 inches of mercury.

Here is the weird part. Disconnect the vacuum hose on the MAP sensor. You would expect the manifold pressure data field to read a constant 30 inches of mercury. What will happen is that the engine will die. When restarted, manifold pressure will read about 12 inches of mercury. When the throttle is snapped, the pressure will rise to nearly 30 inches of mercury then fall to well below 12 when the throttle is allowed to snap shut. Eventually, the manifold pressure would level off at 12 inches of mercury.

Now disconnect the electrical connector and start the engine. The manifold pressure will read about 12 inches of mercury. When the throttle is snapped, pressure will rise to nearly 30 inches of mercury then fall to well below 12 when the throttle is allowed to snap shut. Eventually, the manifold pressure would level off at 12 inches of mercury.

What is displayed in the data field on the scanner is what the computer wishes the MAP sensor was doing. What you are seeing are the default readings based on TPS and rpm. If the MAP sensor circuit is good, these readings are real. If the MAP sensor circuit is defective, these readings will look real. Always use the MAP voltages when troubleshooting.

Manifold Vacuum Reading

This is the reading of manifold pressure displayed in either inches of mercury vacuum or kiloPascals. This, like the coolant temperature data field, is one of those that can be misleading. Imagine that you start the engine and allow the engine to idle. You would expect the manifold vacuum to read about 18 inches of mercury. When the throttle is snapped, you would expect the vacuum to drop to nearly 0 inches of mercury then rise to well above 18 when the throttle is allowed to snap shut. Eventually, the manifold vacuum would level off at 18 inches of mercury.

Here is the weird part. Disconnect the vacuum hose on the MAP sensor. You would expect the manifold vacuum data field to read a constant 0 inches of mercury. What will happen is that the engine will die. When restarted, manifold pressure will read about 18 inches of mercury. When the throttle is snapped, the vacuum will drop to nearly 0 inches of mercury then rise to well above 18 when the throttle is allowed to snap shut. Eventually, the manifold pressure would level off at 18 inches of mercury.

Now disconnect the electrical connector and start the engine. The manifold vacuum will read about 18 inches of mercury. When the throttle is snapped, the inches of mercury will fall to nearly 0 inches of mercury then rise to well above 18 when the throttle is allowed to snap shut. Eventually, the manifold pressure would level off at 18 inches of mercury.

What is displayed in the data field on the scanner is what the computer wishes the MAP sensor was doing. What you are seeing are the default readings based on TPS and rpm. If the MAP sensor circuit is good, these readings are real. If the MAP sensor circuit is defective, these readings will look real. Always use the MAP voltages when troubleshooting.

Manifold absolute pressure 9.0
Manifold vacuum 18.0
Min air flow idle speed 650
Minimum throttle reading 650

The manifold absolute pressure is currently 9.0 inches of mercury pressure. This equals about 1/3 of atmospheric pressure and is a typical idle reading for this data field at an idle. Manifold vacuum is about 18 inches of mercury, which is the reading that would be read on a vacuum gauge connected to the intake manifold. When the automatic idle speed (AIS) control stepper motor is moved to the fully extended position, the computer will be expecting the engine to idle at 650 rpm.

Minimum Airflow Idle Speed

This is the desired idle speed when the automatic idle speed control motor is closed off all the way. The automatic idle speed motor (or AIS) is a stepper motor-controlled valve which the computer moves in order to control the speed of the engine at idle. The AIS can be moved to any one of 256 positions by the computer to ensure the correct idle speed regardless of changes in engine load due to the transmission, power steering, alternator, air conditioning compressor, or anything else. At an idle, the AIS will be at about position 20 with no loads on the engine. As engine loads increase, the rpm will tend to drop. As the rpm drops, the computer steps the AIS to a more open position (higher number). As the rpm decreases, the AIS is stepped in. AIS position is displayed through serial data and is an important piece of troubleshooting information.

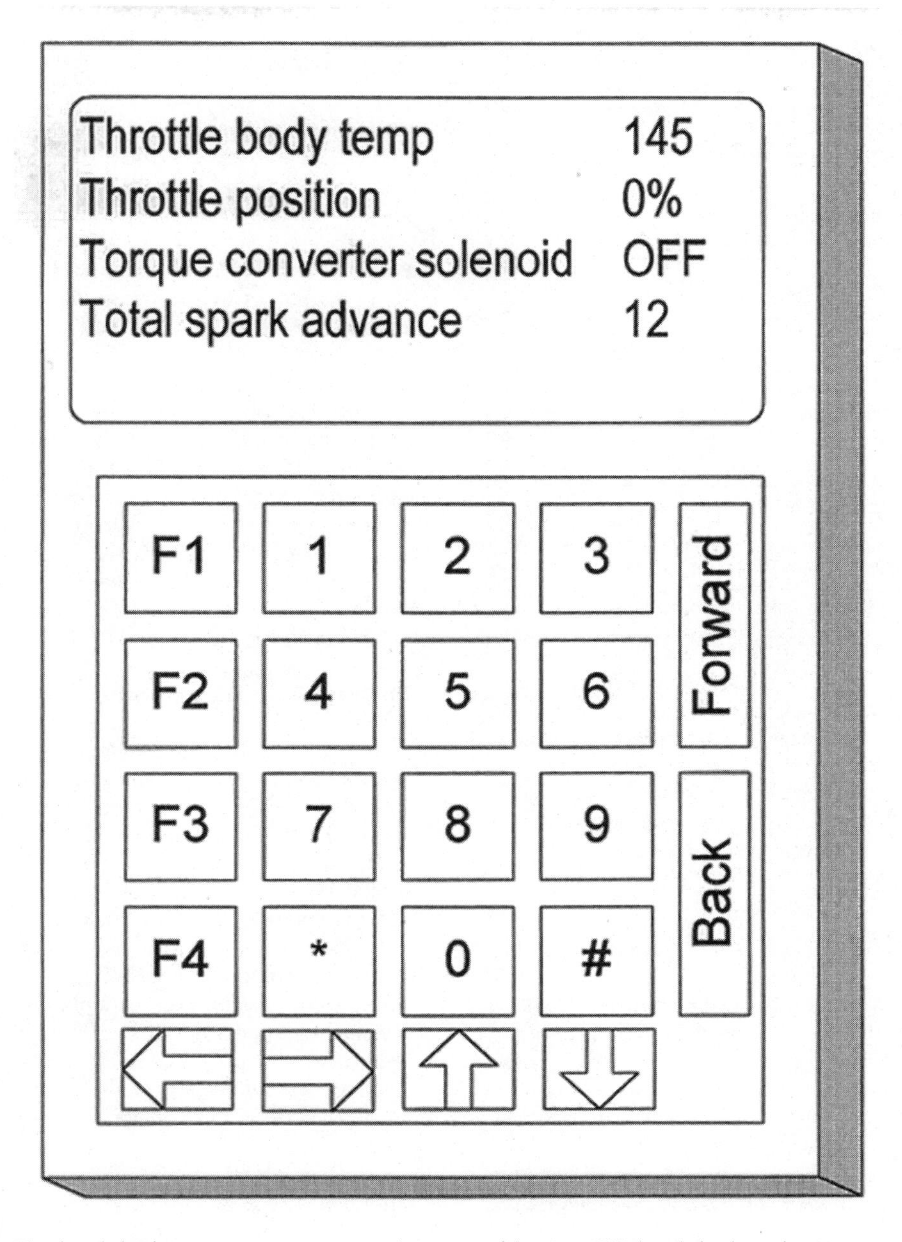

The throttle body temperature sensor is used on many of the post-1985 throttle body applications. These systems use low fuel pressure, and there is a possibility of fuel percolation when the engine is shut off and allowed to go into hot soak. When the driver attempts to restart the engine, the fuel in the throttle body unit will no longer be a liquid but rather a vapor. When the temperature of the throttle body is greater than 179 degrees, the computer will double-pulse the injector while the starter is engaged. This clears the vapor out of the injector quickly to ease start-up. The throttle position sensor is in the 0 percent or fully closed position. This data field will read about 50 percent when the throttle is half open and 100 percent with pedal to the metal. The solenoid that allows oil to flow in the transmission to lock up the torque converter has not been energized by the computer. Total spark advance is 12 degrees beyond initial.

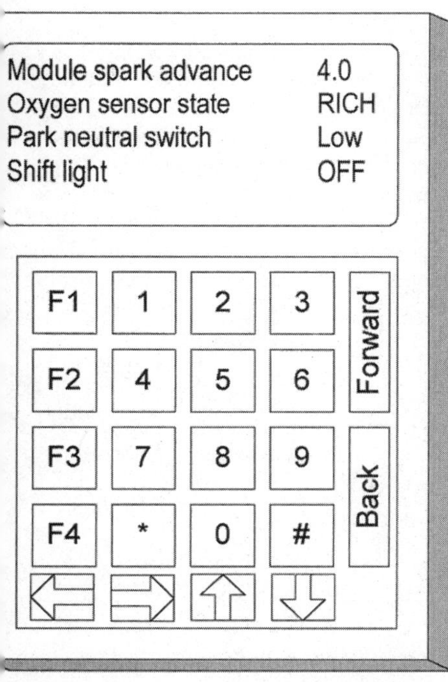

Module spark advance indicates the timing is being advanced 4 degrees. The current average oxygen sensor voltage is above 0.45 volts. The oxygen content of the exhaust gases is low, and therefore the computer believes the engine is running rich. The park/neutral switch is indicating low. This means that the voltage being received by the sensor is low, not that the transmission is in low gear. The shift light is used on manual shift applications to signal the driver that he/she should shift to the next higher gear, thereby maximizing fuel economy.

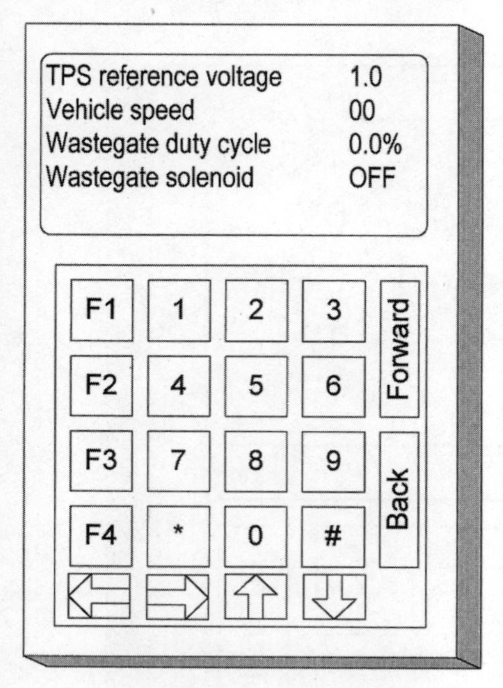

The throttle position sensor (TPS) reference voltage is 1.0 volts. This is a measurement of the minimum voltage seen by the computer from the TPS. The vehicle was currently traveling at 0 miles per hour. The wastegate for the turbocharger is currently not being used to control turbo boost.

Minimum Throttle Reading

This data field shows the lowest voltage seen by the computer from the throttle position sensor. This reading is the reading used by the computer as the closed throttle voltage.

Module Spark Advance

This is the amount of ignition timing advance being provided by the computer.

Oxygen Sensor State

The oxygen sensor produces a voltage that ranges from 100 to 900 millivolts during operation. When the oxygen sensor voltage is averaging below 0.45 volts, the oxygen content of the exhaust is high and the computer assumes that the engine is running lean. This data field will display LEAN. If the oxygen sensor voltage is averaging above 0.45 volts, the oxygen content of the exhaust gases is low and the computer assumes the engine is running rich. The scanner will display RICH in this data field. If the Scanner displays MID or CENTER, then it means that the oxygen sensor is detecting average voltages at very close to 0.45 volts.

Park Neutral Switch

LOW will be displayed in this data field when the vehicle's gear selector is in either park or neutral.

Part Throttle Unlock Solenoid

This relates to the transmission's lock-up converter.

Shift Light

Now this is a technology that I personally truly despise. The shift indicator light is turned on by the computer when you have reached an engine speed/load combination that warrants shift to a higher gear. I hate this data field because it tells me to shift before the valves begin to float. What fun is that?

Throttle Body Temperature

The throttle body temperature signal is used by the computer to determine if the engine is in a hot-soak condition. This sensor is used on many of the low-pressure throttle body injection systems. The reading should be about the same as ambient when the engine is cold. As the engine runs and warms, the reading should rise above ambient but should remain within about 50 degrees of ambient. When the temperature of the throttle body rises above 178 degrees F, it is assumed that the fuel may have boiled in the throttle body. To ease starting, the computer double-pulses the injector to clear out the potentially percolated fuel.

Throttle Position

This data field should read 0 percent when the throttle is closed and 100 percent when the throttle is to the floor. As the throttle is moved forward, the percentage should rise proportionally to the movement from 0 percent to 100 percent.

Torque Converter Solenoid

This relates to the transmission's lock-up converter.

Total Spark Advance

This is the amount of ignition timing advance being provided by the computer.

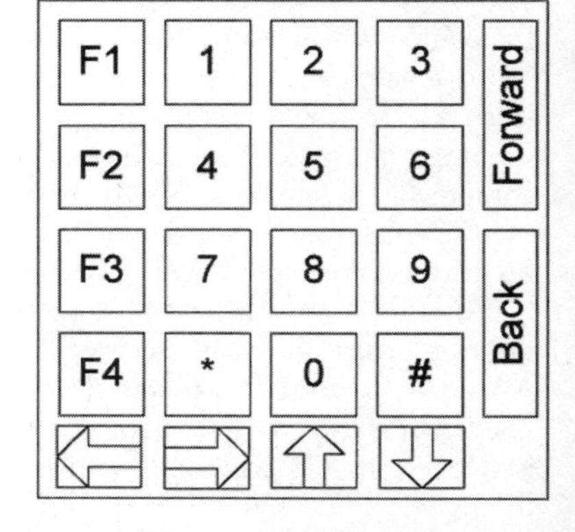

The ambient air temperature sensor and the battery temperature sensor are indicating that the temperature of the air is about half range. The industry standard for minimum temperature to be measured is 40 degrees below zero. The maximum temperature is in the neighborhood of 160 degrees. Since the readings are 2.6 volts, it can be assumed that the ambient temperature is about half way between -40 degrees and 160 degrees, or about 60 degrees F. The battery volts data field relates the computer's perception of battery voltage. The normal reading is often about 1 volt below the actual terminal voltage of the battery. The coolant temperature sensor range of measurement is between -40 degrees and 260 degrees F. On the scanner shown here, the reading is a little below half of the reference voltage. This means that the temperature being detected is about halfway above the minimum measurement of the temperature range. Since the total range is 300 degrees, the temperature being measured is greater than halfway above the mid part of the range. Minus 40 plus 150 is about 110 degrees. Therefore, the temperature of the engine must be a little greater than 110 degrees.

TPS Reference Voltage

This field displays the amount of voltage supplied to the TPS. The reading should be 5 volts.

Vehicle Speed

This is the current vehicle speed.

Wastegate Duty Cycle

This data field corresponds to the amount of wastegate opening being commanded by the computer. As turbo boost increases, the computer needs to limit that boost. If the amount of boost were allowed to go too high, serious damage could occur to the engine. As the amount of boost rises, the computer increases the duty cycle to the wastegate solenoid in an attempt to bypass exhaust gas turbine pressure and limit compressor output pressure. Zero percent means that the solenoid is closed. One hundred percent means that the solenoid is fully open.

Wastegate Solenoid

This is like the wastegate duty cycle but it only returns an "ON" for an open solenoid or an "OFF" for a closed solenoid.

Voltages

Ambient Air Temperature Volts

At 70 degrees F, this data field should read about 2.5 volts. The reading should be a little lower when the temperature is higher and a little higher when the temperature is lower.

Battery Temperature Sensor Volts

At 70 degrees F, this data field should read about 2.5 volts. The reading should be a little lower when the temperature is higher and a little higher when the temperature is lower.

Battery Volts

This should read about the same as the reading on a voltmeter across the terminals of the battery.

Charge Air Temperature Sensor Volts

At 70 degrees F, this data field should read about 2.5 volts. The reading should be a little lower when the temperature is higher and a little higher when the temperature is lower.

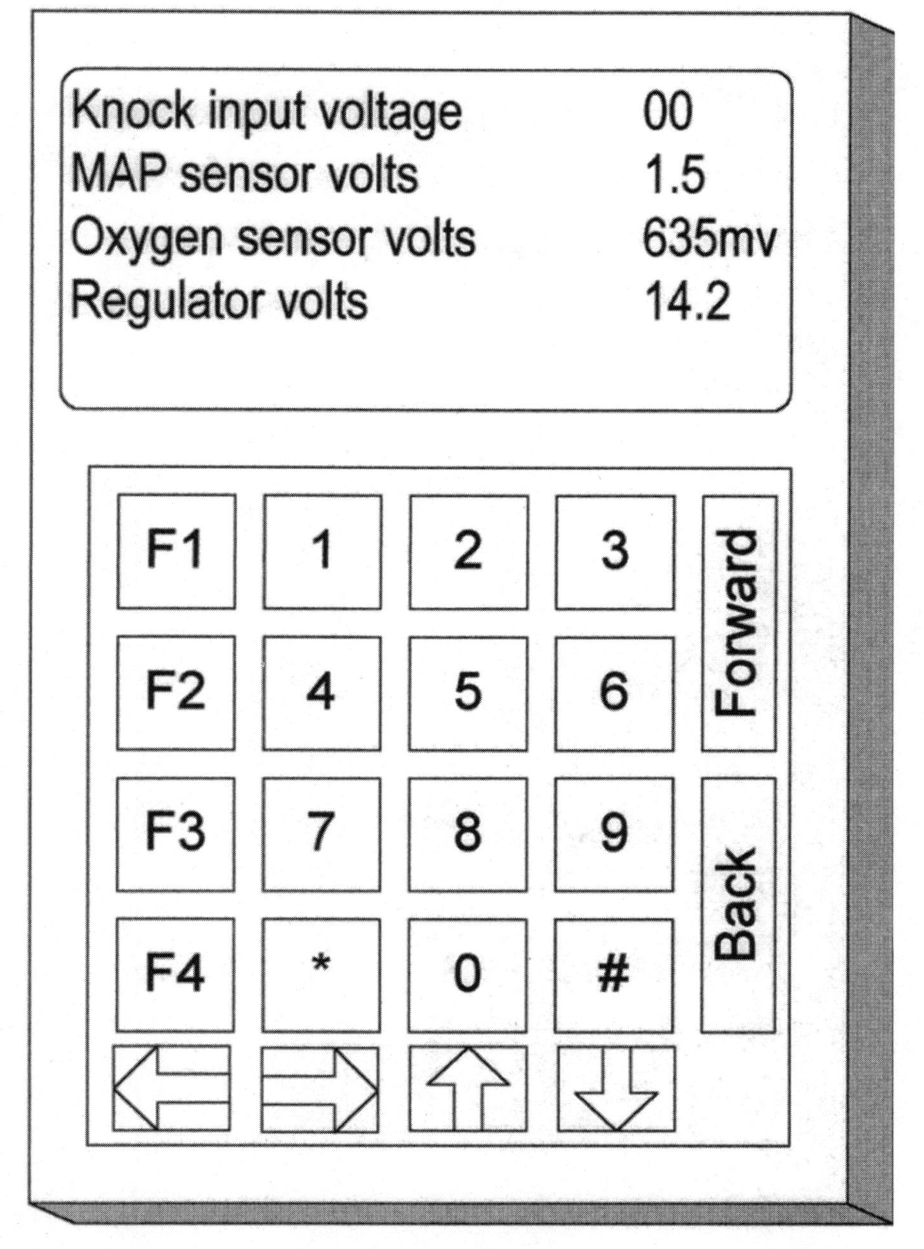

The knock sensor is not detecting any detonations. Therefore, the voltage shown on the scanner is zero. The MAP sensor reading is 1.5 volts. This is about 1/3 of what the reading would have been before the engine is started. When the engine is revved, the voltage will rise. The oxygen sensor voltage should be constantly toggling above and below 0.45 volts when the engine is running and thoroughly warmed up. The regulator volts data field tells the technician the alternator output voltage that the computer's alternator voltage regulator is attempting to achieve.

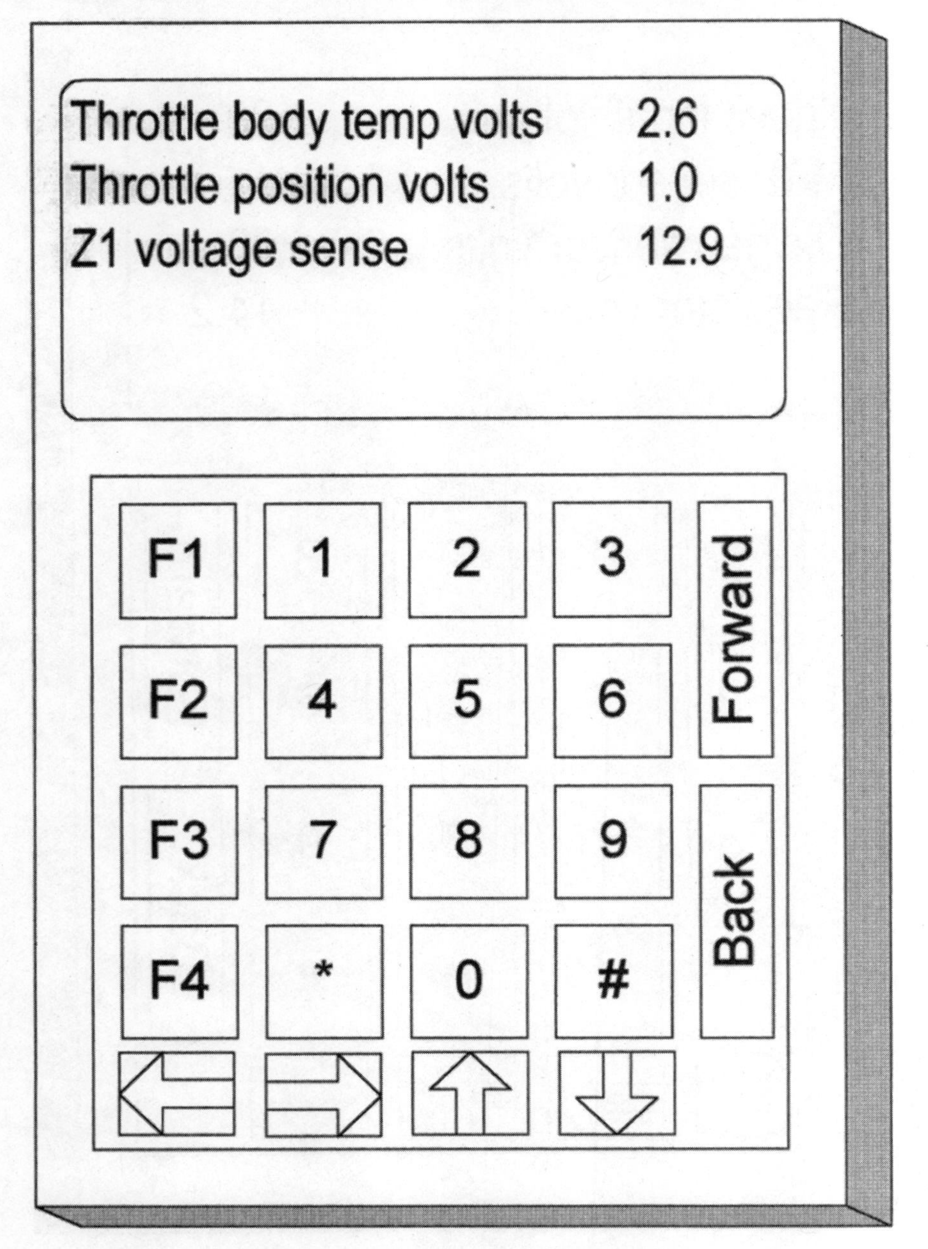

The throttle body temperature sensor range of measurement is between -40 degrees and 260 degrees F. On the scanner shown here, the reading is a little above half of the reference voltage. This means that the temperature being detected is about halfway above the minimum measurement of the temperature range. Since the total range is 300 degrees, the temperature being measured is greater than halfway above the mid part of the range. Minus 40 plus 150 is about 110 degrees. Therefore, the temperature of the throttle must be about 110 degrees. The throttle position sensor is at the normal closed throttle voltage. Z1 volts is the main power supply voltage to the ignition coil, the injector(s), the alternator field, and many other actuators. The reading should be approximately battery terminal voltage.

Coolant Temperature Sensor Volts

The coolant temperature sensor circuit behaves a little differently than the electrical circuits with which most automotive technicians are familiar. A power supply (usually 5 volts) supplies a reference voltage to the circuit. Before leaving the computer, the 5-volt current passes through a fixed value resistor causing a voltage drop. The current then continues through the coolant temperature sensor and on to ground where the voltage is zero. As the resistance of the coolant temperature sensor changes, the voltage on the wire between the fixed value resistor and the coolant temperature sensor will also vary. The computer measures this voltage on the outbound side of the fixed value resistor to determine the temperature.

When the computer sees a comparatively high voltage on the wire to the NTC coolant temperature sensor, it knows that the resistance in the coolant temperature sensor is high; therefore, the temperature of the coolant temperature sensor is low. A low voltage on this wire means that the resistance is low and therefore the temperature must be high.

Chrysler throws a bit of a curve ball on this circuit. Applications after 1986 may have a dual-stage coolant temperature sensor circuit. When the engine is cold, about -40 degrees F, the voltage on the signal wire is close to 5 volts. As the engine warms, the voltage on the signal wire drops rapidly. When the engine temperature rises to about 120 degrees F, the voltage on the signal wire is sitting at about 1.2 volts. At about this point the voltage will suddenly rise to over 3 volts. The Chrysler fuel injection computer uses a two-stage coolant temperature circuit. This trick is performed by placing a second internal resistor in the current path to the coolant temperature sensor. When the engine is cold, the computer uses the current path that contains a 10,000-ohm resistor. When the temperature of the engine reaches about 120 degrees, the computer switches to a 909-ohm resistor. This causes the voltage on the sensor signal wire to rise to a little over 3 volts. Well that is interesting, but what is the reason for doing all this? In a standard coolant temperature sensing circuit, there is realistically 5 volts, actually a little less, in which to measure about 300 degrees of temperature change. By using a two-stage unit, the circuit has effectively 10 volts in which to measure the 300 degrees of change. This means that each volt of change represents about 30 degrees instead of about 60 degrees. This means more accuracy.

Knock Input Voltage

This data field displays the input voltage to the computer from the knock sensor. When the engine is idling, and at other times when the engine is not detonating, the reading should be zero. When a detonation does occur, the voltage should be greater than zero.

MAP Sensor Voltage

The MAP sensor should read about 4.5-4.9 volts key on, engine off. At high altitude, the reading may be lower than 4.5 volts. When the engine is started and allowed to idle, the reading should drop to about one-third of the voltage displayed with the engine not running. If you are working on a turbo application, cut these expected voltage readings in half. A reading of 5.0 or close to 5.0 indicates that there is an open circuit in the wiring harness to or from the MAP sensor.

Oxygen Sensor Voltage

The oxygen sensor might be described as a chemical generator. When it is heated to a minimum of 600 degrees F, it will begin to produce a voltage ranging from 100 to 900 millivolts. Once operational temperature is reached, the sensor will begin to respond to changes in the content of exhaust oxygen. When the oxygen content of the exhaust is high, the computer assumes that the engine is running lean. The design of the oxygen sensor is such that it will produce a low voltage when the exhaust oxygen content is high. Oxygen content in the exhaust gases resulting from a lean combustion will result in a voltage less than 450 millivolts being delivered to the computer. When the exhaust gases result from a rich combustion, the oxygen sensor voltage to the computer will be greater than 450 millivolts. When the oxygen sensor voltage is indicating a lean condition, the computer will respond by enriching the mixture. When the oxygen sensor voltage is high, the computer will respond by leaning out the mixture. In this manner, the computer adjusts for minor errors and variations from the rest of the input sensors and controls the air/fuel ratio at 14.7:1.

Regulator Voltage

This is the target charging system output voltage.

Throttle Body Temperature Volts

At 70 degrees F, this data field should read about 2.5 volts. The reading should be a little lower when the temperature is higher and a little higher when the temperature is lower.

Throttle Position Sensor Volts

If the signal voltage is between .5 and 1 volt, gradually open the throttle and monitor the voltage. The voltage should gradually increase to more than 4 volts. If the increasing voltage hesitates or drops as the throttle is opened, replace the TPS.

Z1 Voltage Sense

This is the signal to the computer that ensures the ASD relay has closed. The signal should be high when the engine is being cranked and when the engine is running.

THE ACTUATOR TEST MODE

CONTENTS

The ATM mode is one of the most unique computer-based test procedures that Chrysler has in its repertoire. This test procedure is initiated when the scanner tells the computer to enter the mode. There is no way to enter the procedure without the use of a scanner. Once the ATM mode is entered, the computer will activate the selected test for several minutes or until the technician tells the scanner to stop. The purpose of this test is to confirm the computer's ability to control its actuators.

Never attempt to run any of the following tests with the engine running!

Ignition Test

This test should be performed when the engine will not start. Remove the coil wire from the distributor cap. Situate the end of the coil wire so that it is about 1/4 inch from a good ground. Initiate the ATM ignition test. The computer should fire the ignition coil once every two seconds. The spark will be weak and yellow. If no spark is detected, connect a test light to the negative terminal of the ignition coil. If the test light pulses, replace the ignition coil. If the light was not illuminated, move the light to the positive terminal of the ignition coil. Is the light lit? If it is, measure the resistance of the ignition coil primary. If the resistance is greater than 10 ohms, replace the ignition coil. If the resistance is less than 10 ohms, inspect the wire that runs from the negative terminal of the ignition coil to ground at the ignition module inside the computer. If this wire is not grounded to the chassis, replace the computer.

If the test light did not light on the positive terminal of the ignition coil, repair the green/black wire that runs between the coil and the ASD.

Note: Wire colors do change from one application to another. While this text does make an effort to discuss the wires in as universal a way as possible, it may be necessary to alter

The computer controls virtually every component and function under the hood. With the use of a scan tool, a technician or enthusiast can request that the computer activate selected actuators.

The first actuator test option that comes up on most scanners is the ignition test. Remove the coil wire from the distributor cap and hold it 1/4-inch from ground.

the aforementioned procedure to compensate for change in wire color.

Injector Test

This test should be performed when the engine will not start. Initiate the ATM injector test. On throttle body applications, just look for pulsing fuel. On multipoint applications, use a mechanic's stethoscope or a piece of heater hose held to the ear and on the injector being tested. The computer should pulse the injector or injectors once every two seconds. If no pulse is detected, connect a test light to the negative terminal of the injector. If the test light pulses, replace the injector. If the light was not illuminated, move the light to the positive terminal of the injector. Is the light lit? If it is, measure the resistance of the injector. If the resistance is greater than 15 ohms, replace the injector. If the resistance is less than 15 ohms, inspect the wire that runs from the negative terminal of the injector to ground at the computer. If this wire is not grounded to the chassis, replace the computer.

If the test light did not light on the positive terminal of the ignition coil, repair the green/black wire that runs between the coil and the ASD.

Note: Wire colors do change from one application to another. While this text does make an effort to discuss the wires in as universal a way as possible, it may be necessary to alter the aforementioned procedure to compensate for change in wire color.

AIS Motor Test

The AIS is either a stepper motor that can extend and retract to any of 256 different positions or it is a DC motor that can extend or retract a plunger to control the position of the throttle. When you enter the AIS ATM test, the computer will extend and retract the plunger. Use a mechanic's stethoscope on the stepper motor type; simply observe the DC motor type. The motor should move the plunger in and out every two seconds. If the stepper motor fails to move, connect a test light to each of the four terminals, one at a time. While watching the test light as it is attached to each terminal, turn power accessories on and off. The accessories will either load the engine directly or load the alternator, which will load the engine. Each of the four terminals should turn the test light on and off. If a terminal does not, inspect the wire for that terminal back to the computer. If the wire is good, disconnect the stepper motor electrical connector. Check the resistance of each of the four stepper motor terminals to ground. Then measure the resistance across pairs of terminals. If the resistance is infinity, replace the computer. If one terminal shows continuity to ground, replace the stepper motor. If you are unable to find two pairs of wires having just a few ohms of resistance, replace the stepper motor. If the resistance checks of the stepper motor show that it is good and the wires to the computer are good, replace the computer.

While in the ignition actuator test mode, all of the components involved in creating the spark are utilized, except the Hall Effects.

It is tempting to blame the ignition coil when the ATM mode for ignition fails to spark the coil wire to ground. Remember, however, that in addition to the coil, problems with the wiring and the computer can affect the operation of the ignition coil.

Alternator Field Test

The easiest way to perform this test is to place a test light on the F2 (field ground) terminal of the alternator. Once in the ATM alternator field test mode, the computer should pulse the test light on and off every two seconds. This indicates that the voltage regulator function of the computer is working.

Remember the code 41 test procedure? The most effective way to troubleshoot a code 41 is with a scanner. However, a second best method would be to attach a frequency counter, tachometer, or dwell meter to the field control wire of the alternator. Which one is the field control wire? It is probably the terminal with the dark green wire connected to it. There are two

On later-model applications equipped with distributorless ignition systems, the scanner has the ability to control the spark through each coil individually. Each coil sends spark to two cylinders simultaneously. An ignition coil in a distributorless application should only be considered a possible cause of a driveability problem when two cylinders are not contributing power.

There is an ATM test for the injector or injectors. On late-model multipoint injection systems, the technician can choose which injector to fire individually.

Each time the computer activates the injector, it is opened for only a few microseconds. As a result, the computer may be left in the ATM injector test mode for several minutes with little risk of flooding the engine or fouling a spark plug.

small wires on the back of the alternator. One of these wires is field power; the other is field control. With the engine running, connect a voltmeter to each of the smaller terminals, first one then the other. One terminal should read almost the exact voltage that is found across the terminals of the battery. The other should be slightly lower. If they are both the same, there is probably an open between the alternator and the computer. Next choose a meter that can detect and confirm a pulsing DC voltage. Connect this meter to the field control wire, the one with the lower voltage. If both terminals had the same voltage, try to detect a pulse on both wires. On the frequency counter, one of the wires should read a frequency greater than zero. On the tachometer, one of the wires should read an "rpm" greater than

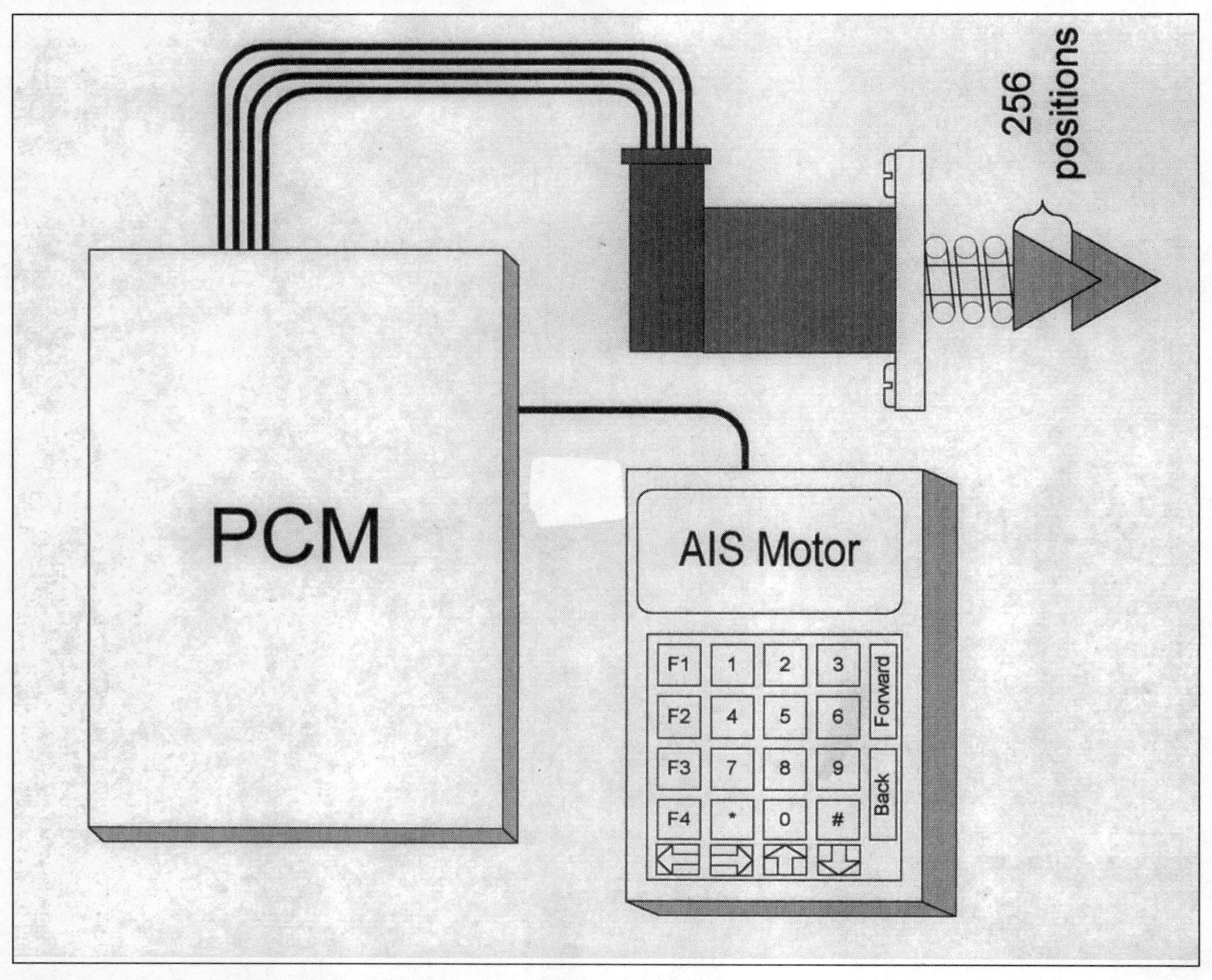

The scanner can also test the computer's ability to control the automatic idle speed (AIS) motor. Shown in this drawing is the stepper motor commonly used on late-model Chrysler applications. The ATM mode will cause the stepper motor to move in and out every two seconds. Usually, you will not be able to hear this movement without the assistance of a mechanic's stethoscope.

zero. On the dwell meter (set on the four-cylinder scale), one of the wires should read a dwell greater than 0 degrees but less than 90 degrees.

If the readings are outside the range of the specifications given above, check the dark green wire to the computer for continuity. If the wire is good, replace the computer.

The repair of, or even the absence of, a code 41 does not mean that the charging system is working, it only means that the computer thinks it has the ability to control the alternator field.

Instead of the above procedure, code 41 could be diagnosed by setting up the scanner to enter the alternator field ATM. Connect a test light to the alternator F2 terminal and instead of using a dwell meter or other pulse detection meter, the frequency of the computer's control of the alternator field has been slowed to the point where the test light can be seen to pulse. Follow the rest of the test procedure as described above.

Tachometer Output Test

Enter the ATM tachometer test mode and the tach on the instrument panel should pulse between zero and some predetermined reading. If the tach fails to pulse, inspect the wire that runs from the computer to the tachometer on the instrument panel. If the wire is in good condition, connect a voltmeter to the wire. If the reading on the voltmeter fluctuates while in the ATM tach mode, replace the tach. If it does not and the wire is good and the connection at the computer is good, replace the computer.

Radiator Fan Relay Test

Use the scanner to place the computer in the radiator fan ATM test mode and the radiator fan should turn on for about half a second every two seconds. This ATM function is very handy for troubleshooting code 35 on

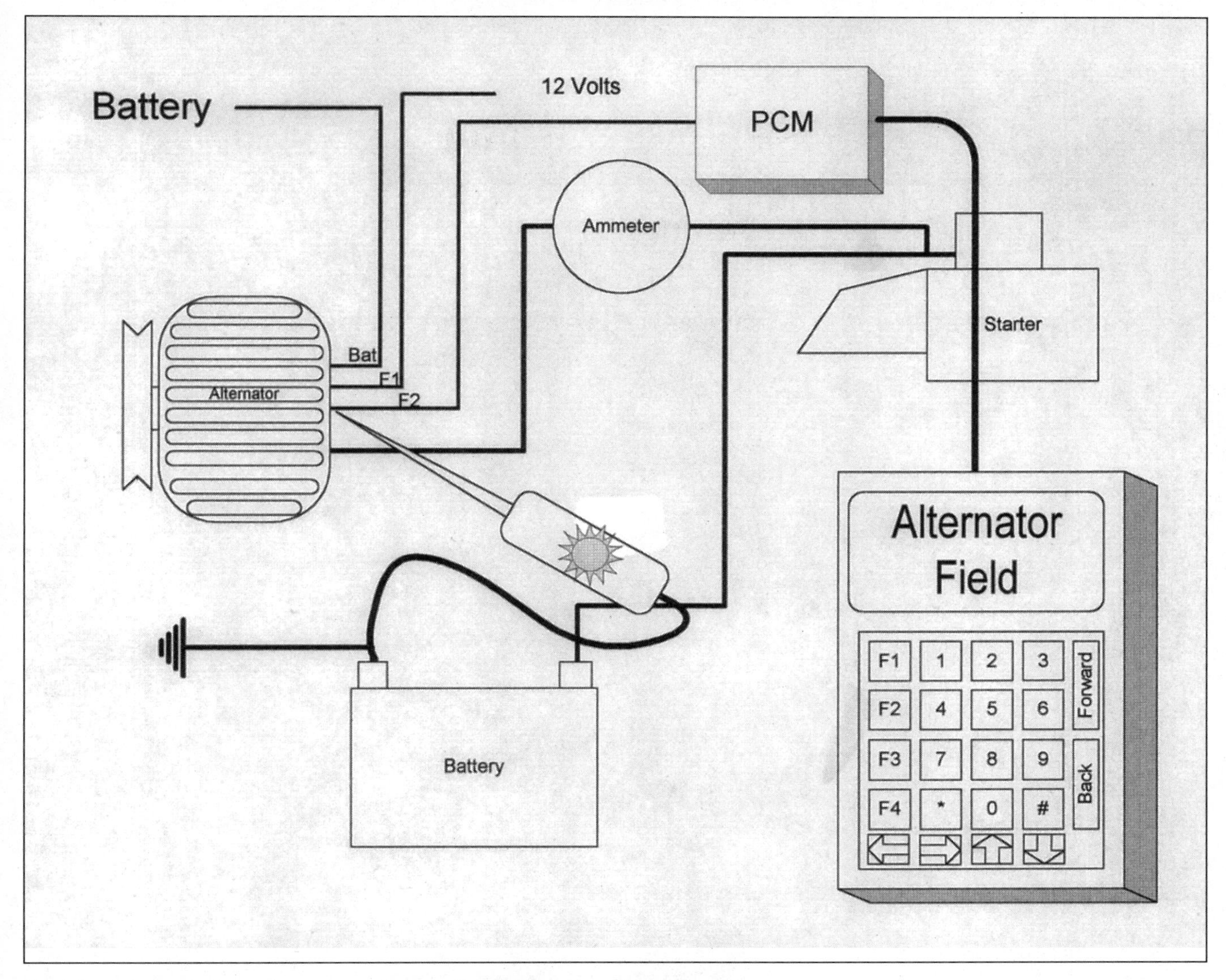

The ATM mode can also be used to test the voltage regulator, which is located inside the computer. Connect a test light to the F2 terminal of the alternator. Enter the ATM alternator test mode. The test light should blink on and off every two seconds.

front-wheel-drive applications. When the computer is placed in the mode and the fan relay does not intermittently spin the fan, then the technician can trace the problem while feeling confident that the computer is supposed to be making the fan work.

A/C Clutch Relay Test

This test mode ensures that the computer can control the air conditioning compressor.

Auto Shutdown Relay Test

This test can be used to specifically test the computer's ability to control the blue/yellow wire that carries the computer's signal to activate the ASD relay.

Purge Solenoid Test

The computer will activate the canister purge solenoid every two seconds. If you cannot feel or hear the solenoid activate, turn the ignition off and disconnect the electrical connector from the canister purge vacuum control valve solenoid. Now turn the ignition switch back on. Connect the black lead of a voltmeter to a good ground. Connect the red lead of the voltmeter to the pink or pink/black wire at the solenoid end. Turn the ignition switch back on. The voltmeter should read approximately battery voltage. If it does not, repair this supply voltage wire.

If the voltmeter read zero, locate the other end of the pink or pink/black wire on the computer. Turn off the ignition switch. Disconnect the connector on the computer and connect the red lead of the voltmeter to the terminal end of the pink or pink/black wire. Turn the ignition on and read the voltage. If the voltmeter reads battery voltage, replace the computer. If it does not read battery voltage, test the resistance through the canister purge solenoid. The resistance should be less than 20 ohms. If greater than 20 ohms, replace the canister purge solenoid. If less than

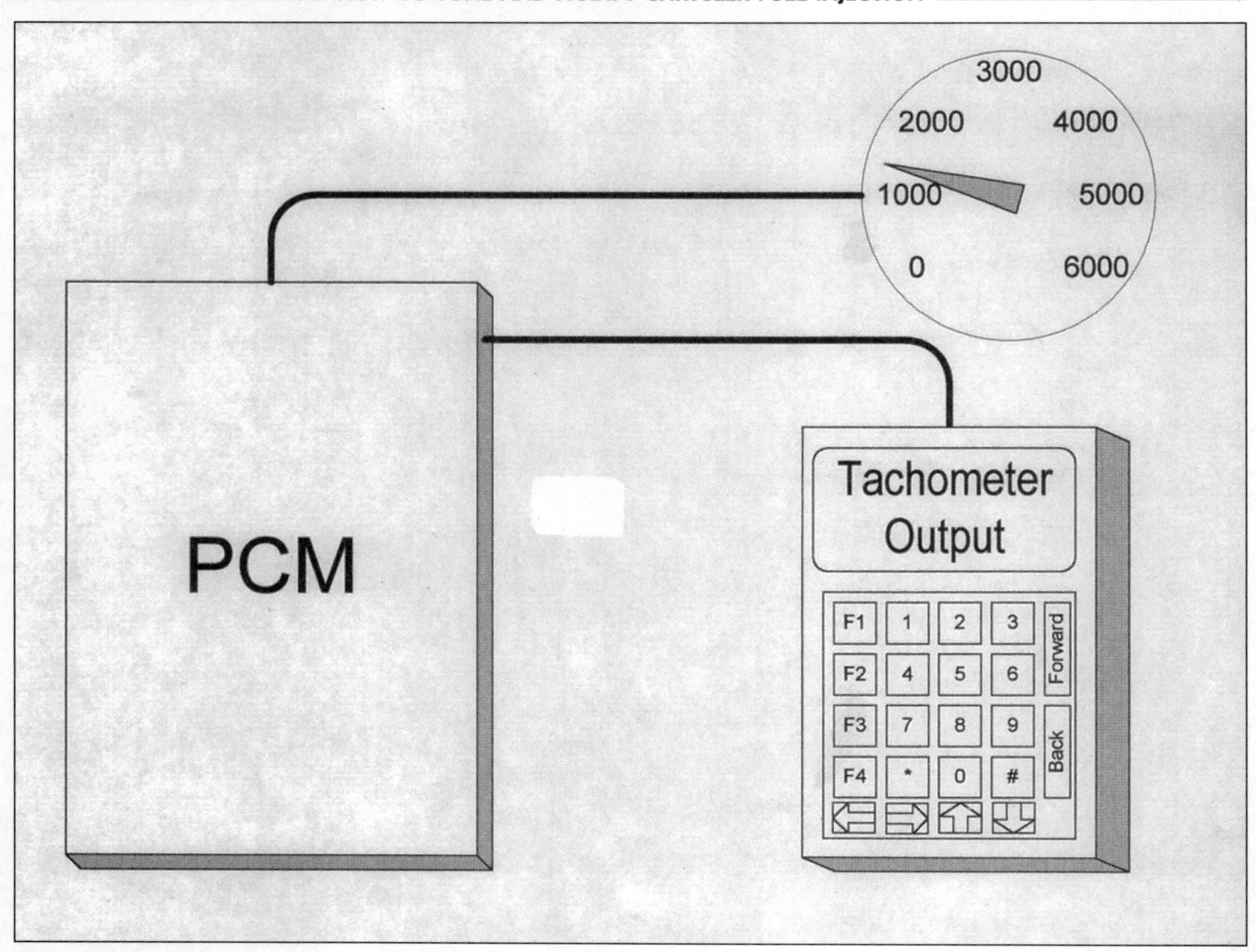

The computer sends the signal to the tachometer to indicate engine speed. This gives the technician equipped with a scanner the ability to test the tachometer and the wiring between the computer and the tachometer. When the scanner is placed in the ATM tachometer test mode, the tachometer should read between 1,000 and 2,000 rpm.

When activated by the ATM test, the radiator fan relay (foreground) should momentarily activate the radiator fan every two seconds. Also in the picture is the ASD relay. When activated during the ATM test, it should provide power to the primary engine actuators (injectors, coil, fuel pump, etc.).

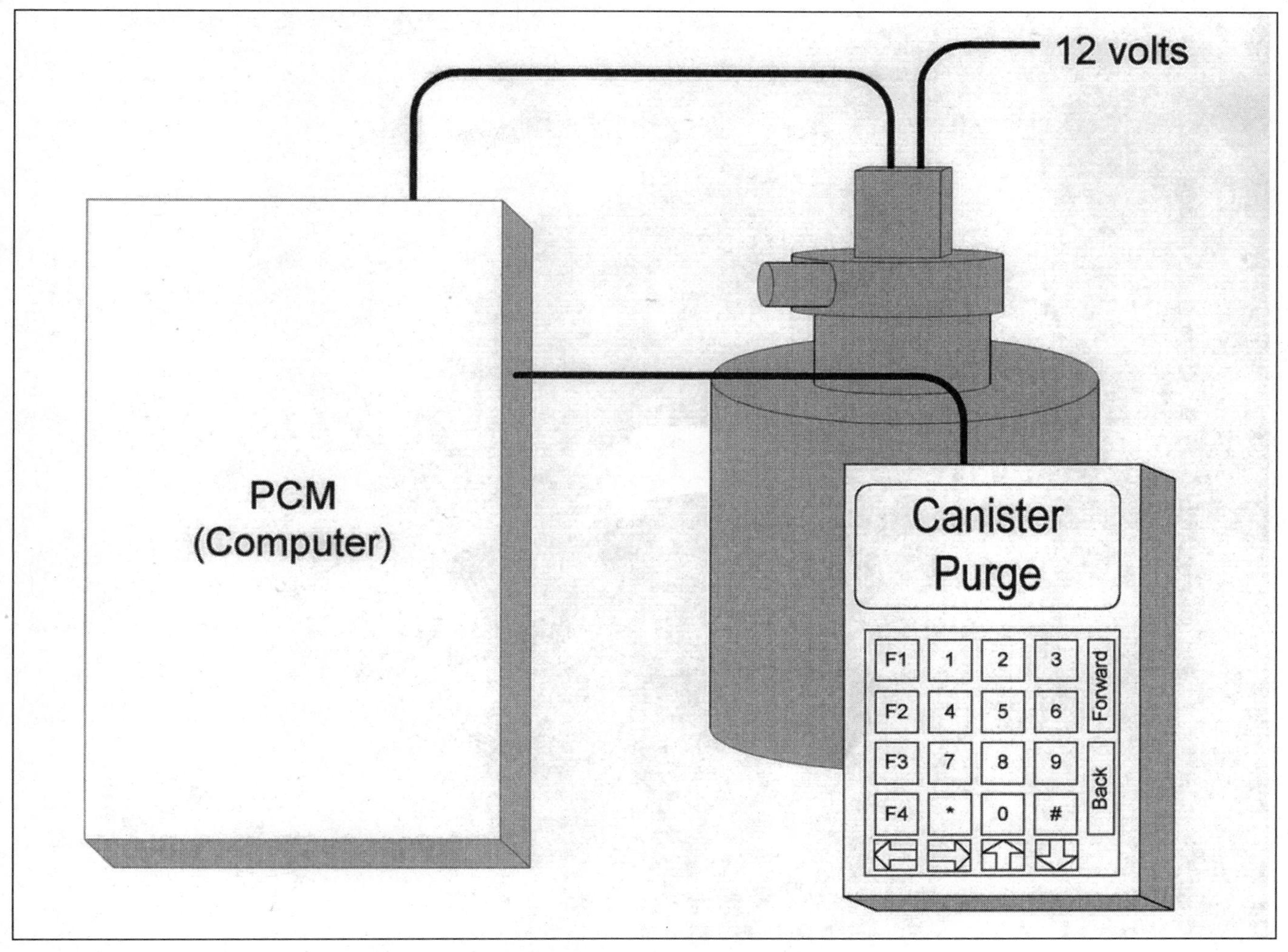

The evaporative canister stores fumes from the fuel tank. When the engine is operating under conditions when extra enrichment is not going to have a great effect on performance or emissions, the computer will open the purge valve to the intake manifold. Upon scanner request in the ATM mode, the computer will open the valve every two seconds. This allows a technician to test its operation.

20 ohms and there was a correct amount of voltage at the power supply terminal, check the resistance of the canister purge solenoid control wire.

Note: Wire colors do change from one application to another. While this text does make an effort to discuss the wires in as universal a way as possible, it may be necessary to alter the aforementioned procedure to compensate for change in wire color.

Speed/Control Servo Solenoid Test

Initiate the Speed/Control solenoids. They are located inside the servo and can probably only be heard with a stethoscope. If they are clicking, no further testing is needed. If they are not, check the fuse. Now verify that there is 12+ volts on the dark blue/red wire. If there is not, repair the defect in the dark blue/red wire between the servo connector and the cruise control fuse. Consult your owner's manual or the diagram on the fuse panel cover to determine which fuse is related to the cruise control. If the fuse is good, turn the ignition switch off. Locate the tan/red wire on the computer that controls the cruise servo. Disconnect the 60-pin connector on the computer. The tan/red wire is usually connected to pin 53. Also locate the light green/red wire that is usually connected to pin 30. Disconnect the cruise servo and place a jumper wire between the tan/red and light green/red wires at the servo end. Now measure the resistance between these two wires at the computer connector end. For practical purposes, the resistance should be zero.

If the resistance is zero, measure the resistance through the servo from where the tan/red wire connects to the servo to where the dark blue/red wire connects to the servo. Repeat this process between the light green/red wire and the dark blue/red wire. In each case, the resistance should be greater than zero, but less than infinity. If these resistances measured through the servo are correct, and the wires are good, replace the computer.

Note: Wire colors do change from one application to another.

The cruise control (speed control) system is also controlled by the computer. Vacuum from the intake manifold is used to control the opening of the throttle. Solenoids control the application and release of this vacuum.

While this text does make an effort to discuss the wires in as universal a way as possible, it may be necessary to alter the aforementioned procedure to compensate for change in wire color.

EGR Solenoid Test

When the EGR solenoid ATM is initiated, the computer should pulse the ignition coil every 2 seconds. If the solenoid fails to pulse, turn the ignition switch on and verify that there is voltage at both terminals of the EGR solenoid. This component, like most of the others controlled by the computer, is grounded by the computer. The EGR will not be requested to work when the engine is idling, much less when it is not running. These facts mean that the voltage on both wires should be approximately battery voltage. If there is battery volts, it means the power supply, the dark blue wire, is in good condition, and the solenoid does not have an open.

The gray/yellow wire between the solenoid and the computer is grounded by the computer to energize the solenoid. Follow this wire to the computer. Turn the ignition switch off. Disconnect the connector containing this wire from the computer. Now connect a voltmeter to the terminal for the gray/yellow wire on the computer harness and turn the ignition switch back on. If the reading is approximately 12 volts, the wire is good and the computer must be replaced.

Note: Wire colors do change from one application to another. While this text does make an effort to discuss the wires in as universal a way as possible, it may be necessary to alter the aforementioned procedure to compensate for change in wire color.

On many applications, the EGR valve is controlled by the computer. This control is accomplished with a signal to a vacuum solenoid. When the computer is put into the ATM EGR solenoid mode, this solenoid will be activated every two seconds.

When the scanner puts the computer into the ATM speed control solenoid test mode, the vacuum control solenoids are activated and deactivated every two seconds.

CHAPTER TEN

MODIFYING FOR PERFORMANCE AND ECONOMY

What Is Legal?

In a few words...not much! The clean air act of 1990 commits the manufacturers of motor vehicles to ensuring that even after the vehicle is out of their control and into the hands of consumers that it will continue to meet the strict emission requirements established by the act until the vehicle is scrapped. This means that the OBD II diagnostic system being introduced at this writing virtually precludes any modification to the injection system, and in fact to the engine as a whole. The OBD II system has therefore been designed to preclude modification.

The OBD I system that has been in effect for several years is not as preclusive to modification as the OBD II system, but nevertheless most modifications are illegal. Basically, you may not modify any emission-related items. This would include, but is not limited to, EGR systems, air pump systems, canister purge systems, timing control systems, and fuel injection systems. Basically, the only functions that are controlled by the fuel injection computer that can be modified are the cruise control and charging system.

In many jurisdictions, internal engine modifications are strictly forbidden. In others, they are permitted on what might be referred to as a "don't ask, don't tell" basis. External engine modifications are never permitted for licensed vehicles unless there is a California Air Resources Board executive order allowing that modification.

Many jurisdictions either have been or are about to be faced with tougher emission testing procedures. These procedures are designed to make sure that even older vehicles meet their design specifications for emissions. Many modifications can therefore cause a vehicle to fail emission tests in jurisdictions where they may have easily passed in previous years.

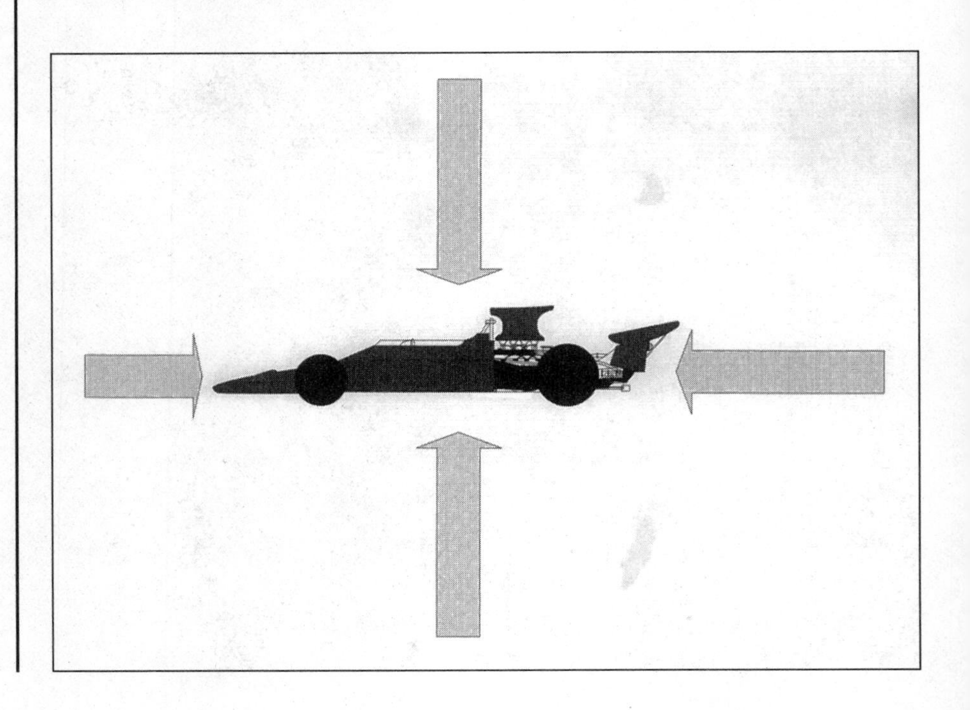

A common misconception of the uninitiated is that you can increase the amount of power from an engine by giving it more fuel. Getting fuel into the engine is easy; a coffee can with a few small holes punched in the bottom and filled with gasoline will accomplish that. The real trick is to get the air surrounding the engine into the engine.

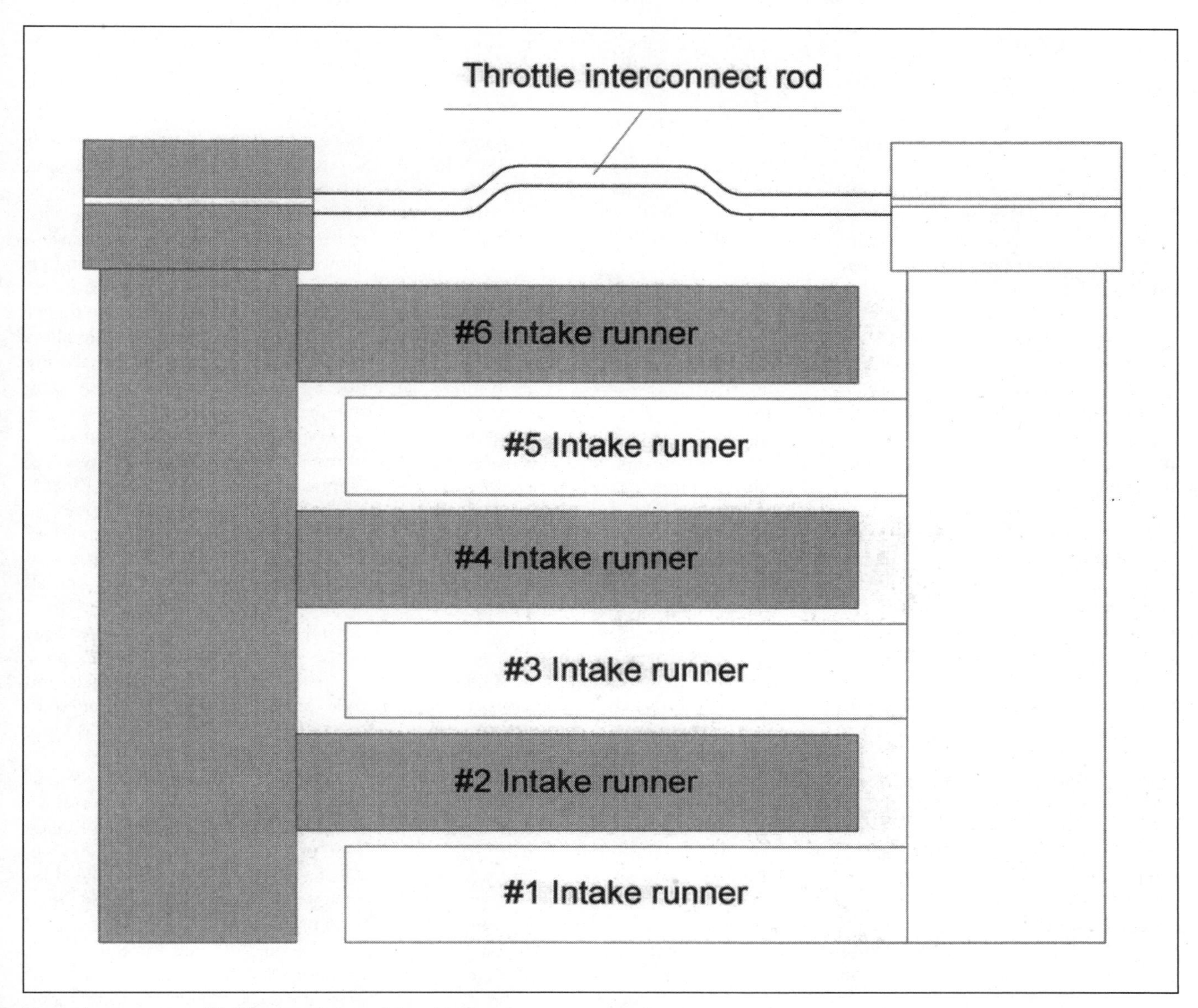

The 3.5-liter engine used by Chrysler in late-model applications uses a cross-ram induction technique pioneered on stock applications in the 1960s. The manifold features two separate air induction systems with two separate throttle plates connected by a common rod. The manifold on the right feeds the left side of the engine, and the manifold on the left feeds the right side of the engine. This increases the length of the runners, and therefore, the velocity of the air traveling through the runners for better torque.

Power Comes from Air, Not Fuel

Power comes from air. Every performance engine builder knows this, yet many people think that power comes from increasing the amount of fuel. Increasing the amount of fuel to increase the amount of power is true only to the extent that the extra fuel can add extra heat to the air to cause the air to expand. Add extra air and you add extra power. The design of the ports in the cylinder head can affect how much air can flow into the engine.

A good cylinder head machinist can increase the power output of the engine by improving air flow through the cylinder heads and across the valves. This procedure may cost several hundred dollars and a couple of weeks of down-time.

Exhaust modifications are probably the avenue of greatest hope for improved performance. There are several California Air Resources Board-approved performance exhaust systems on the market today. The only critical areas here are that the location of the oxygen sensor be the same as in the original exhaust and that the catalytic converter position not change. The manufacturer of the exhaust system component must then seek approval on a vehicle-by-vehicle basis from the California Air Resources Board.

Improving Air Flow (Intake System)

Dynamic Supercharging

When the average person talks about supercharging, one imagines a big blower sitting on top of the engine. One imagines Don Garlits (is my age showing?) sitting behind a monster "jimmy" blower with flames shooting out of the exhaust (yes, my age is definitely reflected in this). In reality, dynamic supercharging involves a much more subtle approach. Dynamic supercharging maximizes the use of the 100-mile column of air to force the air into the cylinders. There are two types of dynamic supercharging: ram pipe supercharging and tuned-intake tube charging.

Ram Pipe Supercharging

In this technology, each cylinder has its own intake manifold usually connected to a common receiver. This design uses the kinetic energy of the gas moving into the cylinder to press the air into the cylinder. In a sense, it uses the air as a mallet to pound the air into the cylinder.

Tuned-Intake Tube Charging

This intake system is just like the ram pipe system except the intake runners are grouped together by ignition interval.

Turbocharging

Turbocharging is a fascinating technology. Essentially, a turbocharger takes energy that would have gone out the tailpipe as heat and converts it into energy. The energy then drives a turbine that spins a compressor. The compressor increases the pressure in the intake manifold. The air molecules become squashed together cramming more oxygen molecules into the cylinders. Since more air in the cylinders translates directly as more power from the engine, this is a very efficient way to increase the power output of an engine. Some turbocharged engines run over 100 psi of boost. This means that the mass of air in the cylinders can be over six times greater than without a turbocharger. This means that the power potential of the engine is many times greater than without the turbocharger. Of course, all good things do have a downside. When the boost pressures are this high, the engines have a very short life expectancy, perhaps seconds.

Supercharging

A supercharger is an engine-driven pump used to force air into the cylinders. Engine-driven superchargers are what people typically think of when the subject of superchargers comes up.

Keeping the Charge in Suspension

When it comes to mixing fuel and air, there are two basic types of intake manifolds: wet manifolds and dry manifolds. Actually, for a gasoline spark-ignition engine, the concept of a dry manifold is not proper. Even in these manifolds the air and fuel are mixed just above the intake valve. Only direct-injected engines, engines where the fuel is injected directly into the combustion chamber, are true dry manifold engines.

This may be where the real trick lies. Wet manifold engines use carburetors or throttle body injection to mix the fuel with the air destined for the combustion chambers. As the air/fuel charge moves through the intake runners, there is a tendency for some of the fuel to drop out of suspension. Imagine a mountain stream for a moment. As the rushing waters come to level ground, they slow; as they slow, rocks begin to drop from the moving water. The slower the waters move, the smaller the rocks that will drop out of suspension. The same thing happens in an intake manifold. Fuel particles will drop out of suspension as the air flow through the intake slows. This drop-out is held to a minimum on multipoint-injected engines. These dry manifold engines mix the fuel with the flowing air only just above the intake valve. As a result, the fuel is more evenly and accurately carried to the cylinder.

Air Cleaner/Filter

The air cleaner has the dubious honor of being the engine's primary line of defense against airborne contaminants and being the first point of serious air restriction in the intake system. Although we seldom really think about it, airborne grit, sand, and other abrasives can destroy an engine very quickly.

There is a story that I heard from some civilian Marine Corps mechanics. I do not know if it is true, but it does paint an excellent picture of the importance of the air filtration system. It seems that there was a training session on the maintenance of diesel engines. The preventative maintenance was done, then, as a treat, the trainees were allowed to drive the vehicles through the desert areas of Twenty-Nine Palms Marine Base. Unfortunately, the trainer failed to emphasize that the air filters needed to be reinstalled before the joy ride. Several engines were destroyed by the sand.

Now I *love* oxymorons. You know, phrases like "military intelligence," "jumbo shrimp," and the like. One of the best oxymorons in the automotive industry is the term "free-flow air filter." Of course, this may be a matter of interpretation. All air cleaners and all air filters should be free-flowing, free-flowing to the maximum requirements of the engine. Those of you that grew up in the world when teenagers and their popular media were limited to drag racing and beach parties, as I did, are very familiar with the idea of monster carburetors. I can actually remember having a serious conversation about whether or not to "put a Holley 1150 cfm" on a 318-cubic inch engine. The amount of air actually inhaled by an engine is usually much less than what pride or desire would like. In general, the air flow rate through the air filtration element should be 15 to 20 times the displacement of one cylinder.

The typical factory air cleaner has provisions for performing three jobs. The typical factory air cleaner filters the air, mixes warm and cold air, and dampens noise. Air filtration will be discussed later, let us look at warm air mixing and noise.

Warm Air Mixing

Okay, people in Minnesota listen up, people in Guam take a snooze. There is no secret to the fact that warm air will hold fuel in suspension better than cold air. There is no doubt that pre-warmed air will also be easier to burn once inside the combustion chamber. Anyone who has ever tried to start a car in Minneapolis in December and then tried in Guam a few days later can confirm this. There is more to this than just the inconvenience of

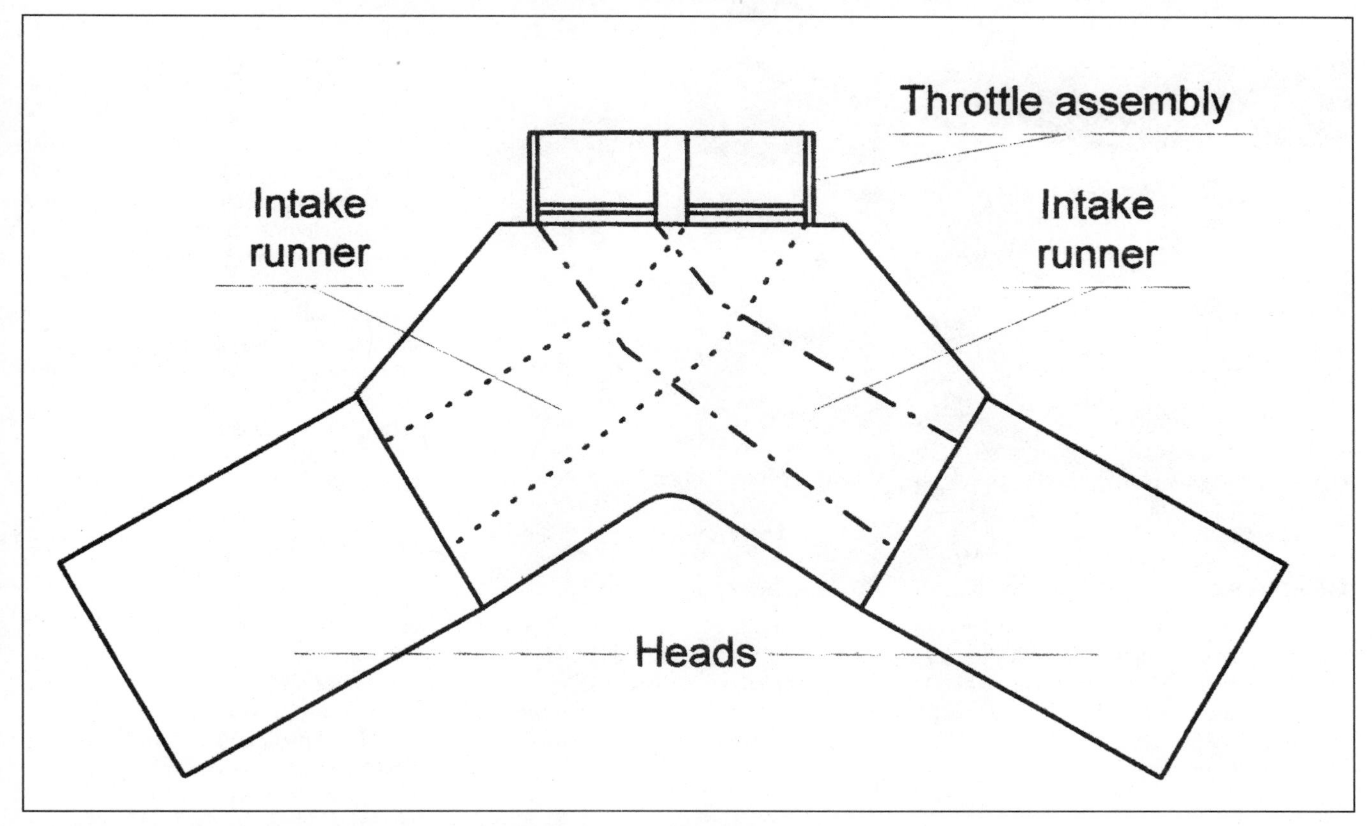

Many of the late-model "V" Chrysler engines with multipoint fuel injection use a high-rise, cross-induction manifold reminiscent of the 1960s.

poor driveability when the engine is cold. Any cylinder in which the air is not warm enough to support proper combustion will not fire properly; any cylinder that is not firing properly will be a heavy polluter.

Once the engine is started, the exhaust manifold will warm up very quickly. Many factory air cleaners take advantage of this heat. The heat is captured through a heat-resistant metal or metalized paper tube. The warmer air coming through this tube is then mixed with air coming through the main air inlet of the air cleaner. At the point where the air from this tube enters the main body of air entering the engine there is a control door. This door is intended to cut off the warm air when the ambient temperature or coolant temperature is above a predetermined point. If the air entering the intake system is warmer than it needs to be, there can be a loss of power. Warm air is air that has been expanded. Therefore, if the temperature of the air is higher than it really needs to be to meet the goal of low emissions of good warm driveability, then there will be a proportional power less.

Noise

In the early 1980s, I began teaching electronic engine controls to automotive fleet maintenance personnel. One of the things I observed early on was the fact that most of the police cars that came in for service had the top of the air cleaner inverted. I thought it odd that intelligent police officers could believe inverting the top of the air cleaner could improve power. For years I sought a more believable answer to this mystery. After years of questioning technicians and police officers, I finally obtained a satisfactory answer. Imagine that you are cruising a dangerous neighborhood, you turn down a dark alley in pursuit of a perpetrator. With the air cleaner lid turned upside down, the air flowing through the carburetor will yield a deep, throaty roar. The hope is that this roar will intimidate the perpetrator.

This story emphasizes the second important job that the air cleaner must do—silence the sound created by several hundred cubic feet of air per minute entering the engine. Although the typical performance-minded individual usually cares little about intake noise, it should be pointed out that the typical manufacturer also does not care much about this noise.

The sound of 300-400 cubic feet of air per minute being pulled through a couple of holes only slightly bigger than a half dollar can be very annoying to the average car owner. Remember the 1980s? Remember yuppies? Can you imagine courting a potential junk-bond client in a car that goes "wonnnnn" every time you step on the accelerator? It would be enough to fry the client's sushi.

The manufacturers expend a great deal of effort trying to reduce the noise created by the induction of air. For big rigs, over-the-road trucks and the like, there are currently Environmental Protection Agency regulations in affect for the noise they generate.

The bottom line concerning sound is that if you decide to replace

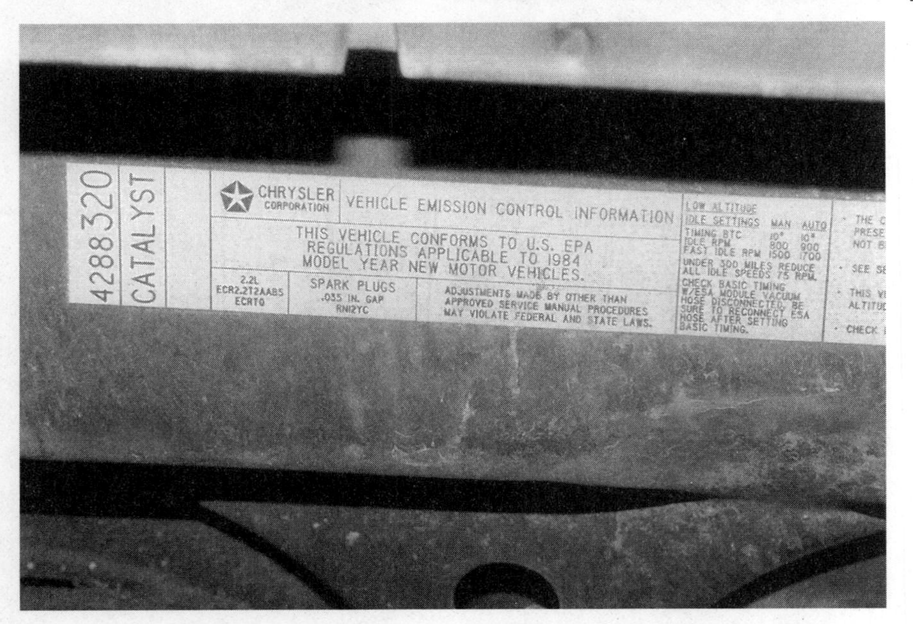

The Environmental Protection Agency, California Air Resources Board, Bureau of Auto Repair, Texas Natural Resources Commission, and other agencies are charged with the responsibility of making sure that a vehicle manufactured during a given year will meet emission requirements. This responsibility does not end when the car is sold. Therefore, the manufacturers are required to design the engine, the fuel injection system, and the emission control systems in such a way as to make tampering difficult.

the air cleaner assembly, expect the noise to increase. When the engineer designed the original air cleaner for this engine in this application, he did not just reach up on the shelf and pull one off. He got out his slide rule and began by estimating the speed of sound in the air where the vehicle is going to be operated. He measured the length of the intake runners. He measured the mean diameter of the intake manifold. He measured the air filter volume. After confirming the filter volume, he calculated the resonance frequency of the intake system.

Emission Laws and the Air Cleaner

Since the 1970s, the air cleaner assembly has been an integral part of the emission control system. In many jurisdictions across the country, replacing the air cleaner with some aftermarket air cleaners may be blatantly illegal. The toughest jurisdiction in the United States is California. There are only a very few street-legal changes allowed for in the State of California Air Resources Board document Reference Number A-92-443, "Modifications to Motor Vehicle Engine and Emission Control Systems Exempted Under Vehicle Code Section 27156." If the vehicle you are modifying is destined for the street, be sure to confirm that the new air cleaner you are purchasing has been approved by your state.

Although you may even be able to prove that the new air cleaner does not affect emissions in the least, you will still have to convince the emission inspector or referee. There will be more about aftermarket air cleaners a little later in this chapter.

Filter Elements

An area where a great deal of improvement can be made without severe legal restrictions is the air filter element. The paper filter used by the manufacturer when the car or truck is assembled usually supplies adequate filtration—just adequate. You may find this hard to believe, but one of the primary issues considered by the manufacturers when they select or design an air filtration element is cost. For some reason, they consider it important to save a few dollars per car and therefore save a few million dollars per year in manufacturing costs. I think that is called capitalism or something. In many cases, the replacement air filter is of lower quality than the original.

To a certain extent, air filtration is a "Catch 22" situation. To provide good filtration, the filter element must filter out all particles capable of damaging the engine. In general, the tinier the particle a given element can filter, the more that filter restricts air flow.

The first car I owned was a Volkswagen with a 1,200 cc engine. This car had an oil bath air filter. Very popular during the first half of the century, the oil bath filter trapped particles in a tub of oil. These are expensive to manufacture and expensive to install on the assembly line.

The most common air filter element in the automotive industry today is the paper air filter. This is one of those areas where it can be said that every filter manufacturer makes the best air filter. It is difficult for me to recommend a particular manufacturer under the purview of this book. Ask your local mechanic or auto parts man which brand he would recommend.

There is a wide range of performance air filters available. These range from paper, to foam, to oil-impregnated foam, to gauze, to combinations of each of these.

My first experience with performance air filters was during the 1970s when the shop I worked in at the time made a specialty of installing Weber carburetors. The air filter we used was a relatively course foam with an oil spray coating. There are several of these on the market currently as replacement cartridges.

The paper, foam, and gauze filters all begin life with good air flow and filtration characteristics. The paper filter, however, suffers a rapid decline in air flow capability as it filters the air. The dust and contamination particles build up rapidly on the surface of the filter. The paper filter is a single-plane filter. The air flow through the filter decreases at the same rate as the trapping of contaminants.

The foam filter features multiple levels of filtration. There are many paths for the air to take through the filter. When a path becomes restricted, there are many other paths for the air to take. These filters are generally "lifetime" filters. They can be cleaned and re-oiled many times.

Like the foam air filter, the gauze air filter features multiple paths for air flow. Generally marketed as "lifetime" filters, gauze filters are much like the foam filters in that they can be cleaned and reused many times.

Aftermarket Air Cleaners

Street-Legal Air Cleaners

As was mentioned earlier, the choices of street-legal aftermarket air cleaners may be extremely limited. The worst case is California. What follows is a complete list of permitted replacement air cleaners in California as per State of California Air Resources Board document Reference Number A-92-443, "Modifications to Motor Vehicle Engine and Emission Control Systems Exempted Under Vehicle Code Section 27156."

Track-Only Air Cleaners

On the track, not withstanding class restrictions, anything goes. Many weekend and professional racers will opt not to have an air cleaner or filtration system at all. It is hard to sit here at this computer keyboard and criticize someone who is blowing away the competition every Friday and Saturday night. Nevertheless, in most environments, there is enough hard particle contamination in the air to do very serious damage to the engine in a very short time frame. Custom air cleaners are available for virtually any carbureted or throttle body-injected engine. Only a few multipoint-injected engines have aftermarket air cleaner assemblies available.

Modifying the Factory Air Cleaner

Most factory air cleaners consist of a shallow tube or pan with a small air inlet cut in the side. A snorkel is usually attached to the air inlet to assist in directing air flow into the air cleaner. Most aftermarket "performance" air cleaners have the walls of the air cleaner tube removed. This enables air flow from 360 degrees around the throttle assembly. The walls of the air cleaner then become the air filter itself. The air cleaner on most cars is capable of flowing more than the actual air flow of the engine on which it is installed during normal operation. If the car is to be used under circumstances where maximum performance is required, such as racing or trailer towing, then the air cleaner assembly can become a restriction to air flow and therefore performance.

In the old days, guys would flip the air cleaner lid upside down. While the effects of doing this were largely psychological, it did bypass the snorkel tube of the air cleaner and greatly increase the potential air flow. The design of fuel injection air cleaners does not make inverting the lid practical or desirable.

Improving Air Flow (Exhaust System)

The range of stock exhaust systems goes all the way from a series of pipes whose sole function is to keep the occupants of the car from becoming asphyxiated or poisoned by carbon monoxide to highly researched performance exhaust systems. It may surprise you to know that many of the more thought-out exhaust systems are found on applications usually thought of as a modest commuter rig. Many of these commuter vehicles are designed to maximize fuel economy; maximizing fuel economy is actually a performance issue. When the exhaust system flows well, less energy is used to help the engine breath, and more energy is used to push the car down the road. When the energy used to flow the exhaust gases is minimized, and the power used to push the car down the road is optimized, the fuel economy improves.

The Manifold

Today's stock exhaust manifolds reflect the technology of the decade in which the engine was designed. In 1957, who would have dreamed that in 1997 we would still be driving around with engines that were designed while the DeSoto was still being sold? I know, every time I am bold enough to make this statement someone always says, "Wait a minute, the P-309 5.1 liter V-10 is a brand new design." Well let us see: overhead valve V layout with the chain-driven camshaft located above the crankshaft, cross-flow heads and hydraulic lifter operating tubular pushrods. Sounds very similar to a Chevy 409 to me.

These older design engines, the 318, the 383, and their children, use an exhaust manifold virtually unchanged in concept since the decade of the ducktail. The cast iron manifold offers low cost, good durability, and reasonably good flow. In applications where they are exposed to limited cooling from outside air and are subjected to vibration, there are cracking problems.

In most cases, the cast iron manifold consists of a tube running the length of the cylinder head with the front exhaust pipe connected at some point. Inherent to this design is the fact that exhaust gas velocity remains quite low and therefore scavenging is kept to a minimum. The best thing that can usually be done with these manifolds is to find a door that needs to be held open and put them to work doing something they are good at.

In the 1970s, and especially in the 1980s, many applications began to use tube steel manifolds. These offered greater flexibility in design, greater immunity to cracking, and a "cooler" look. Rather than each exhaust port in the cylinder head feeding into a short tube that runs immediately into a common tube, these tube steel manifolds allowed the designer to make long runs to the collection point. The longer runs provided for greater exhaust gas velocities and therefore better scavenging.

The Pipes

I have the greatest admiration for those courageous men at Midas. It is a little known fact that when certain alloys oxidize they develop an affinity for saline solutions. As a result, any rust that falls from the exhaust system while changing or working with the pipes or muffler will plot a deliberate vector for your eyes. Additionally, can you imagine spending your entire life with rust flakes between your teeth?

Pipe Design

Single-Wall Design

The single-wall pipe has been the standard of the industry for decades. Back in the days of Eisenhower and "the good life," you could plan on replacing these pipes frequently, in some climates annually. Since the 1950s, the manufacturers and exhaust system suppliers have done many things to improve the life expectancy of these pipes. Today, companies that grew to national size very quickly in the post-war years working exclusively on exhaust systems have had to branch out into other areas of auto repair. This is due in no small measure to the increase in exhaust pipe quality and the decrease in rust.

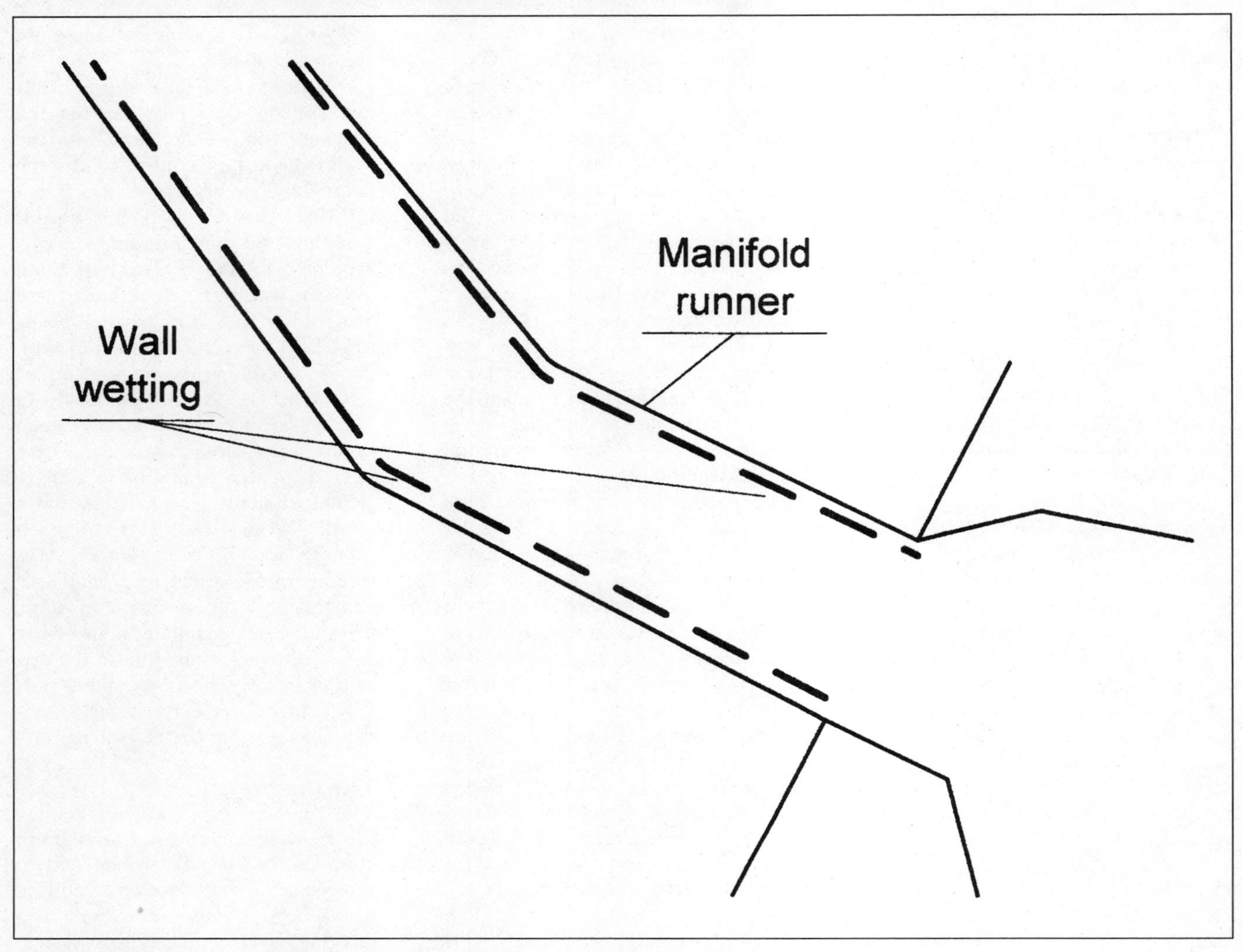

One of the problems of any modern gasoline engine is keeping the air and the fuel mixed. Throttle body-injected engines have little more luck accomplishing this task than do carbureted engines. As the intake manifold pressure changes, fuel being carried toward the combustion chamber tends to fall out of solution. Multipoint injection systems introduce the fuel just above the intake valve. There is little opportunity for the fuel to coat the walls of the intake manifold.

Today's exhaust pipes feature coatings that reduce rust and corrosion to the point where it is virtually non-existent. My dad bought a new 1970 Buick Skylark and, as had been his practice since before the Korean Conflict, immediately made an appointment for exhaust work a year later. The year passed; he had to cancel the appointment, another year, another year, and when the car was traded in 1978, it still had the original exhaust system.

Dual-Wall Design

Reduced performance as a result of exhaust restrictions is not the sole domain of the performance application. In the early 1970s, we began to have cars coming in to the department store auto service center where I worked complaining about reduced power. The answer was always a tune-up. Unfortunately, this answer was not always right. When we checked the exhaust system, we assumed that if the pipe looked good on the outside, then it did not have a rust problem. Unfortunately, we did not, at first, know about dual-wall pipes. On these pipes it was not uncommon for the inner pipe to rust and partially collapse, restricting exhaust flow.

The good part about dual-wall pipes is that it gave me another use for my favorite diagnostic tool—the hammer. When the pipe was tapped on, if you could hear rust falling on the inside, then you knew that the lack of power was probably associated with a collapsed inner pipe.

This whole system is thought of as a restriction to the flow of exhaust gases and is therefore considered to impede performance. There is no doubt that on many applications, and after engine modifications have been performed, the exhaust system may indeed impede flow. Before spending a lot of money on "trick" exhaust systems, however, keep in mind that the total volume of the mufflers is many times greater than the displacement of the engine. In fact, typically, the front

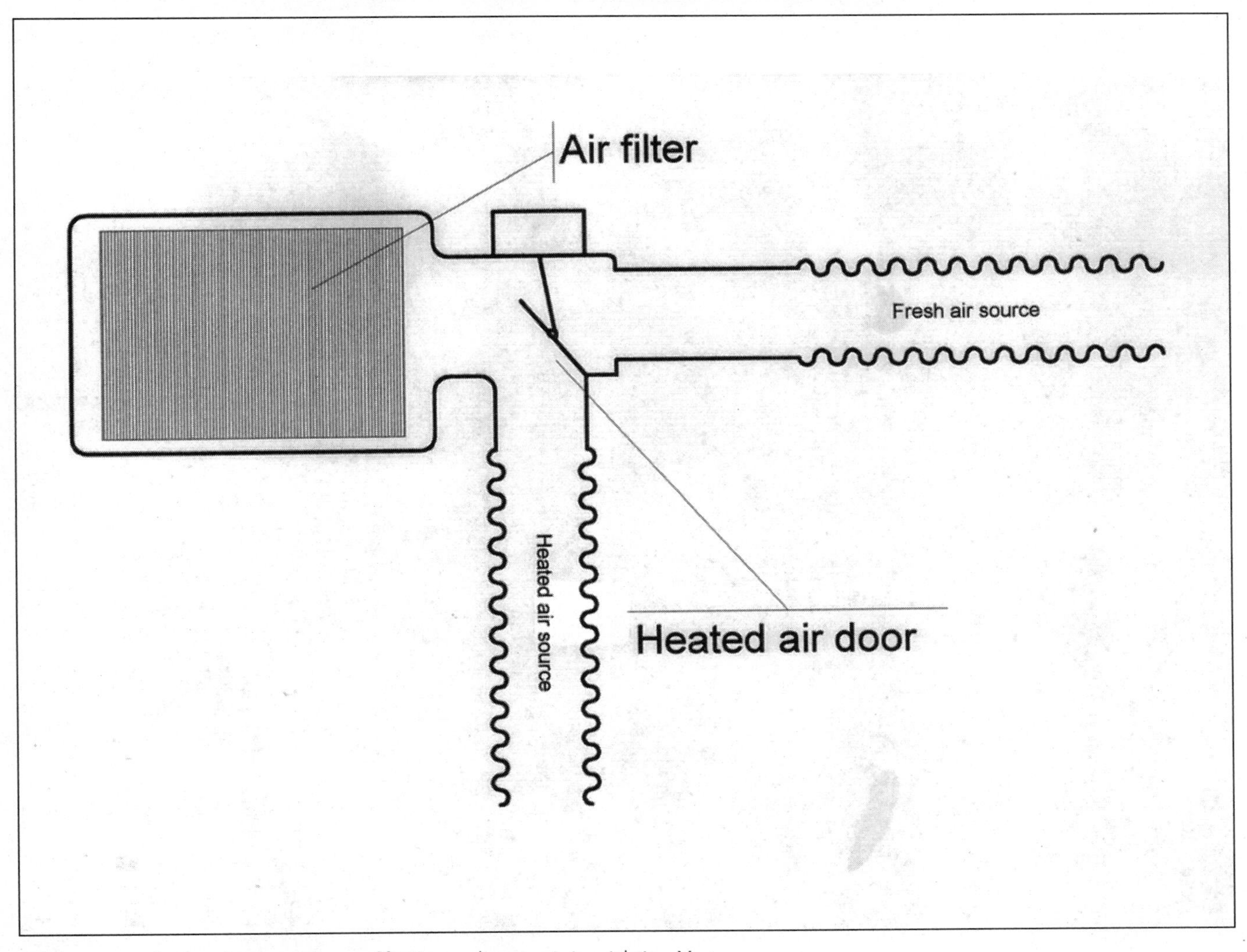

Oddly, even a rather harmless-appearing modification can be an emission violation. Most applications have a door in the air cleaner that routes warm air into the intake when the engine is cold. This door routes warm air from the exhaust manifold to the air cleaner. When this warm air mixes with the air coming from the fresh air source, it raises the temperature of the air destined for the combustion chamber and makes the fuel stay in suspension better. If the door fails to close off the heated air source once the engine is warm, this warmer air can reduce power from the engine.

muffler has 4 to 10 times the swept volume of the engine while the rear muffler has 3 to 8 times. This implies that if backpressure is occurring in an engine, it is not because of restriction but because of resonance waves. The biggest exhaust system, therefore, may not be the best exhaust system.

Heat-Resistant Wrap

I love to go into "speed shops" and ask the benefits of different products they have in the store. For instance, a few years back a heat-absorbing exhaust wrap became available. You would not believe some of the answers I have gotten concerning its benefits. "It protects the plug wires." "It protects the electronics." I could go on. The performance use of this product is to reduce underhood temperatures.

Many performance cars use an air cleaner designed to pick up air from under the hood. In the winter in Fairbanks, Alaska, that is probably a good idea. But in Tuscon, Arizona, the underhood temperature, just caused by the radiant heat of the engine, will be extremely high. Add to that the radiant heat from the exhaust and the air temperature can easily be as high as when under high levels of boost. These exhaust wrapping products reduce the underhood temperatures, allowing the intake system to utilize cooler, denser air. The denser the air going into the intake the more horsepower the air will yield.

Catalytic Converter

When the catalytic converter was introduced in 1975, it meant the beginning of the end for leaded gasoline. These early oxidizing catalytic converters consisted of a platinum/palladium coating over an aluminum oxide substrate. The platinum was not compatible with tetra-ethyl lead

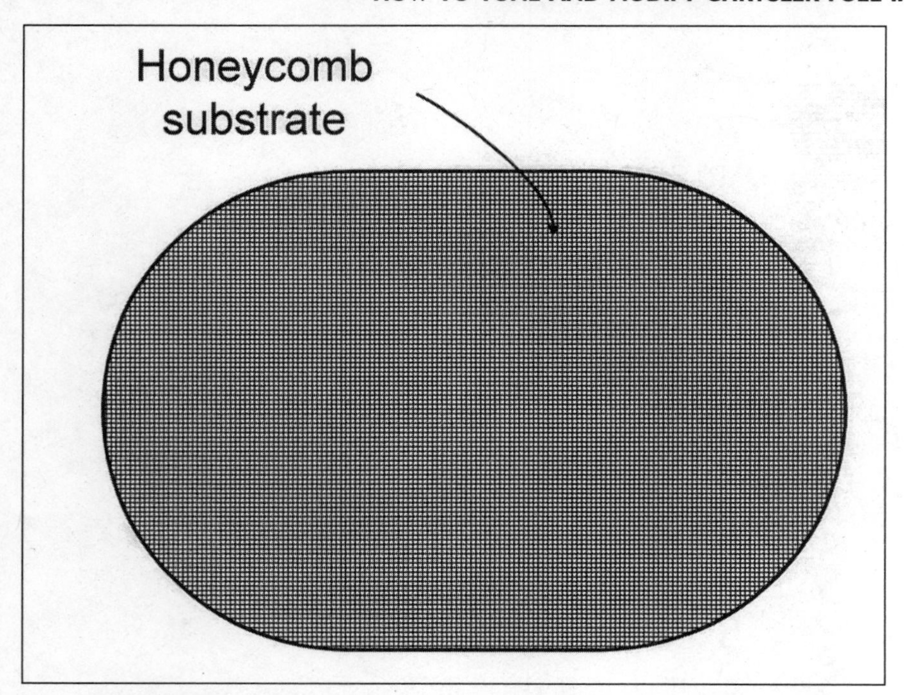

The catalytic converter is often blamed for having reduced automobile performance. In actuality, the cross-section, and therefore, the flow potential of the catalytic converter is greater than that of the exhaust pipes leading to and from the catalytic converter.

requiring that it not be present in the fuel used in catalyst-equipped vehicles. The job of the catalyst is to provide an environment where enough heat can be generated to allow further combustion of the HC and CO to occur. The converter is heated by a chemical reaction between the platinum and the exhaust gases. The minimum operating, or light-off, temperature of the converter is about 600 degrees F with an optimum operating temperature of about 1,200 to 1,400 degrees. At a temperature of approximately 1,800 degrees, the substrate will begin to melt. This means the range of temperatures at which the catalytic converter can properly operate is very narrow. These excessive temperatures can be reached when the engine runs too rich or is misfiring.

A few years ago I attended a meeting of the Society of Automotive Engineers where a representative of a company that built the converter substrates stated that a 25-percent misfire (one cylinder on a four-cylinder engine) for 15 minutes was enough to begin an irreversible self-destruction process in the converter. This type of catalytic converter requires plenty of oxygen to do its job. This means that the exhaust gases passing through it must be the result of an air/fuel ratio of 14.7:1 or leaner.

Oxides of nitrogen are created in the combustion chamber anytime the combustion temperature exceeds 2,500 degrees F. Between 1978 and 1982, there was a gradual introduction of the dual-bed converter. This converter adds a second rhodium catalyst, known as the reducing section, ahead of the oxidizing section. The rhodium coats an aluminum oxide substrate and reacts with the NOx passing through it. When heated to more than 600 degrees F, the nitrogen and oxygen elements of the NOx passing through it will be stripped apart. Although only about 70 to 80 percent efficient, when coupled with the EGR valve it does a dramatic job of reducing NOx. Since the job of the reducing catalyst is to strip oxygen away from nitrogen, it works best when the exhaust gases passing though it are the result of an oxygen-poor air/fuel ratio of 14.7:1 or richer.

The only air/fuel ratio that will permit both sections of the converter to operate efficiently is an air fuel ratio of 14.7:1. The job of the oxygen feedback fuel injection systems being used today is to control the air/fuel ration at 14.7:1 as often as possible.

On many applications, the air from the air pump, after completing its role of preheating the catalytic converter, will be directed between the front reducing section and the rear oxidizing section of the converter. This is to supply extra oxygen to improve the efficiency of the oxidizer.

Resonators

There are many types of mufflers in use today. These range from simple resonance chambers to highly sophisticated electronic devices. To paraphrase the *Automotive Handbook (2nd Edition)* published by Robert Bosch Corporation, "...the resonance chamber, or resonator, by reflecting sound back toward the sound source and by multiplying the number of sound emission points." There are tubes with small holes, or slits, in the resonator. These holes break the sound up into smaller units. While each of the smaller units individually has the same decibel potential as the exhaust gases coming down the exhaust pipe from the cylinder head, the fact that they are smaller allows the sound to dissipate more readily, and therefore a quieter exhaust is achieved. In effect, it is easier to deaden a bunch of little sounds than it is to deaden one big sound.

Forcing the big sound to break up into a bunch of little ones requires energy. This energy must come from the exhaust gases themselves, and since the energy in the exhaust gases originates with the engine, resonators do affect the power output of the engine.

Additionally, the resonator will usually contain 2 to 7 "acoustic elements." Acoustic elements are abrupt changes in pipe cross-section or direction. Anytime the exhaust gases are forced through a smaller cross-section, energy is required; the ultimate source for this energy is the engine, to compact the gases. Also, a change in direction means that kinetic energy must overcome inertial force and more power is lost.

The location of the resonator is critical to the way sound is dampened and the amount of energy the resonator diverts from pushing the car down the road. In the resonator, exhaust gases are forced against the outer walls. When the resonator is located close to the engine, this will result in the loss of a great deal of heat energy, and therefore a loss of power. Additionally, the outside of the resonator can become extremely hot. Depending on location, this heat can be transferred to the underbody and eventually to the passenger compartment. Many resonators feature a double outer wall with insulating material to reduce the transferal of heat to the outside of the resonator.

If the resonator is located near the end of the exhaust, it has less effect in deadening the sound of the exhaust. When located at the rear of the car, the sound-deadening ability will be greatly affected by the length of the tailpipe. Therefore, if the purpose of the exhaust system modification is strictly to sound "cool" (which *is* an honorable desire), then a single resonator at the end of the exhaust system with a selection of various length tailpipes would allow you to tune the sound of your exhaust to match the occasion. A short tailpipe to "impress" your girlfriend with the throatiness of your rod, and a long tailpipe with which to depart the wedding ceremony.

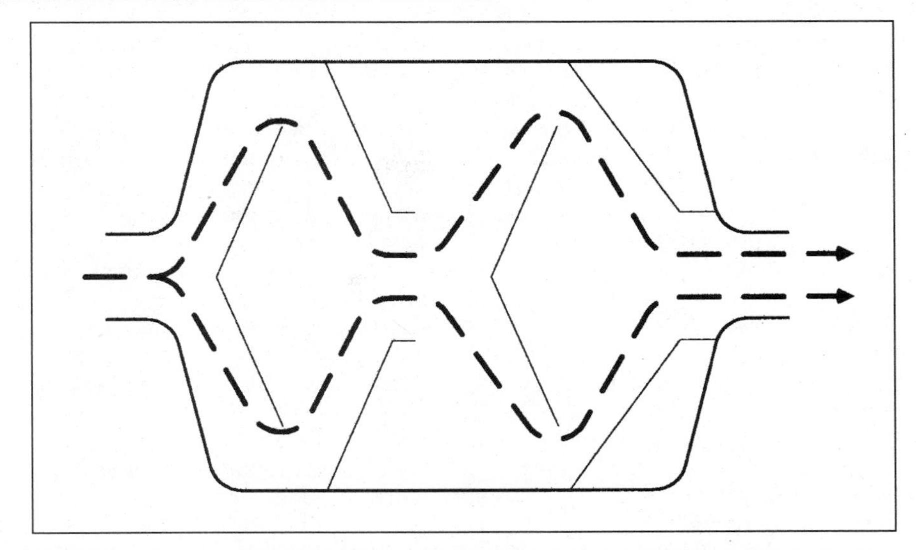

There are many new muffler designs that have been introduced in the past decade. Installation of one of these new muffler designs might be the simplest and cheapest legal way to improve performance.

The Muffler

Absorption Mufflers

Absorption mufflers use a sound-deadening material to reduce engine noise. The desire of every teenage male during the 1960s who did not have a pocket protector, a loop on the back of his shirt, or a slide rule on his hip was to have a set of "glass packs" on his car. Imagine this: A 1959 Plymouth Belvedere with push-button automatic, a smoking V-8, and a 17-year-old at the wheel; a car so fine that when the owner visited used car lots, he would leave the keys in the door in hopes they would sell it. Yet, add a set of glass packs and the car was magically transformed into something they would have to ban at Indy out of fairness to the competition.

"Glass pack" absorption mufflers were considered for a long time to be the ultimate in performance exhaust mufflers. Essentially, the exhaust is directed through a pipe in the muffler. Inside the muffler, the pipe has a series of holes in it. The exhaust gases are directed through the holes into a heat-resistance dampening material such as basalt wool. The advantage of this type of muffler is that it permits the exhaust gases a straight path through the muffler. The absence of bends and turns would logically reduce the back-pressure in the exhaust and increase power. Unfortunately, we still do not live in that ideal world discussed in previous chapters. As the gases pass across the holes in the tube, the holes set up eddy currents in the gases. These eddy currents tend to impede the flow of the exhaust gases. A well-thought-out muffler will reduce the resistance to gas flow by carefully spacing the holes.

Acoustic Tuning Noise-Dampening Principles:

Pipes Perforated with Holes

Actually this principle has been discussed already. Although a physicist would probably disagree with this analogy, the principle is as follows. Think about the flute carried by Kwai Chang Caine of the old "Kung Fu" television series. This flute is basically a bamboo tube with several holes drilled in it. Each of the holes is placed at a different length from the end of the tube. When Caine blew through the end of the tube with all the holes uncovered, a tone was generated. As he progressively placed his fingers over the open holes, the sound frequency would increase. In the perforated pipe acoustic exhaust tuning system, there is a series of holes located along the length of the pipe. These holes filter out and dissipate the various frequencies that make up the sound aggregate that comes through the exhaust ports of the cylinder head. These holes also form sound emission points which tend to cancel one another and create a canceling effect on the frequencies in the exhaust gases. To put it a little more directly, a perforated pipe takes one big sound and makes a bunch of little sounds out of it.

Helmholtz Resonator (Ladder Filter)

The Helmholtz resonator is based on a mathematical theory of the motion of air at the end of an organ pipe, which was proposed by Hermann Ludwig Ferdinand Von Helmholtz in 1859. Essentially what he said was that if you move air to the end of the tube, the reaction of the air in the end of the tube will amplify some frequencies of sound when dampening others. Helmholtz resonators are therefore used within an exhaust system to subdue frequencies within the range of hearing while actually increasing non-offensive frequencies, such as those outside the range of human hearing or those deemed desirable by pubescent males.

Reflection Orifices

Imagine a pipe traveling through a series of chambers. In each chamber, the pipe has been perforated with a series of holes. Entering from the other end of the series of chambers is a second pipe, also perforated with a series of holes. Since the holes occur at different lengths along the two pipes, they will filter out different frequencies. The ends of the chambers act as baffles to further reduce the sound of the exhaust.

This type of muffler, though popular for many decades, tends to be extremely heavy and cause a great deal of backpressure.

Venturi Nozzles

A venturi nozzle is simply a narrowing of the pipe used to decrease the intensity of low-frequency sounds. This is obviously a restriction to gas flow.

Internal Engine Modifications

In preparing this section on performance modification, it comes to mind that there is one major difference between modifying 1960s engines and today's engines—emission laws! Different jurisdictions will have different mandates concerning emission control systems and what modifications are legal. As a whole, throughout the United States, it can be safely said that it is illegal to remove or modify any emission control device on a car which will be licensed for street use. There is, however, a small amount of latitude permitted in some states for the replacement of some components like intake manifolds, throttle bodies, and camshafts with specific tested and approved parts.

Most people who get involved in a "port and polish" job do not really have a good understanding of flow dynamics. The result is that many performance modifiers end up paying a lot of money or spending a lot of time to accomplish very little or to even decrease performance. The typical port and polish job consists of enlarging the intake and exhaust ports of the cylinder head and then polishing the enlarged surface to a high-gloss. While this may look impressive, the ports are basically large enough for the camshaft applications found in common street use, and often the high gloss shine reduces turbulence which can actually decrease the ability of the fuel to remain atomized in the air/fuel charge and therefore reduce performance. By the same token, for most "street" applications, big valves can have an adverse affect on air flow in and out of the combustion chamber.

The area that needs the most work or concentration is the valve pocket. This is the area just in front of the valve in the cylinder head. The valve guide boss can restrict air flow; on many engines this boss is much larger than it needs to be. Additionally, there are casting ridges which can be removed. Flatten and enlarge the radius where the air passage turns down toward the valve.

Shop around a little for a qualified performance machinist if you decide not to do the work yourself. As you do, remember that the best person may not always have the biggest reputation in town, and who your buddy says is the best in town may not be. If performance is really your goal, spend a couple of days at the local drag strip or circle track; ask who has the most consistently fast car. Some of the fastest cars may not be consistent performers; inconsistency may be a clue that the performance comes from "bells and whistles" rather than quality workmanship. Once you have identified the consistent performers, find out who does their machine work; this is the guy you want working for you.

Valve Action

I want to remind you at this point that in this section we are talking about street performance. And while there can be some tweaking such as cylinder heads, camshafts, and intake modifications, there are other performance parts that do not justify their expense on street applications by improved torque or horsepower. One of these areas is in the valvetrain. Where valvetrain modifications can provide an all-out race car with 1/4-mile, tenth-of-a-second improvement, these improvements would only be seen between the ears on a street rod.

Valvetrain modifications include ratio'ed rocker arms, high-tech push rods, performance lifters, and heavy duty valve springs. Engine rebuilders disagree on the necessity of replacing these components when the camshaft is replaced. Personally, I feel that when a new piece of metal rubs with a force of up to 300,000 psi against another piece of metal with only a thin film of oil in between, and we are going to expect it to do this for 100,000 to 150,000 miles, I would prefer that the piece of metal the new piece is rubbing against also be new.

The bottom line of all of this is that if you decide to replace the camshaft, then spending a few extra bucks on high-performance valvetrain parts that should be replaced anyway makes good sense, otherwise put your money into areas where you can get more bang for the buck.

Performance Computers

All I have to say here is to beware of those selling snake oil. Ask the supplier of the performance computer for test documentation and information about California Air Resources Board approval before you buy.

CHAPTER ELEVEN

LEGAL ASPECTS OF MODIFICATION

The basic rule when making internal engine modifications for vehicles that are in smog check areas or are in areas that are likely to be smog check areas in the future is: don't! The law in California, which we are forced to use as the strictest standard, permits no internal modifications for domestic cars after 1966 and for foreign cars after 1968. There is the temptation to hide the fact that high-compression pistons have been installed in the engine. After all, when the emissions are tested, they are not going to take the engine apart to determine if the stock pistons are installed. While that may be technically accurate, it is very likely that the emissions may be changed enough to fail emissions.

In Fairbanks, Alaska, there is a severe problem with CO emissions during the winter. Hydrocarbon (HC) and oxides of nitrogen (NOx) emissions are not a problem. In this jurisdiction, it is legal to make any internal engine modifications you desire. But even in the great white north there is no immunity to federal intervention. At any point in the future, the federal EPA may come in and require tighter emission standards. The best policy when modifying an engine destined for street use is to have an honest heart-to-heart talk with the I/M referee for your jurisdiction. Most of these people are motorheads anyway. They love to bench race with the best of them. I sympathize in advance with those who end up with a grumpy referee.

In many jurisdictions, the depth of the regulation is well beyond what the referee can handle. There is still an avenue for those who have a great deal of time, money, and an entrepreneurial spirit. Let me explain the procedure. Let us say that we have a 1995 Dodge light-duty truck that we wish to convert to natural gas. When converted to natural gas, the emissions will be lower. These lower emissions come at the expense of power. To boost the power back to about where it was, it would be advisable to increase the compression ratio. Logic and science dictate that the use of high-octane fuel like natural gas in a high-compression engine should not create high levels of emissions. Carbon monoxide, hydrocarbons, and oxides of nitrogen should all remain within their legal limits. California Air Resources Board is open to that concept. All you have to do is prove it. Modify about four vehicles of the exact type that you wish to modify. Have each of the four vehicles CARB certified. This procedure is much cheaper than it used to be. A typical test will only cost you about $1,500 to $2,000. So, for about $6,000 to $10,000 (not counting the cost of the vehicles), you can have your engine modification certified.

Unlike most scientific regimens, the law has subtle twists, turns, and traps. This chapter is not intended to be legal advice; it is intended to point out some of the issues involved in the high-performance modification of late-model cars. The bulk of the information that follows came from California's Bureau of Auto Repair (BAR) and the California Air Research Board (CARB). The laws on the books in many states are very similar to those of California. California has the longest record of stringent enforcement, which is why I am using its information.

There are three categories of replacement pars recognized by the CARB.

Category 1

Category 1 items are not considered by the BAR or CARB to be of any concern as long as the required emission controls are not tampered with.

- PCV air bleeds
- Air cleaner modification
- Air conditioner cut-out systems
- Anti-theft systems
- Blow-by oil separators and filters
- Electronic ignition systems retro-fitted to vehicles originally fitted with point/condenser systems as long as the original advance controls are maintained
- Engine shut-off systems
- Ignition bridges and coil modifications
- Throttle lock-out systems
- Intercoolers for OEM turbocharger
- Under-carburetor screens
- Vapor/steam/water injectors

Category 2

Category 2 addresses allowable replacement parts.

- Headers on non-catalyst cars
- Heat stoves for allowed headers
- Intake manifolds for non-EGR vehicles must allow for the installation and proper functioning of the OEM emission controls
- Approved aftermarket catalytic converters
- Carburetors marketed as "emission replacement"
- Replacement fuel fill pipe restrictors
- Replacement gas caps

You can see from this list that for catalyst-equipped fuel-injected cars that there are no performance replacements that are allowable without type approval from the CARB. Today's cars are EPA inspected as an integrated system; disturbing even the minutest portion of the emission control package would constitute a violation.

Category 3 parts must have verification of acceptability. If you are replacing a part in Category 3, ask for and retain a copy of the verification of acceptability for that product. It may prove handy later on, even if you live in an area that is not currently strictly controlled.

Category 3

Category 3 includes:

- Carburetor conversions
- Carburetors that replace OEM fuel injection
- EGR system modifications
- Replacement PROMs (computer chips)
- Electronic ignition enhancements for computerized vehicles
- Exhaust headers for catalyst vehicles
- Fuel injection systems that replace OEM carburetors
- Superchargers
- Turbocharger

PIN AND WIRE IDENTIFICATION

Throttle Body Injection

Logic module pin identification 1984 and earlier (Connector 1)

Pin Number	Wire Color	Wire Function
1	Black	Sensor input ground
2	Black/blue	Engine sensor common ground
3	Orange	8 volts from power module
4	Violet	5-volt supply MAP sensor
5	Violet/yellow	Injector information input
6	Violet	None
7	Yellow	Power module timing information
8	Dark Gray	None
9	Yellow/red	None
10	None	None
11	Gray	Distributor RPM information
12	White/orange	Vehicle speed input
13	Blue/black	Fuel gauge display information
14	Light green	"Hold" function on the test box
15	Pink	Test box fault code information
16 and 17	Dark blue	Fused power from power module
18	None	None
19	None	None
20	Orange/white	5-volt TPS power supply
21	Brown/red	Battery voltage for memory

Logic module pin identification 1984 and earlier (Connector 2)

Pin Number	Wire Color	Wire Function
1	Brown/yellow	Park/neutral switch input
2	Light blue/yellow	Rear defrost input
3	Dark green/red	None
4	Dark green/red	MAP sensor input
5	Orange/dark blue	TPS input

6	Dark green	Battery voltage from auto shut down relay
7	Brown	A/C clutch input
8	Tan	Coolant temperature sensor input
9	Dark blue/yellow	A/C switch input
10	Black/red	Charge temperature sensor input
11	Dark blue/orange	Wide open throttle relay output
12	Gray/red	Automatic idle speed control output
13	White/tan	Brake switch input
14	Brown	Automatic idle speed control output
15	Light blue/red	Ground for logic module
16	Light blue/red	Back-up ground for logic module
17	Black/orange	Ground for power loss light
18	Gray/yellow	EGR ground
19	Pink	Ground for purge solenoid
20	None	None
21	Black	Oxygen sensor input

Power module 10-pin identification 1984 and earlier

Pin Number	Wire Color	Wire Function
1	Black/yellow	Coil ground
2	Dark blue	J2 feed from ignition switch
3	Dark blue	Fused J2 power
4	Tan	Pulse width signal to injector
5	White	Injector feedback signal
6	None	None
7	Dark blue/yellow	Auto shut down relay ground
8	Dark green	Spark and fuel power supply
9	Black	Power module ground
10	Black	Back-up power module ground

Power module 12-pin identification 1984 and earlier

Pin Number	Wire Color	Wire Function
1	Violet/yellow	Injector input from logic module
2	Black/light blue	Sensor "noise" ground
3	None	None
4	None	None
5	None	None
6	None	None
7	Gray	Distributor signal input
8	None	None
9	Yellow/red	None
10	Yellow	Logic module advance input
11	Violet	None
12	Orange	8-volt output for Hall pickup and logic module

Logic module pin identification 1985-1987 (Connector 1)

Pin Number	Wire Color	Wire Function
1	Wire inside computer	5-volt
2	Violet/yellow	Injector control
3	None	None
4	None	None
5	Dark green/orange	Alternator field control
6	Yellow	Dwell control
7	Dark blue	Fused J2
8	Dark blue	Fused J2
9	None	None
10	Gray	Distributor reference
11	Pink	Scanner interface
12	Gray/light blue	Tachometer signal
13	Light green	Scanner interface
14	Light blue	Fuel monitor
15	Orange/light green	Shift indicator light
16	Violet/black	Automatic idle speed motor
17	Dark blue/yellow	Automatic shut down relay control
18	Gray/red	Automatic idle speed motor open
19	None	None
20	Violet	Automatic idle speed motor
21	Dark blue/pink	Radiator fan relay
22	Brown	Automatic idle speed motor close
23	Orange	7.5-volt input
24	Black	Signal ground
25	Black/light blue	Sensor return

Logic module pin identification 1985-1987 (Connector 2)

Pin Number	Wire Color	Wire Function
1	Orange/white	TPS 5-volt
2	Red/white	Battery standby
3	Dark blue/orange	A/C cut-out relay
4	Black/orange	Power loss light
5	Pink	Surge solenoid
6	None	None
7	Light blue/red	Power ground
8	Light blue/red	Power ground

9	None	None
10	None	None
11	Brown	A/C clutch
12	Brown/yellow	Park/neutral switch
13	White/tan	Brake switch
14	White/orange	Distance sensor
15	None	None
16	None	None
17	None	None
18	Black	Oxygen sensor
19	Wire inside computer	MAP sensor signal
20	Red/black	Battery temperature signal
21	Orange/dark blue	TPS
22	Red/white	Battery voltage sensor
23	Tan	Coolant sensor
24	None	None
25	Black/red	Injector temperature sensor

Power module 10-pin identification 1985-1987

Pin Number	Wire Color	Wire Function
1	Black/yellow	Coil negative terminal
2	Dark blue	J2 feed from ignition switch
3	Dark blue	Fused J2 power
4	Tan	Pulse width signal to injector
5	White	Injector feedback signal
6	Dark green/black	Switched battery
7	White	Injector
8	Dark green	Alternator field
9	Black	Power ground
10	Black	Power ground

Power module 12-pin identification 1985-1987

Pin Number	Wire Color	Wire Function
1	Violet/yellow	Injector control unit
2	Black/light blue	Signal ground
3	Red/black	Battery temperature sensor
4	None	None
5	Dark blue/yellow	Auto shut down relay control
6	Red/white	Battery sense and standby
7	None	None
8	None	None
9	None	None
10	Yellow	Dwell control
11	Dark green/orange	Alternator field control
12	Orange	8-volt output

SMEC 14-pin identification 1988

Pin Number	Wire Color	Wire Function
1	Orange	8-volt output
2	Black/white	Ground
3	Dark blue/white	FJ2 output
4	Dark blue	12 volts
5	Gray/white	Injector control #2 (rear-wheel-drive only)
6	Black	Ground
7	Black	Ground
8	Violet/yellow	Injector control
9	White	Injector driver #1
10	Tan	Injector driver #2 (rear-wheel-drive only)
11	Dark green/orange	Voltage regulator signal
12	Black/yellow	Ignition coil driver
13	Yellow	Anti-dwell signal
14	Dark green	Regulator control

SMEC 60-pin identification 1988 front-wheel-drive

Pin Number	Wire Color	Wire Function
1	Dark green/red	MAP sensor
2	None	None
3	Tan/white	Coolant sensor
4	Black/light blue	Sensor return
5	Black/white	Signal ground
6	None	None
7	White/light green	Speed control resume
8	Yellow/red	Speed control on/off
9	Brown/red	Speed control set
10	White	B1 input
11	None	None
12	Dark blue/white	FJ2
13	Violet/white	5-volt supply
14	Dark green/orange	Alternator field control
15	Light blue/red	Ground
16	Light blue/red	Ground
17	Brown/white	Automatic idle speed-1
18	Yellow/black	Automatic idle speed-2
19	Gray/red	Automatic idle speed-3
20	Violet/black	Automatic idle speed-4
21	Black/red	Throttle body temperature sensor
22	Orange/dark blue	TPS
23	Black/dark green	Oxygen sensor
24	None	None
25	None	None
26	None	None
27	None	None
28	None	None
29	White/pink	Brake switch
30	Brown/yellow	Park/neutral switch
31	Light green	Scanner receive
32	None	None

33	Violet/yellow	Injector control #1
34	Yellow	Dwell control
35	None	None
36	Light blue/black	Fuel monitor
37	None	None
38	None	None
39	None	None
40	Gray/yellow	EGR solenoid
41	Red	Direct battery
42	None	None
43	None	None
44	None	None
45	Brown	A/C clutch input
46	None	None
47	Gray/black	Reference pickup
48	White/orange	Vehicle distance pickup
49	None	None
50	Gray/light blue	Tachometer signal
51	Pink	Scanner transmit
52	Orange	8-volt input
53	Tan/red	Speed control vacuum solenoid
54	Pink/black	Purge solenoid
55	Orange/black	Lock-up torque converter
56	Dark blue/orange	A/C wide open throttle cut-out
57	Dark blue/pink	Radiator fan relay
58	Dark blue/yellow	Auto shut down relay
59	Black/pink	Check engine light
60	Light green/red	Speed control vent solenoid

SMEC 60-pin identification 1988 rear-wheel-drive pickup and van

Pin Number	Wire Color	Wire Function
1	Dark green/red	MAP sensor
2	None	None
3	Tan/white	Coolant sensor
4	Black/light blue	Sensor return
5	Black/white	Signal ground
6	None	None
7	None	None
8	None	None
9	None	None
10	Dark green/black	Z1 input
11	None	None
12	Dark blue/white	FJ2
13	Violet/white	5-volt supply
14	Dark green/orange	Alternator field control
15	Light blue/red	Ground
16	Light blue/red	Ground
17	Brown/white	Automatic idle speed-1
18	None	None
19	Gray/red	Automatic idle speed-3
20	None	None
21	Black/red	Throttle temperature sensor (5.2 liter)
22	Orange/dark blue	TPS
23	Black/dark green	Oxygen sensor
24	None	None
25	None	None
26	None	None
27	None	None
28	None	None
29	White/pink	Brake switch
30	Brown/yellow	Park/neutral switch
31	Light green	Scanner receive
32	Gray/white	Injector control #2
33	Violet/yellow	Injector control #1
34	Yellow	Dwell control
35	None	None
36	None	None
37-53	Not Used	
54	Pink/black	Purge solenoid
55	Orange/black	Lock-up torque converter or shift light
56	Dark blue/orange	A/C wide open throttle cut-out
57	None	None
58	Dark blue/yellow	Auto shut down relay
59	Black/pink	Check engine light
60	None	None

Logic module 10-pin identification 1985-1987

Pin Number	Wire Color	Wire Function
1	Black/yellow	Coil negative terminal
2	Dark blue	J2 feed from ignition switch
3	Dark blue	Fused J2 power
4	Tan	Pulse width signal to injector
5	White	Injector feedback signal
6	Dark green/black	Switched battery
7	White	Injector
8	Dark green	Alternator field
9	Black	Power ground
10	Black	Power ground

Multipoint Fuel Injection
Logic module pin identification 1984 (Connector1)

Pin Number	Wire Color	Wire Function
1	Black	Sensor "noise" ground
2	Black/blue	Sensor common ground
3	Orange	8-volt from power module

4	Violet	5-volt supply MAP sensor
5	Violet/yellow	Injector information input
6	None	None
7	Yellow	Timing information input
8	Gray/white	Injector on signal
9	Yellow/red	Knock sensor output
10	None	None
11	Gray	Distributor input signal
12	White/orange	Vehicle speed input
13	Light blue	Fuel gauge display information
14	Light green	"Hold" function on the test box
15	Pink	Test box fault code information
16 and 17	Dark blue	Fused power input
18	Dark blue/pink	Radiator fan relay
19	Tan/yellow	Distributor pickup signal
20	Orange/white	5-volt TPS power supply
21	Brown/red	Battery voltage memory

Logic module pin identification 1984 (Connector 2)

Pin Number	Wire Color	Wire Function
1	Brown/yellow	Park/neutral switch information input
2	Light blue/yellow	Rear defrost input
3	Dark green/red	MAP sensor input
4	Dark green/red	MAP sensor input
5	Orange/dark blue	TPS input
6	Dark green	Battery voltage from auto shut down relay
7	Brown	A/C clutch input
8	Tan	Coolant temperature sensor input
9	Dark blue/black	A/C switch input
10	Black/red	Charge temperature sensor input
11	Dark blue/orange	Wide open throttle relay output
12	Gray/red	Automatic idle speed control output
13	White/tan	Brake switch input
14	Brown	Automatic idle speed control output
15	Light blue/red	Ground for logic module
16	Light blue/red	Logic module back-up ground
17	Black/orange	Ground for power loss light
18	Gray/yellow	EGR solenoid ground
19	Pink	Ground for purge solenoid
20	None	None
21	Black	Oxygen sensor input

Power module 12-pin identification 1984

Pin Number	Wire Color	Wire Function
1	Violet/yellow	Injector input from logic module
2	Black/light blue	Sensor "noise" ground or knock sensor ground
3	None	None
4	Black/light green	Knock sensor input
5	None	None
6	None	None
7	Gray/white	Injector on control
8	Tan/yellow	Distributor synchronization input
9	Yellow/red	Knock sensor input
10	Yellow	Logic module spark advance input
11	Violet	None
12	Orange	8-volt output for Hall pickup and logic module

Logic module pin identification 1985-1987 (Connector 1)

Pin Number	Wire Color	Wire Function
1	Violet/white	5-volt power supply
2	Violet/yellow	Injector control
3	Gray/white	Injector on signal
4	Black/pink	Power loss light
5	Green/orange	Alternator field control
6	Yellow	Dwell control
7	Dark blue	Fused J2
8	Dark blue	Fused J2
9	White/orange	Distance sensor
10	Gray	Distributor reference
11	Pink	Scanner interface
12	Gray/light blue	Tachometer signal
13	Light green	Scanner interface
14	Light blue	Fuel monitor
15	Light blue	Barometric solenoid
16	Black/yellow	Automatic idle speed motor—Turbo II
17	Dark blue/yellow	Automatic shut down relay control
18	Gray/red	Automatic idle speed motor open
19	Light green	Wastegate solenoid
20	Violet/black	Automatic idle speed motor—Turbo II
21	Dark blue/pink	Radiator fan relay
22	Brown	Automatic idle speed motor close

23	Orange	7.5-volt input
24	Black	Signal ground
25	Black/light blue	Sensor return

Logic module pin identification 1985-1987 (Connector 2)

Pin Number	Wire Color	Wire Function
1	Orange/white	TPS 5-volt
2	Red/white	Battery standby
3	Dark blue/orange	A/C cut-out relay
4	Black/pink	Power loss light
5	Pink	Surge solenoid
6	Gray/yellow	EGR solenoid
7	Light blue/red	Power ground
8	Light blue/red	Power ground
9	Brown/red	Speed control set
10	White	Speed control resume
11	Brown	A/C clutch
12	Brown/yellow	Park/neutral switch
13	White/tan	Brake switch
14	White/orange	Distance sensor
15	Yellow/red	Speed control on/off
16	None	None
17	Tan/yellow	Distributor switch
18	Black	Oxygen sensor
19	Dark green/red	MAP sensor signal
20	Red/black	Battery temperature signal or battery sensor
21	Orange/dark blue	TPS
22	Red/white	Battery voltage sensor
23	Tan	Coolant sensor
24	Black/light green	Detonation sensor
25	Black/red	Charge sensor

Power module 10-pin identification 1985-1987

Pin Number	Wire Color	Wire Function
1	Black/yellow	Coil negative terminal
2	Dark blue	J2 feed from ignition switch
3	Dark blue	Fused J2 power
4	Pink	Direct battery
5	White	Injector feedback signal
6	Dark green/black	Switched battery
7	Tan	Injector
8	Dark green	Alternator field
9	Black	Power ground
10	Black	Power ground

Power module 12-pin identification 1985-1987

Pin Number	Wire Color	Wire Function
1	Violet/yellow	Injector control unit
2	Black/light blue	Signal ground
3	Red/black	Battery temperature sensor
4	None	None
5	Dark blue/yellow	Auto shut down relay control
6	Red/white	Battery sensor and standby
7	None	None
8	Gray/white	Injector control
9	None	None
10	Yellow	Dwell control
11	Dark green/orange	Alternator field control
12	Orange	8-volt output

Power module 10-pin identification 1988

Pin Number	Wire Color	Wire Function
1	Black/yellow	Ignition coil trigger
2	Dark blue	J2 feed from ignition switch
3	Dark blue	Fused J2 power
4	Tan	Injector 3 and 4 output
5	White	Injector 1and 2 output
6	None	None
7	Dark blue/yellow	ASD relay ground
8	Dark green	Power supply form power module
9	Black	Power module ground
10	Black	Power module back-up ground

SMEC 14-pin identification 1988

Pin Number	Wire Color	Wire Function
1	Orange	8-volt output (9-volt on 3.0 liter)
2	Black/white	Ground
3	Dark blue/light blue	FJ2 output
4	Dark blue	12 volts
5	Yellow/white	Injector control #2 (rear-wheel-drive only)
6	Black	Ground
7	Black	Ground
8	Violet/yellow	Injector control #1
9	White	Injector driver #1
10	Tan	Injector driver #2 (rear-wheel-drive only)
11	Dark green/orange	Voltage regulator signal
12	Black/yellow	Ignition coil driver
13	Yellow	Anti-dwell signal
14	Dark green	Voltage regulator alternator field control

SMEC 60-pin identification 1988

Pin Number	Wire Color	Wire Function
1	Dark green/red	MAP sensor
2	Black/light green	Detonation sensor (excluding 3.0 liter)
3	Tan/white	Coolant sensor
4	Black/light blue	Sensor return
5	Black/white	Signal ground
6	None	None
7	White/light green	Speed control resume
8	Yellow/red	Speed control on/off
9	Brown/red	Speed control set
10	Dark green/black	Z1 input
11	None	None
12	Dark blue/white	FJ2
13	Violet/white	5-volt supply
14	Dark green/orange	Alternator voltage regulator
15	Light blue/red	Ground
16	Light blue/red	Ground
17	Brown/white	Automatic idle speed-1
18	Yellow/black	Automatic idle speed-2
19	Gray/red	Automatic idle speed-3
20	Violet/black	Automatic idle speed-4
21	Black/red	Charge temperature sensor
22	Orange/dark blue	TPS
23	Black/dark green	Oxygen sensor
24	None	None
25	None	None
26	Tan/yellow	Fuel synchronizer pickup (excluding 3.0 liter)
27	None	None
28	None	None
29	White/pink	Brake switch
30	Brown/yellow	Park/neutral switch
31	Light green	Scanner receive
32	None	None
33	Violet/yellow	Injector control #1
34	Yellow	Dwell control
35	None	None
36	Light blue/black	Fuel monitor
37	None	None
38	None	None
39	Light green/black	Wastegate solenoid (excluding 3.0 liter)
40	Gray/yellow	EGR solenoid (3.0 liter)
41	Red	Direct battery
42	None	None
43	None	None
44	None	None
45	Brown	A/C clutch input
46	None	None
47	Gray/black	Reference pickup
48	White/orange	Vehicle distance pickup
49	Tan/yellow	High data rate pickup
50	Gray/light blue	Tachometer signal
51	Pink	scanner transmit
52	Orange	8-volt input (9-volt input 3.0 liter)
53	Tan/red	Speed control vacuum solenoid
54	Pink/black	Purge solenoid
55	Orange/black	Lock-up torque converter (3.0 liter only)
56	Dark blue/orange	A/C wide open throttle cut-out
57	Dark blue/pink	Radiator fan relay
58	Dark blue/yellow	Auto shut down relay
59	Black/pink	Check engine light
60	Light green/red	Speed control vent solenoid

RESOURCES

Alldata Corporation
9412 Big Horn Blvd.
Elk Grove, CA 95758l

Diacom
Rinda Technologies
5112 N. Elston Ave.
Chicago, IL 60630

Hypertech, Inc.
2104 Hillshire Cr.
Memphis, TN 38133

Mitchell On-Demand
P.O. Box 26260
San Diego, CA 92126-7030

OTC Monitor SPX Aftermarket Tools and Equipment Group
655 Eisenhower Drive
Owatonna, MN 55060

Snap-On Scanner
Kenosha, WI 53141-1410

Fluke Automotive Tools
P.O. Box 9090
Everett, WA 98206

INDEX